Advances in Sustainability Science and Technology

Series Editors

Robert J. Howlett, Bournemouth University, Poole, UK;
KES International, Shoreham-by-sea, UK

John R. Littlewood, School of Art & Design, Cardiff Metropolitan University, Cardiff, UK

Lakhmi C. Jain, University of Technology Sydney, Broadway, NSW, Australia

The book series aims at bringing together valuable and novel scientific contributions that address the critical issues of renewable energy, sustainable building, sustainable manufacturing, and other sustainability science and technology topics that have an impact in this diverse and fast-changing research community in academia and industry.

The areas to be covered are

- Climate change and mitigation, atmospheric carbon reduction, global warming
- Sustainability science, sustainability technologies
- Sustainable building technologies
- Intelligent buildings
- Sustainable energy generation
- Combined heat and power and district heating systems
- Control and optimization of renewable energy systems
- Smart grids and micro grids, local energy markets
- Smart cities, smart buildings, smart districts, smart countryside
- Energy and environmental assessment in buildings and cities
- Sustainable design, innovation and services
- Sustainable manufacturing processes and technology
- Sustainable manufacturing systems and enterprises
- Decision support for sustainability
- Micro/nanomachining, microelectromechanical machines (MEMS)
- Sustainable transport, smart vehicles and smart roads
- Information technology and artificial intelligence applied to sustainability
- Big data and data analytics applied to sustainability
- Sustainable food production, sustainable horticulture and agriculture
- Sustainability of air, water and other natural resources
- Sustainability policy, shaping the future, the triple bottom line, the circular economy

High quality content is an essential feature for all book proposals accepted for the series. It is expected that editors of all accepted volumes will ensure that contributions are subjected to an appropriate level of reviewing process and adhere to KES quality principles.

The series will include monographs, edited volumes, and selected proceedings.

More information about this series at http://www.springer.com/series/16477

Robert J. Howlett · John R. Littlewood ·
Lakhmi C. Jain
Editors

Emerging Research in Sustainable Energy and Buildings for a Low-Carbon Future

 Springer

Editors
Robert J. Howlett
'Aurel Vlaicu' University of Arad
Arad, Romania

Bournemouth University
Poole, UK

KES International
Shoreham-by-sea, UK

Lakhmi C. Jain
Liverpool Hope University
Liverpool, UK

University of Technology Sydney
Sydney, NSW, Australia

KES International
Shoreham-by-sea, UK

John R. Littlewood
School of Art & Design
Cardiff Metropolitan University
Cardiff, UK

ISSN 2662-6829   ISSN 2662-6837   (electronic)
Advances in Sustainability Science and Technology
ISBN 978-981-15-8774-0   ISBN 978-981-15-8775-7   (eBook)
https://doi.org/10.1007/978-981-15-8775-7

Preface

There is a great awareness of the urgent need to eliminate carbon emissions and improve the operational energy efficiency of the built environment in order to reduce the harmful effects on the ecosystem of human economic development and mitigate climate change reality. This has led to a huge growth in research around the science and technology of sustainable and resilient development.

The series *Advances in Sustainability Science and Technology (ASST)* was created by Springer Nature and KES International to respond to the need for a publication channel for the latest high-quality research on a broad range of sustainability topics.

Emerging Research in Sustainable Energy and Buildings for a Low-Carbon Future is the first volume published in the ASST series. It contains an introduction and 20 studies of recent research in the area of sustainable and resilient buildings, built environment infrastructure and renewable energy.

The book is directed to engineers, scientists, researchers, practitioners, academics and all those who are interested in developing and using sustainability science and technology for the betterment of our planet and humankind.

Thanks are due to the authors and reviewers for their expertise and time. The assistance provided by Springer during the development phase of this book is gratefully acknowledged.

Shoreham-by-sea, UK
Cardiff, UK
Sydney, Australia

Robert J. Howlett
John R. Littlewood
Lakhmi C. Jain

Contents

Part III

Part IV

Part V

Part VI

Editors and Contributors

About the Editors

Dr. Robert J. Howlett is the Executive Chair of KES International, a non-profit organization that facilitates knowledge transfer and the dissemination of research results in areas including intelligent systems, sustainability and knowledge transfer. He is a Visiting Professor at 'Aurel Vlaicu' University of Arad, Romania, and Bournemouth University in the UK. His technical expertise is in the use of intelligent systems to solve industrial problems. He has been successful in applying artificial intelligence, machine learning and related technologies to sustainability and renewable energy systems; condition monitoring, diagnostic tools and systems; and automotive electronics and engine management systems. His current research work is focused on the use of smart microgrids to achieve reduced energy costs and lower carbon emissions in areas such as housing and protected horticulture.

Dr. John R. Littlewood graduated in Building Surveying, holds a Ph.D. in Building Performance Assessment, and is a Chartered Building Engineer. He is Head of the Sustainable and Resilient Built Environment group in Cardiff School of Art & Design at Cardiff Metropolitan University (UK). He coordinates three Professional Doctorates in Art & Design, Engineering and Sustainable Built Environment, plus contributing to teaching in Architectural Design & Technology. John's research is industry focused, identifying and improving fire and thermal performance in existing and new dwellings, using innovative materials and construction and also improving occupant quality of life and thermal comfort. He has authored and co-authored 150 peer-reviewed publications and was also Co-editor for the 'Smart Energy Control Systems for Sustainable Buildings' book published in June 2017.

Dr. Lakhmi C. Jain Ph.D., M.E., B.E. (Hons), Fellow (Engineers Australia), is with the University of Technology Sydney, Australia, and Liverpool Hope University, UK. Professor Jain founded the KES International for providing a professional community the opportunities for publications, knowledge exchange, cooperation and teaming. Involving around 5,000 researchers drawn from universities and companies worldwide, KES facilitates international cooperation and generates synergy in teaching and research. KES regularly provides networking opportunities for professional community through one of the largest conferences of its kind in the area of KES.

Contributors

Douglas Aghimien Department of Construction Management and Quantity Surveying, University of Johannesburg, Johannesburg, South Africa;
SARChi in Sustainable Construction Management and Leadership in the Built Environment, Faculty of Engineering and the Built Environment, University of Johannesburg, Johannesburg, South Africa

Lerato Aghimien SARChi in Sustainable Construction Management and Leadership in the Built Environment, Faculty of Engineering and the Built Environment, University of Johannesburg, Johannesburg, South Africa

Clinton Aigbavboa Cidb Centre of Excellence, Faculty of Engineering and the Built Environment, University of Johannesburg, Johannesburg, South Africa;
Department of Construction Management and Quantity Surveying, University of Johannesburg, Johannesburg, South Africa;
SARChi in Sustainable Construction Management and Leadership in the Built Environment, Faculty of Engineering and the Built Environment, University of Johannesburg, Johannesburg, South Africa

John Aliu Cidb Centre of Excellence, Faculty of Engineering and the Built Environment, University of Johannesburg, Johannesburg, South Africa

Maria Benítez Instituto Municipal de La Vivienda E Infraestructura Habitacional (I.M.V. e I.H.), Municipalidad de Villa María, Córdoba, Argentina

Giangiacomo Bravo Social Studies/Centre for Data Intensive Studies & Applications, Linnaeus University, Växjö, Sweden

Hercilia Brusasca Instituto Municipal de La Vivienda E Infraestructura Habitacional (I.M.V. e I.H.), Municipalidad de Villa María, Córdoba, Argentina

Aantonella Caballero Instituto Municipal de La Vivienda E Infraestructura Habitacional (I.M.V. e I.H.), Municipalidad de Villa María, Córdoba, Argentina

Mouatassim Charai Mechanics and Energy Laboratory, Mohammed First University, Oujda, Morocco;
CERTES, Université Paris-Est, Créteil Cedex, France

Joanna Clarke SPECIFIC, Swansea University, Swansea, UK

Danae Conti Municipal, Municipal Institute of Housing and Municipal Housing Structure (I.M.V.e I.H.) from Villa Maria, Villa Maria, Cordoba, Argentina

Soledad Cormick Instituto Municipal de La Vivienda E Infraestructura Habitacional (I.M.V. e I.H.), Municipalidad de Villa María, Córdoba, Argentina

Lucas Daher Instituto Municipal de La Vivienda E Infraestructura Habitacional (I.M.V. e I.H.), Municipalidad de Villa María, Córdoba, Argentina

A. I. Dayneko Irkutsk National Research Technical University, Irkutsk, Russia

D. V. Dayneko Irkutsk National Research Technical University, Irkutsk, Russia; Irkutsk Scientific Center SB RAS, Irkutsk, Irkutsk, Russia

V. V. Dayneko Irkutsk National Research Technical University, Irkutsk, Russia

Youssef Errami Laboratory: Electronics, Instrumentation and Energy—Team: Exploitation and Processing of Renewable Energy, Department of Physics, Faculty of Science, University Chouaib Doukkali, Eljadida, Morocco

Pablo Gonzalez Instituto Municipal de La Vivienda E Infraestructura Habitacional (I.M.V. e I.H.), Municipalidad de Villa María, Córdoba, Argentina

E. Hale Neath, UK

Robert J. Howlett 'Aurel Vlaicu' University of Arad, Arad, Romania; Bournemouth University, Poole, UK; KES International Research, Shoreham-by-sea, UK

Oriabure Ijieh The Federal University of Technology Akure, Akure, Nigeria

Matthew Ikuabe Department of Construction Management and Quantity Surveying, University of Johannesburg, Johannesburg, South Africa

Lakhmi C. Jain University of Technology Sydney, Australia

Mustapha Karkri CERTES, Université Paris-Est, Créteil Cedex, France

John R. Littlewood Cardiff Metropolitan University, the Sustainable & Resilient Built Environment Group, Cardiff, UK

Krushna Mahapatra Department of Built Environment & Energy Technology, Linnaeus University, Växjö, Sweden

Abbas Mahravan Razi University, Kermanshah, Iran

Brijesh Mainali Department of Built Environment & Energy Technology, Linnaeus University, Växjö, Sweden

Carlos Mateu-Royo Researcher, ISTENER Research Group, Mechanical Engineering and Construction Department, Universitat Jaume I, Castelló, Spain

Ahmed Mezrhab Mechanics and Energy Laboratory, Mohammed First University, Oujda, Morocco

V. L. Moorhouse Cardiff Metropolitan University, the Sustainable & Resilient Built Environment Group, Cardiff, UK

Milad Moradibistouni Victoria University of Wellington, Wellington, New Zealand

Adrián Mota-Babiloni Postdoctoral Researcher, ISTENER Research Group, Mechanical Engineering and Construction Department, Universitat Jaume I, Castelló, Spain

Ibrahim Motawa Belfast School of Architecture and the Built Environment, Ulster University, Londonderry, UK

Joaquín Navarro-Esbrí Full Professor, ISTENER Research Group, Mechanical Engineering and Construction Department, Universitat Jaume I, Castelló, Spain

Farnush Nazipov School of Engineering and Digital Sciences, Nazarbayev University, Nur-Sultan, Kazakhstan

Yambenu Ngaj Department of Construction Management and Quantity Surveying, University of Johannesburg, Johannesburg, South Africa

Abdellatif Obbadi Laboratory: Electronics, Instrumentation and Energy—Team: Exploitation and Processing of Renewable Energy, Department of Physics, Faculty of Science, University Chouaib Doukkali, Eljadida, Morocco

Olanrewaju Ogunniyi The Federal University of Technology Akure, Akure, Nigeria

Ayodeji Oke SARChi in Sustainable Construction Management and Leadership in the Built Environment, Faculty of Engineering and the Built Environment, Department of Construction Management and Quantity Surveying, University of Johannesburg, Johannesburg, South Africa;
The Federal University of Technology Akure, Akure, Nigeria

Michael Oladokun Heriot Watt University, Edinburgh, UK

Georgios Pardalis Department of Built Environment & Energy Technology, Linnaeus University, Växjö, Sweden

Thanuja Ramachandra Department of Building Economics, University of Moratuwa, Moratuwa, Sri Lanka

Lucia Rodriguez Instituto Municipal de La Vivienda E Infraestructura Habitacional (I.M.V. e I.H.), Municipalidad de Villa María, Córdoba, Argentina

Smail Sahnoun Laboratory: Electronics, Instrumentation and Energy—Team: Exploitation and Processing of Renewable Energy, Department of Physics, Faculty of Science, University Chouaib Doukkali, Eljadida, Morocco

Romina Sangoy Municipal, Municipal Institute of Housing and Municipal Housing Structure (I.M.V.e I.H.) from Villa Maria, Villa Maria, Cordoba, Argentina

Haitham Sghiouri Mechanics and Energy Laboratory, Mohammed First University, Oujda, Morocco

Edgardo Suarez Instituto de Sustentabilidad Edilicia (ISE), Colegio de Arquitectos Provincia de Córdoba, Córdoba, Argentina

Carolyn Thomas Nottingham Trent University, Nottingham, England, UK

Wellington Thwala SARChi in Sustainable Construction Management and Leadership in the Built Environment, Faculty of Engineering and the Built Environment, University of Johannesburg, Johannesburg, South Africa

Serik Tokbolat School of Engineering and Digital Sciences, Nazarbayev University, Nur-Sultan, Kazakhstan

Brenda Vale Victoria University of Wellington, Wellington, New Zealand

Michael S. J. Walter Ansbach University of Applied Sciences, Ansbach, Germany

Achini Shanika Weerasinghe Department of Building Economics, University of Moratuwa, Moratuwa, Sri Lanka

Stefan Weiherer Ansbach University of Applied Sciences, Ansbach, Germany

Lukas Wildner Ansbach University of Applied Sciences, Ansbach, Germany

P. Wilgeroth Cardiff Metropolitan University, the Sustainable & Resilient Built Environment Group, Cardiff, UK

Yangang Xing Nottingham Trent University, Nottingham, England, UK

S. V. Zykov National Research University Higher School of Economics, Moscow, Russia

Chapter 1
An Introduction to Emerging Research in Sustainable Energy and Buildings for a Low-Carbon Future

Robert J. Howlett, John R. Littlewood, and Lakhmi C. Jain

This book contains an introduction and 20 chapters, each describing a recent research investigation in the area of sustainable and resilient buildings, built environment infrastructure and renewable energy. Contributions are from many different countries of the world and on a range of topics, representing a sample of research within the 'sustainable energy and buildings' field.

Part 1: The first part of the book looks at the sustainable design of buildings.

Chapter 2, 'Designing Active Buildings', discusses the evolution and validation of a design guide for Active Buildings. This was developed to aid in building design to enable the Active Building concept to be adopted by the construction industry, contributing to reducing the energy consumption of buildings and aligning with the UK Government's industrial strategy to at least halve the operational energy consumption of all new buildings by 2030.

Chapter 3, 'Sustainable Housing Solutions', describes design based on experience gathered over more than 20 years in social housing construction in Argentina for vulnerable families. The design principles incorporate the axioms of sustainability in its three dimensions: economic, social and environmental. The design premise

R. J. Howlett (✉)
'Aurel Vlaicu' University of Arad, Arad, Romania
e-mail: rjhowlett@kesinternational.org

Bournemouth University, Poole, UK

KES International, Shoreham-by-sea, UK

J. R. Littlewood
School of Art & Design, Cardiff Metropolitan University, Cardiff, UK
e-mail: jlittlewood@cardiffmet.ac.uk

L. C. Jain
University of Technology Sydney, Broadway, NSW, Australia
e-mail: jainlakhmi@gmail.com

is based on reducing the energy demands of the housing, considering bioclimatic conditions appropriate to the location and incorporating the active participation of the families from the very beginning.

Part 2: The next four chapters describe issues relating to the renovation, restoration and reconstruction of existing buildings or in one case a railway wagon.

Chapter 4, 'Future Energy-Related House Renovations in Sweden: One-Stop-Shop as a Shortcut to the Decision-Making Journey', discusses the attitudes of owners of detached houses in Sweden towards future renovations and considers their opinions of a one-stop-shop to provide deep renovation services, based on an online survey.

Chapter 5, 'Crisis of Institutional Change: Improving Restoration and Reconstruction Methods for Estate Cultural Heritage', presents the problem of institutional changes in Russian Urban planning. The chapter discusses institutional problems in the sphere of cultural heritage. The necessity for the research is justified, and methods and tools for evaluation of the effectiveness of the institutional are suggested. A programme of urban development is proposed, and tested methods for estate objects of cultural heritage protection, which are to be implemented further in Russia considering climatic, seismic and ecological peculiarities of the regions, are suggested.

Chapter 6, 'Greening Existing Garment Buildings: A Case of Sri Lanka', considers green retrofitting as a solution for contemporary issues such as global warming, resource depletion and greenhouse gas emissions which have arisen due to the conventionally built environment. Building owners may be unwilling to invest in green retrofits due to their perceptions of the first cost and payback period implications of the green retrofit. Therefore, the chapter assesses the first costs and life cycle saving implications of fourteen energy and water-efficient retrofits incorporated into four garment buildings in Sri Lanka to find the retrofit options which are financially sound.

Chapter 7, 'Sustainable Cultural Wagon', describes the conversion of a redundant railway wagon into a cultural and creative space for community use, responding to a local need in an area of Argentina. The cultural wagon was equipped with a range of features, including ramps for easy access, thermal-acoustic insulation, a photovoltaic panel system, indoor LED lighting, cross ventilation and a community accessible hot-water pump station powered by solar panels. This resulted in a public project that was sustainable, encouraging energy production through renewable sources, contributing to the reduction of energy consumption and the generation of clean energy with significant economic and environmental benefits.

Part 3: This section contains two chapters that consider barriers or impediments to low- or zero-carbon buildings.

Chapter 8, 'Unearthing the Factors Impeding Sustainable Construction in Developing Countries—A PLS-SEM (Partial Least Square Structural Equation Modelling) Approach', presents the result of a case study of the factors impeding the sustainability of construction projects in developing countries using Nigeria as an example. The study sought responses from construction managers, project managers and quantity surveyors from the six different regions of the country. Data gathered was analysed using factor analysis and structural equation modelling. The findings revealed that issues surrounding regulation and policy, information and management, sustainability knowledge and the availability of sustainable materials and technology are significant reasons for the poor record of sustainability in the country's construction industry.

Chapter 9, 'Barriers to the Adoption of Zero Carbon Emissions in Buildings: The South African Narrative', continues this theme with a study that evaluated the barriers to the adoption of zero carbon emissions in occupied buildings with a view to proffering ways to mitigate such practices. A comprehensive review of relevant literature was done which aided the identification of the barriers. Data for the study was elicited through questionnaire surveys from built environment professionals. Methods of data analysis used were percentage, mean item score and principal component analysis while Cronbach's alpha was used in testing the reliability of the questionnaire. Findings from the study revealed several factors hampering the adoption of zero carbon emissions in buildings. Recommendations were made to encourage the adoption of zero carbon emission processes in building operations and activities by its occupants.

Part 4: The next two chapters are on policy and certification.

Chapter 10, 'System Dynamics Analysis of Energy Policies on Buildings Performance', describes an investigation which aimed to develop a dynamic model to analyse the impact of energy policies on buildings performance in the UK. The principle of socio-technical systems was adopted as an approach to the modelling. A system dynamics model was developed to simulate the intrinsic interrelationship between the dwellings, occupants and environment systems. This chapter considered the impact of various policy scenarios on energy consumption in building towards achieving the UK national targets, namely improvements in the uptake of dwelling insulation measures, occupants' behavioural changes and policy change on energy prices. An integrated scenario was also assumed to combine the effect of the first three ones. The main findings indicate that it is unlikely for any one scenario alone to meet the required binding reductions unless an integrated solution is adopted. The developed model considers various qualitative conditions which are not usually simulated using the traditional regression-based forecasting of energy use in buildings. The developed model can be used to test various policies in other than UK context considering various datasets of the model variables.

Chapter 11, 'Investigating the Application of LEED and BREEAM Certification Schemes for Buildings in Kazakhstan', describes a study that looks at the application of rating and certification schemes on Kazakhstan and investigates some of the first pioneering buildings that have undergone certification process in that country. The construction industry in Kazakhstan has started adopting widely recognised environmental assessment certification schemes such as the Leadership in Energy and Environmental Design (LEED) and the Building Research Establishment Environmental Assessment Method (BREEAM). Up to the present time, more than 50 buildings, especially from rapidly expanding cities such as Nur-Sultan and Almaty, have obtained LEED and BREEAM certificates and have been recognised as green buildings. This study investigates the adoption of these methods in the context of Kazakhstan with the aim of understanding the driving factors of such application, characteristics of the certified buildings and potentials of promoting the certification schemes at a wider scale.

Part 5: There then follow four chapters on various topics related to sustainable buildings.

Chapter 12, 'Examining Undergraduate Courses Relevant to the Built Environment in the 4IR Era: a Delphi Study Approach', discusses a research project carried out in South Africa looking at undergraduate courses that will be relevant to the built environment in the imminent future. A qualitative Delphi approach was adopted to validate these courses as institutions of higher learning prepare students for this latest wave of innovation. Fourteen experts completed a two-stage iterative Delphi study process and reached consensus on all 29 courses identified. This study found that courses such as data analytics, artificial intelligence, computer programming, computer coding and data mining should be integrated into the curricula of universities to ease the transition of students from the lecture room to the world of work.

Chapter 13, 'An Appraisal of the Level of Awareness and Adoption of Insurance Policies on Sustainable Construction', describes a study, the purpose of which is to appraise the level of awareness and adoption of insurance policies on the sustainability of construction projects in Nigeria. The study involved a questionnaire survey method that was self-administered to a range of professionals in the construction industry. The data collected was analysed using Statistical Package for Social Sciences (SPSS). The findings of the study indicate the types of insurance policies available to the construction industry and concluded that insurance policies have positive impacts highly beneficial to the performance of construction projects.

Chapter 14, 'To What Extent is Biophilia Implemented in the Built Environment to Improve Health and Well-being?—State-of-the-Arts Review and a Holistic Biophilic Design Framework', is on an investigation of the application of biophilia in building design practices for improved health and well-being. Biophilic theoretical frameworks developed by leading biophilic experts were examined and compared to health and wellness performance certifications such as WELL Building and Living Building Challenge (LBC) standards. Then, a holistic biophilic framework inspired by Kellert and Calabrese was elaborated to assess the biophilic features in the built environment. Multiple case studies were studied during the project. The findings revealed that the biophilic applications linked to direct experiences of nature were implemented inefficiently and lacked a holistic approach to improve health and well-being. The authors argue that biophilia needs to be included holistically to maximise the benefits of nature's experiences.

Chapter 15, 'Thermal Conductivity Characterisation of Industrial Small-Sized Building Materials: Experimental and Simulation Study', describes a new experimental procedure for determining the thermal conductivity of small-sized building materials using the boxes method and an approach that does not require additional sensors. The measurement is based on the permutation of sensors and then the interpretation of the steady-state heat balance of samples. Local earthen blocks from eastern Morocco were developed. The measured thermal conductivity values were compared with those obtained by an accurate transient hot disc method. The comparison shows a good agreement and verifies the performance of the permutation approach. The chapter goes on to describe an evaluation of the thermal performance of the developed building materials and the results obtained. The building performance analysis clearly proved that earth walls can play a great role in mitigating the cooling demand and improving the thermal comfort of buildings in summer.

Part 6: This section contains three chapters relating to renewable energy.

Chapter 16, 'Coordinated Control Strategy to Improve Performance of Permanent Magnet Synchronous Generator Wind Power Systems', is intended to aid the reader in their understanding of the principles of control for wind energy conversion systems based on the permanent magnet synchronous generator. Modelling of the conversion system is described, including models of the wind turbine, the permanent magnet generator and the power converters. Then, the controls of the machine side converters and grid side converter are introduced. The proposed control methods were used to maximise the generated power from wind turbine generators, to keep a constant DC bus voltage for the grid side converter and to control the power fed to the grid.

Chapter 17, 'High-Temperature Heat Pumps for a Sustainable Industry', discusses developments in heat pump technology applicable to the decarbonisation of industry. This chapter gives a comprehensive overview of the current status and future possibilities of high-temperature heat pump technology based on the working principles, working fluids, configurations, existing prototypes and the possibility of reversible operation. Many working fluids are available with low global warming potential, but none of them are perfect as they come with their own disadvantages. The selection must be based on many factors. As well as various working fluid options, there are several configurations. Working fluids and configurations are described and explained.

Chapter 18, 'Replacing Fossil Fuels by On-Site Sources of Energy in a Residential Building in Chalus, Iran', has as its aim the evaluation of the possibility of replacing fossil fuels by renewable sources of energy in a residential building in Chalus, an Iranian city with a high potential for using on-site sources of energy. This study first investigated the annual operating energy used by households in Chalus. To do this, the type of appliances and average time of their use, including heaters and air conditioning units, is extracted from official reports together with a local field study. The ability of renewable on-site sources of energy to supply this load was calculated by considering the specific characteristics of the region. The result shows these sources have the potential to provide approximately 98% of the annual energy the household consumes.

Part 7: This part of the book contains two chapters with a sustainable transport theme, one relating to electric vehicles and the other about a sustainable road infrastructure.

Chapter 19, 'Simulation of an Adsorption Machine with Auxiliary Heater for CO_2-Neutral Air-Conditioning of Electric Utility Vehicles', looks at air-conditioning for electric vehicles, which is important for driver comfort. In conventional battery-operated commercial vehicles, the energy required to operate the air conditioning system is used from the battery which can reduce the range of the vehicle by up to 30%. To reduce energy consumption due to air-conditioning, it is preferable to utilise a technology that is independent of the battery. There are a number of options for controlling the temperature of a moving vehicle, but only a limited number that is CO_2-neutral. This chapter focuses on adsorption chiller technology in combination with an auxiliary heater based on bioethanol. To understand the advantage of an adsorption machine, a simulation model can provide useful data on scaling and ease of use and thus be the basis for design and assembly of a prototype system. Therefore, a mathematical model of the adsorption technology is described which is combined with the known dimensional parameters of electric vehicles, and the results are presented in the form of a simulation model.

Chapter 20, 'Sustainable Road Infrastructure in Rural Areas in South Africa—A Preliminary Study', describes a study that assessed the possible measures for attaining sustainable road infrastructure within rural communities in Limpopo province, South Africa. The study sought answers from rural dwellers and construction workers within the study area through a questionnaire survey. Data gathered was analysed using percentage, mean item score, standard deviation and one-sample t-test. The reliability of the questionnaire was also tested using Cronbach's alpha which gave an alpha value of 0.948 which indicates the questionnaire used was reliable. Findings of the study revealed that the most significant measures for attaining more sustainable road constructions within the rural areas include the following: using quality materials that will last the expected lifespan of the road, having planned maintenance, proper investment on road projects and using contractors and skilled workers with the right experience in road construction. It is believed that the findings of this study will help increase the delivery of sustainable road projects within the rural areas in a bid to provide a better standard of living for rural dwellers.

Part 8: The final chapter is on the manufacture of sustainable building components for the UK housing sector.

Chapter 21, 'Optimising Offsite Manufacturing of Timber-Frame Roof Trusses for UK Housing', describes a study involving sustainable offsite manufacturing and modern methods of construction. This was undertaken in response to drivers from the Welsh Government to increase the number of houses, and their quality, to contribute to meeting Wales' low to zero carbon agenda, launched in March 2019. The chapter presents a case study undertaken in one of the largest manufacturers of timber-frame construction systems in Wales (one of the countries that make up the UK). The project involved time and motion and value stream mapping studies to evaluate optimisation opportunities in the offsite manufacture of roof components, such as trusses. The preliminary results are presented and highlight opportunities for quick win refinement to the company's operational processes with the aim of increasing production efficiency, reducing waste and closing the performance gap. By this means, increased quality and thermal performance of offsite manufactured timber-frame buildings were achieved, resulting in reduced operational energy usage and therefore minimising greenhouse gas emissions.

The chapters that make up this volume are from a range of diverse topics within the sustainable energy and buildings area. Hence, it is not claimed that they represent a comprehensive record of research in the field. However, this diversity does indicate the breadth of research that is being undertaken with the aim of reducing carbon emissions and achieving energy efficiency in sustainable building and renewable energy.

Part I

Chapter 2
Designing Active Buildings

Joanna Clarke

Abstract This paper discusses the development and validation of an Active Building Design Guide being developed to enable the design of Active Buildings. The definition of Active Buildings and the key principles to be considered when designing them are discussed and illustrated. The background to the research project is discussed, which includes addressing the UK Government's aim to at least halve the energy consumption of all new buildings by 2030, and how the author's experience in designing Active Buildings and developing a Design Guide to enable others to design Active Buildings could help achieve this aim. The author has identified a need for some design guidance to enable the Active Building concept to be adopted by the construction industry, contributing to reducing the energy consumption of buildings, aligning with the UK Government's Industrial Strategy. The paper presents progress on the Design Guide development, testing and refinement, including the considerations for ensuring it is a document that architectural designers of Active Buildings will genuinely find useful and will use to ensure reductions in energy use and associated carbon emissions through the design of such buildings.

Keywords Architecture · Buildings · Energy · Construction · Design · UK government's industrial strategy · Transforming construction · Low carbon · Net zero carbon

1 Introduction

An Active Building is one that *'supports the energy network by intelligently integrating renewable energy technologies for heat, power and transport'* [1]. The purpose of this paper is to discuss and illustrate the first Design Guide for Active Buildings, intended for use by Architects and others to aid the design of Active Buildings. This supports the role of the Active Building Centre (ABC), Swansea

J. Clarke (✉)
SPECIFIC, Swansea University, Swansea SA1 8EN, UK
e-mail: joanna.r.clarke@swansea.ac.uk

University, UK, in enabling the UK construction industry to meet the goal to halve the energy consumption of all new buildings by 2030 as set out in the UK Government's Industrial Strategy [2] and to reduce pressures on the wider UK energy networks that connect buildings. The energy networks in the UK were originally designed to supply power and heat to buildings in a distributed way from a centralised supply [3]. However, with the increase in renewable energy generation across the UK as reported since 2016 [4], the UK energy networks are now more dynamic, presenting both technical and commercial issues for the grid networks [3]. The UK electricity grid was not designed or installed to respond to dynamic electrical energy generation and supply, where at certain peak renewable energy generation across the UK, the grid has too much power and therefore must curtail the generation of renewable energy, and at times of peak demand, the grid struggles to supply the electricity demanded across the UK.

Active Buildings could offer a solution to the above peak electricity demand/supply of renewable energy generation. Using intelligent controls and energy storage, it is possible to manage import and export of energy from a building such that buildings could support the grid, rather than adding to its current constraints [1]. This concept is currently being tested on a building designed by the author whilst she was employed at SPECIFIC Innovation and Knowledge Centre (IKC) at Swansea University [5], prior to the establishment of the ABC, where she is currently employed as Head of Design. Known as the Active Office [6], this building acts as a demonstrator to trial different modes of interaction with the UK electricity grid, utilising a combination of renewable energy generation, energy storage, intelligent control strategies and smart electric vehicle charge points. Linking the Active Office's energy strategy to control algorithms that monitor the carbon intensity of the grid, the current energy tariffs, weather predictions and occupancy patterns, the building is capable of managing its import and export of energy, depending on different factors [7]. For example, the control system may choose to import energy when the carbon intensity of the grid is low, maximising the use of renewable energy generation, and to export energy when the carbon intensity is high, that is when the energy mix contains a high proportion of fossil fuel sources [8].

In September 2018, the ABC was established [9] to develop solutions to enable the UK construction industry to investigate how to adopt the Active Building concept, supporting the UK Government's ambition to Transform Construction, by at least halving the energy consumption of buildings by 2030 [2]. To date (2020), no specific guidance has been developed to aid those wishing to implement the Active Building concept. The Active Building Design Guide (ABDG) currently under development will be the first source of guidance produced to aid the design of Active Buildings and one of the first outputs from the ABC. Figure 1 illustrates the crucial role of the ABDG within the context of the ABC.

This paper sets the context to the development of the ABDG, describes the development to date (2019) and discusses the methods used to test the ABDG.

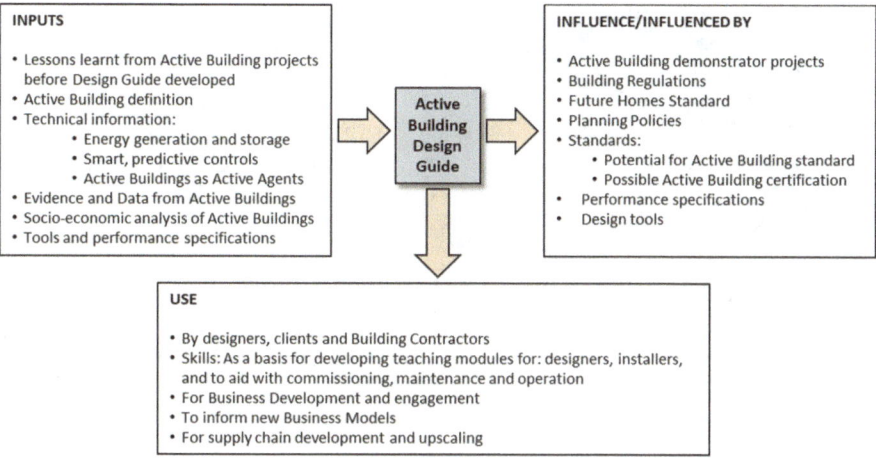

Fig. 1 Inputs, influencers and use of the active building design guide (ABDG)

2   Background and Related Work

Before the ABC was established in 2018, the author worked as the Building Integration Manager for SPECIFIC IKC [5], where her role was to encourage UK construction industry stakeholders to adopt the Active Building concept on building projects, utilising renewable energy technologies developed by SPECIFIC and their industry partners. The author's experience as an architect helped her to identify the best way to achieve this as being to develop demonstrator buildings to showcase the Active Building concept, which was at the time (2014) referred to as *'Buildings as Power Stations'*—buildings that generate, store and release their own energy. The first building designed by the author was a garden office building, known as the Pod [10], which demonstrated the *'Buildings as Power Stations'* concept on an off-grid and self-sustaining building (see Fig. 2). The aim of the Pod was not only to demonstrate the concept to the UK construction industry stakeholders and potential building

Construction
Structurally Insulated Panels (SIPs)
Innovative Technologies
Building Integrated Photovoltaic (BIPV) Roof
Glazed Solar Air Collector
Battery Storage
Novel Resistive Heating System
Bespoke Control System

Fig. 2 Pod [10]

Construction
Novel interlocking steel framed panels
Innovative Technologies
BIPV Roof
Transpired Solar Collector (TSC)
Air Source Heat Pump (ASHP)
Mechanical Ventilation and Heat Recovery System (MVHR)
Battery Storage
Novel Resistive Heating System
Bespoke Control System

Fig. 3 Active classroom [11]

owners and clients, but also to provide a building that the researchers at SPECIFIC could relate to as a home for the innovative technologies they were developing.

The Pod became the catalyst for a new opportunity for SPECIFIC (to focus on demonstrating generate, store and release technologies on buildings, rather than simply developing individual technology prototypes), and soon after, the author was asked to design a second building—the Active Classroom [11], shown in Fig. 3.

The name Active Classroom was chosen as it highlights that the building envelope is 'activated'. Rather than the facades and roof being 'passive' elements to simply provide shelter for the building occupants, the technologies embedded within the fabric generate heat and electricity, for use in the building. Hence, the term *'Active Buildings'* was created [11]. The Active Classroom has been successful in raising the profile of SPECIFIC within the Welsh and UK construction industries, winning several awards, including the prestigious 'Project of the Year' at the Royal Institute of Chartered Surveyors (RICS) Wales Awards in 2018 [12].

In designing and delivering the Active Classroom, the author worked with the technical team of scientists and engineers at SPECIFIC, who developed control systems to ensure the building operated effectively, and that data collected from a range of sensors within the building could be used to learn about what worked well and what did not work so well—lessons that could be used in the design of further Active Buildings. The Active Classroom demonstrated that it was possible for a building to produce more energy than it consumed over an annual period, described as being *'energy positive'* [11]. However, during the process of designing the next building—the Active Office [6], a two-storey office building (shown in Fig. 4)—it became clear that achieving *'energy positive'* is more challenging the more storeys a building incorporates. With experience, the SPECIFIC team also realised that the capability of a building to interact intelligently with the energy grid, controlling when power is imported to or exported from (made possible through the use of energy storage and smart controls), has the potential to truly transform both the energy and building sectors and is hence of more value than simply being energy positive over a year.

Construction
Modular – 12 modules
Innovative Technologies
BIPV Roof
Battery Storage
Combines Solar Thermal and PV
(PV-T) System
Thermal Store
Bespoke Control System

Fig. 4 Active office [6]

These Active Buildings are fully electric, operating without consuming gas for heating and hot water, aligning with the UK Government's target to decarbonise heat in buildings, as outlined in their Clean Growth Strategy [9].

2.1 Professional Doctorate—Sustainable Built Environment

The ABDG will form part of a Professional Doctorate in Sustainable Built Environment (D.SBE) [13] research project which the author commenced in April 2017, after completion of the Active Classroom [11], based on her work to date on designing Active Buildings [14]. The progress of the author's doctorate project has included taking a multi-methods approach after Creswell and Plano Clark [15] including a critical review of both academic and professional publications; engaging with UK construction industry stakeholders and architectural students (delivering workshops and Continuous Professional Development (CPD) seminars and hosting innovation visits to the Active Building demonstration projects); developing and testing a Design Guide for Active Buildings; presenting at various events; and contributing to articles in publications [16, 17, 18, 19, 14], Premier [20, 21].

In addition, performance data is being collected from demonstration buildings the author has either designed or influenced the design of. This data is analysed by the technical teams at SPECIFIC and the ABC and used to detect faults, optimise performance of building systems and develop predictive control strategies for Active Buildings. Data collection is essential to enable the ABC to build an evidence base that can be used to support the transformation of the UK construction industry to meet the goals set out in a report entitled *Construction 2025* [22] and the UK Government's Industrial Strategy [2]. Ambitious targets for the UK construction industry, including lowering carbon emissions from the built environment by 50% by 2025 [22], indicate the need for innovative solutions to reducing energy consumption and carbon emissions from buildings, which the Transforming Construction Challenge [23] and the Active Building approach [1] set out to deliver.

Before embarking on the development of the Active Building Design Guide (ABDG), the author undertook literature reviews on the current status of the UK construction industry and the UK Government's ambitions to reduce the energy consumption of buildings [2, 22]. The evaluation of these publications was discussed and presented in a paper at Eco-Architecture 2018: 7th International Conference on Harmonisation between Architecture and Nature [17].

The ABDG under development is an enabler to the adoption of the Active Building concept in the UK, setting out the six core principles and the process for meeting these, as summarised in Fig. 5. As Active Buildings are a relatively new concept in the UK and there is currently (2020) limited data available from Active Building projects, the principles are deliberately not too prescriptive which may otherwise reduce clients and developers' consideration for future projects. As data and evidence are collected from Active Building projects, the minimum performance specifications will be developed, which will prescribe target energy values to be achieved.

In developing the ABDG, the author has reviewed existing information available to Architects on these topics and referenced the most appropriate documents within the Design Guide, for example, environmental design [24] and building physics [25] to reduce the energy consumption of buildings, which relate to the first principle of an Active Building. The intention is that section one (energy reduction) will act as an

Fig. 5 Core principles of an active building [1]

aide-memoire to those reading and using the guide on fundamental passive design principles without repeating information described in detail in existing literature [26, 25, 24, 27].

To further support the development of the ABDG, existing guidance documents used by Architects in the UK will be reviewed, including the Approved Documents used to provide guidance on compliance with UK Building Regulations [27] and guidance documents produced by other bodies, such as Local Authorities (e.g. [28]).

Feedback from a pilot project undertaken as part of the author's D.SBE between February and June 2019 and documented in a paper entitled *'Active Buildings in Practice'* was presented at the 11th International Conference on Sustainability in Energy and Buildings [18]. This paper (ibid) and the feedback from the presentation are being used to refine the ABDG in development. In the pilot project, the author tested a first version of the ABDG with Architects in UK practice and feedback suggested that a mixture of small amounts of text, images, diagrams and tables is preferred to a document consisting of main text. The ABDG will be developed to align with existing design guidance used by UK-based Architects and feedback from participants of the pilot project.

3   Methods to Test the Active Building Design Guide (ABDG)

Qualitative research methods are deployed in this D.SBE project in order to test, evaluate and refine the ABDG using semi-structured interviews and focus groups, following research methods described by Saunders et al. [29]. As mentioned above in Sect. 2.1, before commencing development of version two of the ABDG, the author undertook a pilot project, which took place in two stages:

Stage 1. Identified the main challenges facing UK construction industry stake-holders when introducing innovative technologies and practices into construction projects, between March and November 2018. The results from the analysis of this stage influenced the decision to develop an ABDG to address some of the challenges identified;

Stage 2. Development of a draft ABDG which was tested with UK-based Architects in professional practice between January and March 2018 and also with architectural students studying at one of the two Schools of Architecture in Wales during the same time period, January–March 2018. The analysis of the feedback led to version two of the ABDG.

Final testing and refinement of version two of the ABDG will be undertaken between January and June 2020. This final testing and evaluation will use *Action Research* [29], more specifically a *design-decision research* approach described by Groat and Wang [30], where the researcher is embedded in the research process. Figure 6 shows the strategy for developing and testing the different versions of the ABDG.

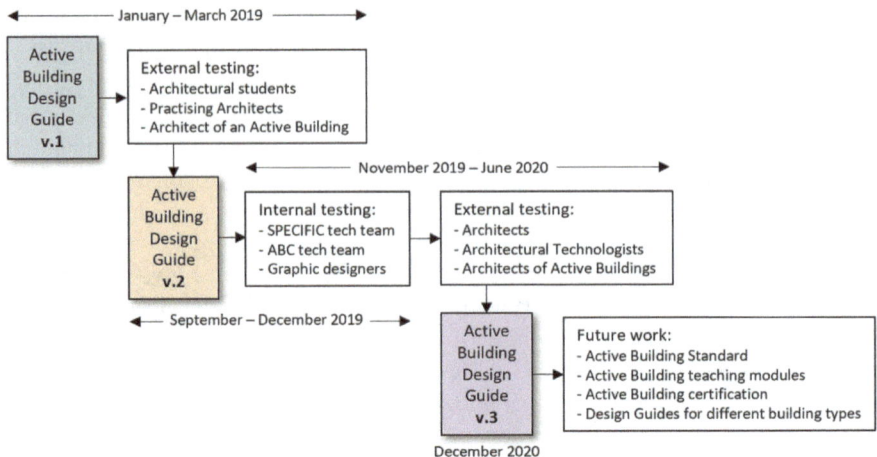

Fig. 6 ABDG development and testing

A *probabilistic* or *random* sampling strategy [30] was used in the pilot project, with the aim of achieving a sample representative of the larger population of Architects Participants included those who signed up to workshops, rather than selecting individual participants. As climate change and the need to reduce the energy consumption of buildings are high on the UK Government agenda and that of construction industry stakeholders, the author has been approached by Architects known to her and bodies, such as the Royal Society of Architects in Wales (RSAW) [31], requesting workshops on the design of Active Buildings. This suggests there is a desire within the UK construction industry to seek ways to reduce the energy consumption of buildings, potentially by incorporating the Active Building concept, and an appetite for some guidance on how to do this. Table 1 illustrates the proposed sampling plan for testing the ABDG in the final phase of the D.SBE project [18].

As illustrated in Table 1, before testing the ABDG externally to the ABC with UK construction industry stakeholders, the author will obtain validation from the internal teams at both the SPECIFIC [10] and the ABC [1]. After any refinement to version two of the ABDG from the internal testing, testing will take place with external stakeholders, primarily Architects in professional practice, and students of both architecture and architectural technology, who are the main target audience for the Design Guide. Feedback will be sought on aspects such as technical content accuracy; level of technical content provided; usability of the ABDG, including accessibility, intelligibility, visual appearance, structure and graphical representation; and use of Active Building case studies.

An *Active Building Overlay* to the RIBA Plan of Work [32] has also been developed to form part of the ABDG, as a way of ensuring Active Building elements are considered at every stage in the design and delivery of Active Building projects. Figure 7 illustrates how this will be presented within the ABDG.

Table 1 Sampling plan for testing the ABDG

Participant group	Method
Internal testing	
SPECIFIC technical team	Workshops
Expert knowledge of technical aspects of Active Buildings	Individual meetings
ABC technical team	Workshops
Expert knowledge of technical aspects of Active Buildings	Individual meetings
	Use on Active Building projects
Graphic designers (SPECIFIC & ABC)	Discussions
Expert knowledge on presentation of documents	Individual meetings
	Individual meetings
External testing	
Architects	Workshops
Expert knowledge of architectural design, aesthetics and building fabric	On Active Building projects
Architectural technologists	Workshops
Expert knowledge of detailing building designs, including building fabric and integration with building services	
Building services engineers	Workshops
Expert knowledge of building services design	On Active Building projects
Building contractors	Workshops
Expert knowledge in construction and installation of services	On Active Building projects
British standards institution	Development of a scoping document to determine potential
Expert knowledge of building standards	
BREEAM	Workshop to discuss a scoping exercise
Expert knowledge of environmental assessment methods and criteria for building projects	

4 Results

The results of the pilot project were discussed in a paper entitled *Active Buildings in Practice*, which was presented at the eleventh International Conference on Sustainability in Energy and Buildings in 2019 and published in Smart Innovation, Systems and Technologies, Volume 163 [18].

These results are summarised in Table 2.

The results from the pilot project will be used to refine the ABDG in the next phase. From the results, the author has determined the need for separate documents

Stage	Core Objectives	Active Building Centre Input
0 **Strategic Definition**	Identify client's Business Case and Strategic Brief and other core project requirements	Define Active Building (AB) requirements: Key considerations and Design Guide
		Provide advice on passive design, such as siting and orientation
		Set objectives for Project Delivery Team (PDT) on Active Building elements
		Identify key stakeholders, e.g. building owners, facility management team (FMT)
		Review data and lessons learnt from previous AB projects to inform PDT
		Additional duties: Ensure AB requirements and access to data in accordance with AB monitoring specifications are included in contractual documents
1 **Preparation & Brief**	Develop Project Objectives, including Quality Objectives and Project Outcomes, Sustainability Aspirations, Project Budget, other parameters or constraints and develop Initial Project Brief, undertake Feasibility Studies and review of Site Information	Provide feasibility study on proposed site(s) incorporating constraints - financial, site, other
		Provide integrated PDT advice/support
		Provide high level data monitoring and performance specifications
2 **Concept Design**	Prepare Concept Design, including outline proposals for structural design, building services systems, outline specifications and preliminary Cost Information along with relevant Project Strategies in accordance with Design Programme. Agree alterations to brief and issue Final Project Brief	Develop initial low complexity concept model on early design scheme(s)
		Provide report on early design(s) and recommendation for steps to enable AB elements
		Provide information to support early LCC assessment to support the aim to reduce whole life costs
		Provide information on AB technologies to support Design and Access Statement (DAS) for planning application
3 **Developed Design**	Prepare Developed Design, including coordinated and updated proposals for structural design, building services systems, outline specifications, Cost Information and Project Strategies in accordance with Design Programme	Provide advice on renewable energy technologies using evidence from concept model
		Assist the development of the energy strategy in conjunction with the design team
		Provide performance specifications for an AB to PDT
4 **Technical Design**	Prepare Technical Design in accordance with Design Responsibility Matrix and Project Strategies to include all architectural, structural and building services information, specialist subcontractor design and specifications in accordance with Design Programme	Develop detailed building physics/dynamic thermal model
		Review design information and Mechanical, Electrical & Plumbing (MEP) strategy
		Review specifications developed by PDT to ensure Active Building elements included
		Provide monitoring specifications
		Provide naming schema for Building Management System (BMS)
		Provide information on AB technologies to support design development
5 **Construction**	Offsite manufacturing and onsite Construction in accordance with Construction Programme and resolution of Design Queries from site as they arise	Provide/deliver toolbox talks for on-site inductions, management and commissioning
		Undertake site visits/inspections to support the delivery of the AB
		Work with BMS installer to ensure naming schema used properly and monitoring in place
		Work with MEP installers to ensure correct installation of equipment
		Advice on commissioning and testing to ensure all systems working properly before building is signed off and handed over
6 **Handover**	Handover of building and conclusion of Building Contract	Provide information for Operation & Maintenance (O&M) manuals
		Review the whole design and construction process and capture lessons learnt
		Undertake post-project review workshop with all stakeholders
		Support handover workshops with building owners/occupiers/FMTs
7 **In Use**	Undertake In Use services in accordance with Schedule of Services	Assist with monitoring of building performance
		Capture data in National Active Building Evidence Base (NABEB) and assess optimised performance of systems
		Provide updated information for O&M manuals as necessary
		Assist the PDT with Life Cycle Cost reporting
		Support Post Occupancy Evaluation (POE) with occupants and FMT

Fig. 7 Active building overlay to the RIBA plan of work

for students and designers. In addition to the ABDG for Architects in practice, she will also develop an Active Building Student Guide, which will contain more detailed information than the ABDG for Architects in practice. Version 2 of the ABDG is currently underway and is being developed as a more visual document, with more diagrams and less text than version 1. The author is also developing a detailed case study of the Active Classroom, which includes considerations at each of the RIBA Plan of Work stages [32]. This will provide a supplementary document to the ABDG.

Table 2 Summary of findings from pilot project

1. Architecture students (10 participants)	• Found the document helpful • Understood the definition and what constitutes an Active Building
2. Architects in practice (10 participants)	• Understood what an Active Building is • Found content was good but would prefer less text & more graphics • Wanted to see case studies included • Wanted information on costs and comparisons of technologies to advise their clients
3. Architect of an active building project (1 participants)	• Understood what an Active Building is • Found content was good but would prefer less text & more graphics • Wanted to see case studies included

5 Discussion

To date (2020), the author has determined that design guidance is needed to enable the UK construction industry to adopt the Active Building concept. Testing of the ABDG in the pilot project and the literature review undertaken by the author have provided a strategy for developing the ABDG, based on a combination of short paragraphs of text, diagrams, images, tables and case studies, all based on the Active Building principles. The Action Research approach [29] which is being adopted enables the ABDG to be tested in its various versions (two thus far), reviewed, updated and tested again, which is one output of the author's D.SBE project. Based on the pilot project, the interest in the Active Building approach the author has experienced since designing the first Active Building (the Pod) in 2014 [10], and the proposed building demonstrator projects that the author in her role at the ABC will have design input into (e.g. Active Homes Neath [33]), and it is anticipated that there will be plenty of opportunities for testing the ABDG further.

Figure 6 sets out the steps to the completion of both the testing and publication of the ABDG. Table 3 sets out the steps to the completion of the author's D.SBE project.

Table 3 Steps to completion of D.SBE project

Date	Activity
January–March 2020	Refine ABDG and other supporting documents and publish ABDG v2
January–June 2020	Test ABDG and other supporting documents with architectural designers and architecture students
June–December 2020	Develop ABDG v3
December 2020	Publish ABDG v3
January 2021 onwards	Develop further ABDG documentation

The recommendations from the completion of the work documented in this paper are that use of the ABDG and the associated Active Buildings designed using the ABDG will enable the construction industry to deliver low energy buildings that incorporate energy generation, storage and intelligent controls to support local and national grid networks, by managing the building's interactions with the grids. Deploying the Active Building concept, as set out in the ABDG, on building projects will help reduce the energy consumption of buildings in the UK and their associated carbon emissions, contributing to mitigating climate change.

The limitations to the completion of the work documented in this paper have been the ability to recruit participants to test the developing ABDG with. Time and geographical constraints have led to the decision by the author to restrict the testing of the ABDG to Architects and architectural students in Wales.

6   Conclusions

This paper has discussed the development of an ABDG which will help enable the construction industry to adopt the Active Building concept on building projects, helping to achieve the UK Government's mission to halve the energy use of all new buildings by 2030, as set out in the Transforming Construction Challenge [23]. Halving the energy consumption of buildings will also help meet the targets set out in the 2019 amendment to the Climate Change Act (2008) [34] to reduce the UK's greenhouse gas emissions (GHGs) by 100% below 1990 levels by 2050. The author believes that the ABDG provides one enabler to the adoption of the Active Building concept by the construction industry.

The ABDG developed by the author provides an opportunity to collate all the information available to date on Active Buildings together in one reference document that can be utilised by Architects and other building design professionals in the design and development of Active Building projects. It will provide guidance on the elements that should be considered when designing Active Buildings and examples of how these might be achieved. Once complete, the ABDG can also be used as a basis for further work, such as developing training modules, standards and certification schemes. Further work could also include developing individual design guides for different building typologies—residential, commercial, educational, industrial, leisure and healthcare—and also for retrofit.

The next steps are to refine the ABDG; develop additional associated documents, such as the Active Building Student Guide; test these documents with architectural students and Architects in practice; and use feedback from testing to refine the developing documents further as an output to the D.SBE project.

One limitation of this project is the ability to test the ABDG with a large population of architectural practitioners. During the next stage of the project, the author will develop a sampling strategy to determine an appropriate number of participants needed for a meaningful study, depending on the population of Architects in Wales. It will not be possible to test in the whole of the UK, due to time and geographical constraints. Another limitation is that the author has identified a number of documents that would be useful to develop to support the roll-out of the Active Building concept, which will not be able to be developed during this research project. These are listed as future work in Fig. 6.

Acknowledgements The author wishes to acknowledge the support of UK Research and Innovation (UKRI) who fund the ABC, as part of the 'Transforming Construction' Challenge within the Industrial Strategy Challenge Fund (ISCF). SPECIFIC is part-funded by the European Regional Development Fund (ERDF) through the Welsh Government, and by Innovate UK and EPSRC. The author wishes to acknowledge the support of all funders. In addition, she wishes to thank her Director of Studies and supervisor from Cardiff Metropolitan University: Dr John Littlewood and Prof George Karani, and advisers and mentors at the ABC and SPECIFIC at Swansea University.

References

1. ABC (2019) Active building definition. Cited at: https://www.activebuildingcentre.com/active-buildings/. Accessed 16th Jan 2020
2. BEIS (2017) Industrial strategy: building a britain fit for the future. Cited at: https://www.gov.uk/government/publications/industrial-strategy-building-a-britain-fit-for-the-future. Accessed 13th Oct 2019
3. Shaw R, Attree M, Jackson T (2010) Developing electricity distribution networks and their regulation to support sustainable energy. Energy Policy 38(10):5927–5937. Cited at: https://doi.org/https://doi.org/10.1016/j.enpol.2010.05.046.Accessed 15th Jan 2020
4. Energy UK (2016) Electricity charging arrangements report. Cited at: https://www.energy-uk.org.uk/publication.html?task=file.download&id=5903. Accessed 13th Oct 2019
5. Clarke J (2019) From SPECIFIC to active building centre. Cited at: https://designingactivebuildings.blog/2019/10/03/from-specific-to-active-building-centre/. Accessed 20th Jan 2020
6. SPECIFIC (2018)[1] The active office. Cited at: https://www.specific.eu.com/assets/downloads/casestudy/Active_Office_Case_Study.pdf. Accessed 22nd Feb 2019
7. Searle J (2019) The active office: one year on… Cited at: https://www.specific.eu.com/the-active-office-one-year-on/. Accessed 16th Jan 2020
8. The Solarblogger (2017) The carbon intensity of UK Grid Electricity. Cited at: https://www.solarblogger.net/2017/11/the-carbon-intensity-of-uk-grid.html. Accessed 15th Jan 2020
9. BEIS (2018)[1] News Story: Swansea receives £36 million UK government funding for its clean energy tech breakthrough. Cited at: https://www.gov.uk/government/news/swansea-receives-36-million-uk-government-funding-for-its-clean-energy-tech-breakthrough. Accessed 12th Dec 2018
10. SPECIFIC (2019)[1] The pod. Cited at: https://www.specific.eu.com/assets/downloads/casestudy/CaseStudy_POD.pdf. Accessed 13th Oct 2019
11. SPECIFIC (2018)[2] The active classroom. Cited at: https://www.specific.eu.com/assets/downloads/casestudy/Active_Classroom_Web_Case_Study.pdf. Accessed 22nd Feb 2019
12. Business News Wales (2018) RICS awards 2018–winners announced. Cited at: https://businessnewswales.com/rics-awards-2018-winners-announced/. Accessed 13th Oct 2019
13. Cardiff Metropolitan University (2020) Doctor of sustainable built environment (D.SBE). Cited at: https://www.cardiffmet.ac.uk/research/Pages/CSAD-DSBE.aspx. Accessed 10th Jan 2020
14. Morgan J, Littlewood JR, Wilgeroth P, Jones P (2017) Testing and validation of building as power station technologies in practice, to maximise energy efficiency and user comfort and minimise carbon emissions. Sustain Energy Build Res Adv J 6(1):20–28

15. Creswell JW, Plano Clark VL (2017) Designing and conducting mixed methods research. 3rd edn. Cited at: https://us.sagepub.com/en-us/nam/designing-and-conducting-mixed-methods-research/book241842. Accessed 10th Jan 2020
16. Clarke J (2018) Buildings as power stations. A case study: the active classroom. Delta T. pp 28–29
17. Clarke J Littlewood J Wilgeroth P Jones P (2018) Rethinking the building envelope: building integrated energy positive solutions. WIT Trans Built Environ 183:151–161. Cited at: https://www.witpress.com/elibrary/wit-transactions-on-the-built-environment/183/37068. Accessed 15th Jan 2020
18. Clarke J, Littlewood JR, Jones DP, Worsley D (2019) Active buildings in practice. Smart Inn Syst Technol 163(47):555–564
19. Ijeh I (2015) The appliance of science. Building. Issue 05:38–40
20. Premier Construction (2018) The active classroom premier construction. Issue 25(5):68–70
21. Smit, J (2019) Carbon control goes back to the office drawing board. Cited at: https://www.ribaj.com/products/government-limits-zero-carbon-plans-industry-takes-sustainability-initiative. Accessed 18th April 2019
22. HM Government (2013) Construction 2025: industrial strategy: government and industry in partnership. Department for business, innovation and skills. Cited at: https://www.gov.uk/government/publications/construction-2025-strategy. Accessed 17th Feb 2015
23. BEIS (2018)[2] The grand challenge missions. Cited at: https://www.gov.uk/government/publications/industrial-strategy-the-grand-challenges/missions. Accessed 12th Dec 2018
24. Pelsmakers S (2015) The Environmental design pocketbook, 2nd edn. RIBA Publishing, London, UK
25. Evans HMA (2016) How buildings work. RIBA Publishing, RIBA Enterprises, Newcastle-upon-Tyne, UK
26. Clegg P, Bradley K, Fielden R, Gething B (2007) The environmental handbook. Right Angle Publishing Ltd., London, UK
27. Roaf S (2013) Ecohouse: a design guide, 4th edn. Routledge, Oxon, UK
28. City and County of Swansea (2014) Places to live: residential design guide. Cited at: https://issuu.com/swanseacitycouncil/docs/residential_design_guide_-_adopted. Accessed 25th Sept 2019
29. Saunders M, Lewis P, Thornholl A (2012) Research methods for business students. Pearson Education Limited, Harlow, UK
30. Groat L, Wang D (2013) Architectural research methods. Wiley, Hoboken, New Jersey, USA
31. RIBA (2019) Royal society of architects in Wales. Cited at: https://www.architecture.com/my-local-riba/rsaw. Accessed 20th Oct 2019
32. RIBA (2013) RIBA plan of work 2013. Cited at: https://www.ribaplanofwork.com/Download.aspx. Accessed 20th Oct 2019
33. Pobl (2019) Active homes neath. Cited at: https://www.poblgroup.co.uk/activehomes/. Accessed 16th Jan 2020
34. Legislation.gov.uk (2019) The climate change Act 2008 (2050 Target Amendment) Order 2019. Cited at: https://www.legislation.gov.uk/ukdsi/2019/9780111187654. Accessed 16th Jan 2020
35. SPECIFIC (2019)[2] Active buildings. Cited at: https://www.specific.eu.com. Accessed 13th Oct 2019

Chapter 3
Sustainable Housing Solutions

Pablo Gonzalez, Lucia Rodriguez, Soledad Cormick, Lucas Daher, Edgardo Suarez, Maria Benítez, Hercilia Brusasca, and Aantonella Caballero

Abstract Based on our experience of more than 20 years in housing construction for vulnerable families, from the IMV and IH, we have assumed the commitment to project integral housing solutions, incorporating the principles of sustainability in its three dimensions: economic, social and environmental. Our design premise is based on reducing the energetic demand in the houses, projecting in bioclimatic conditions which correspond to the convenient site for the facilities, incorporating the active participation and the multidisciplinary accompaniment to the families from the very beginning. The Institute of Building Sustainability (ISE) has evaluated the first houses in their energetic and environmental performance, allowing us to learn about the functioning of the homes and be able to meet the minimum requirements in terms of comfort.

1 Introduction

From the public policy and referring to the previous municipal, provincial and national housing construction programmes, a physiognomy of identical housing units is observed in constructions which do not respond to the economic, environmental and social dimensions, stigmatizing the social welfare dwelling as a repetitive element, without considering the participant's own identity. Thus, families receive a problem and not a solution to their housing issue.

P. Gonzalez (✉) · L. Rodriguez · S. Cormick · L. Daher · M. Benítez · H. Brusasca · A. Caballero
Instituto Municipal de La Vivienda E Infraestructura Habitacional (I.M.V. e I.H.), Municipalidad de Villa María, Córdoba, Argentina
e-mail: estropablo@gmail.com

E. Suarez
Instituto de Sustentabilidad Edilicia (ISE), Colegio de Arquitectos Provincia de Córdoba, Córdoba, Argentina

© The Author(s), under exclusive license to Springer Nature Singapore Pte Ltd. 2021
R. J. Howlett et al. (eds.), *Emerging Research in Sustainable Energy and Buildings for a Low-Carbon Future*, Advances in Sustainability Science and Technology,
https://doi.org/10.1007/978-981-15-8775-7_3

Our work is carried out within the Eva Perón Housing Program with the incorporation of an interdisciplinary perspective, reconsidering social housing as an integral dwelling solution for families located in high-risk communities.

In terms of economic and environmental dimensions, it is necessary to submit the result of this process to the building sustainability assessment entity, the eSe protocol of the Building Sustainability Institute (CAPC), in order to allow continuity to the improvement process.

This model shift seeks to influence future housing policies, oriented towards the construction of sustainable cities, and resilient to climate change.

The modern concept of "sustainability" implies the development of socio-ecological systems reconfiguring the three dimensions (environmental, economic and social) of the so-called sustainable development.

2 Objectives

- Develop a quantitative approach to the housing deficit from an integral conception of housing as a process and as habitat, incorporating alternative constructive variants and systems, applying sustainable criteria in the design and development and involving social actors in the different stages and production processes of habitat.
- Overcome traditional ways of conceiving housing policies based on assumptions of homogeneity of their recipients and centralized implementation, advancing in the construction of methodological instruments that facilitate the recognition of specific and heterogeneous needs and situations, as well as the strengthening local community management and social connection.
- Strengthen the community by means of providing tools and strategies (work co-ops, home gardens, job training) which diminish their work and food vulnerability.
- Diagnose the final performance of the use of resources, particularly of the primary energy demanded and promote the improvement of their use, using design tools first and technological ones later. (The figure shows the assigned order of strategies and their impact.)

Costs and impacts of efficiency strategies (Fig. 1).
MITIGATION Renewable energies to reduce the impact of fossil fuels.
EFFICIENCY Devices/support systems to improve efficiency/reduce consumption.
Enclosure/passive design to establish efficiency and reduce demand.

Costs and impacts of efficiency strategies

Fig. 1 Efficiency strategies impacts

3   Background Information and Related Works

The Housing Program Eva Perón (Programa de Viviendas Eva Perón) is based on the reality of social housing which affects a considerable group of families located in different city neighbourhoods and is part of precarious places living under poor conditions.

Since 2016, other types of houses have been considered, where families become part of the project together with the development of the design alongside the architect, as well as each dwelling being sustainable and much better at saving energy.

This paradigm shift makes each member of the families active characters of their own change, helping them step out of the shadows of being a mere recipient of public policies.

Once the work was completed, an audit of building performances of the two finished houses was carried out, consisting of a diagnosis of the 33 variables (18 calculation engines), which make up the eSe/building sustainability labelling protocol, developed by the iSe of the College of Architects of the province of Córdoba.

The eSe protocol of the iSe is an ordered set of objectives, requirements and strategies that, as an operational instrument, allows the improvement and sustainable integral efficiency of the design, manufacture and effective use of buildings. The environmental factor and economic and social aspects are especially taken into account in this system, in order to find a better balance in the framework of sustainable development.

4 Project Two Houses in Los Olmos Neighbourhood

Its main aim is to be energy efficient by taking advantage of natural resources, fostering environmental care, promoting waste reduction and generating the possibility of developing an economic activity within the same architectural work. The main principles to be mentioned are housing with universal accessibility, DVH openings, cross ventilation and natural light, eaves for solar radiation control, home/community gardens and vertical gardens, a compost of organic waste for the home gardens, water heating solar tank, energy-efficient LED lighting, Russian stove, grey water recovery, rainwater collection and storage to use when potable water is not necessary and double-flushing toilets.

The IMV and IH subsequently perform the verification of results related to the adaptation of families to the principles of sustainable housing.

Then the performance building diagnosis is implemented, applying the "eSe" diagnostic protocol, aimed at the final improvement of the environmental performance of both buildings and the micro-sector, where members can apply their skills in successive possible interventions (Fig. 2).

4.1 Work and Sustainability Cooperatives

Work co-ops participated in the construction and development of the houses, generating genuine job opportunities, incorporating working habits through the learning of trades.

Co-op groups are intrinsically a sustainable and hands-on company model. They encourage job safety and improved working conditions; they foster democratic principles and knowledge and social inclusion. Therefore, they are well equipped

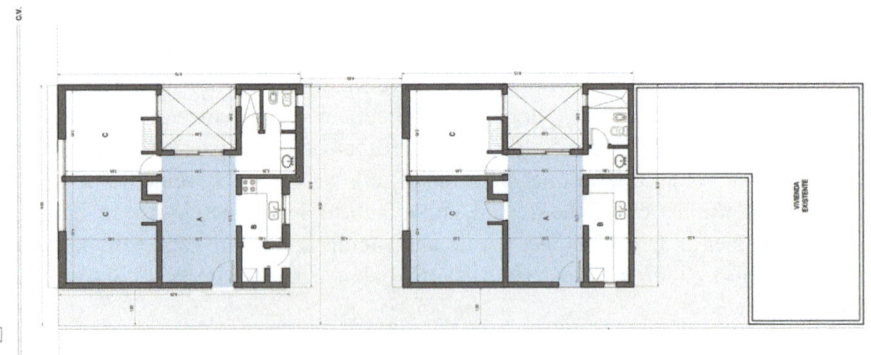

Fig. 2 General plant plan

to contribute to the triple balance of economic, social and environmental objectives of sustainable development, among other reasons because they are companies committed to achieving the economic progress of their partners, while attending to their socio-cultural interests and protecting the environment.

5 Economic Dimension

We owe ourselves the responsibility of defining the economy as a set of integral and inclusive production models that take into account environmental and social variables. This definition is the "green economy" that produces low carbon emissions, uses resources efficiently and is socially inclusive.

The social cost is composed of: the alternative cost of the resources used in the production of a good, plus the loss of well-being and the increase in costs that the production of that good may cause to any other productive activity. To lessen the impact, different alternatives were sought in food generation, health management and the decrease in the demand for the use of resources, particularly water and energy.

The IMV and IH decided to incorporate said economy in support of all their projects, since it, in addition to recognizing and demonstrating the value of natural capital, invests in it and seeks to increase it in order to give society sustainable economic progress.

With respect to the projects carried out, in the construction of the first two sustainable homes, the amount paid for each of the homes was € 53,077.11. (€ 780.55 per square metre).

This cost represents a 23% increase in price when compared to the construction cost of more traditional dwellings.

6 Results

Within this context, the results of the audit of building performances were the following:

6.1 Potable Water Demand

Audit: 239 lt/person/day. With strategies for improving device replacement, a 44% reduction in demand is achieved: 136 lt/person/day, below the sustainable reference base (100 lt/person/day).

6.2   Effluent Septic Systems

Audit: 958 lt/day impulsion of sewerage effluents. Considering the potential of reduction in the potable water demand, the quantity of sewerage liquid poured to the septic system can be reduced an average of 413 l/day (43% saving).

6.3   Management Rainwater System/Sigall

Each house has a usable recollection capacity of 10,000 lt per month. There is a tank system of 3000 lt, on deck. The resource provision for the efficient irrigation is proposed (trickle irrigation) suggested by the *Cobertura Vegetal Equivalente/Cve (Equivalent Vegetable Cover)*.

6.4   Equivalent Vegetable Cover/Cve

The audit detected that the equivalent environmental correction generated by the project is 170 m^2 equivalent to 25 large shrubs, generating an equivalent return of 70%.

6.5   Superficial Reflectance Index/Irs

The final objective of this index is to quantify and then reduce the impact of overheating of the built environment, associated with the heat island effect (ICU). The IRS index of the existing housing is 35.75 BAD. With IRS improvement strategies, such as designing "cold" covers, which involve the use of "cold" materials or high reflectance index, an improvement to 74.52% is achieved: GOOD.

6.6   Greenhouse Gas Emissions/Gei

Audit: 26, 11 kg CO_2/m^2 year, the integration with renewables represents a 0, 02 kg CO_2/m^2 (2%) saving, because of the solar water heating system use.

6.7 *Energetic Demand of Primary Energy/Ipe*

Modelling and simulation of energetic demand and hygrothermal adaptive comfort ran under the calculation engine of the transient dynamic regime, ECOTEC@, a system developed by Western Australia University, where there is register of the thermal balance of the building, and of each specific area, without restrictions related to the building geometry or the number of internal thermal areas which can be analysed in a simultaneous way in multi-stationary regime. The mathematical protocol of thermal analysis used is the admittance method of Chartered Institute of Building Services Engineers (CIBSE Admittance Method).

The real environmental and context conditions were established from a private climatic data file (according to the geographic location of the audit location), corrected from the climate base EPW (Energy Plus Weather Format). The model is constructed with a general orientation and precise environmental conditions to develop a close to the real bioclimatic evaluation.

The dynamic simulations (8760 h) are carried out considering zoned models by indoor areas (thermal areas) which allow to shape the monthly and annual energetic demand, establishing levels of seasonal thermal comfort (in correspondence to the local regulations).

The thermal areas are defined as such among those facilities which present divisions limited by walls, partitions and carpentries or watertight compartments. The non-thermal areas are determined likewise, and they include obstacles that directly influence the percentage of shadow over each building (Fig. 3).

In the first diagnosis made, there is an attempt to emulate the existing infiltration conditions and air movement. In this case, the calculation base is established as 0, 5 ACH (air changes per hour) and an air sensibility speed of 0, 25 m/s.

Afterwards, in the process of the improvement proposal, these values are arranged to simulate conditions of air renewal for convective ventilation systems, not only for the traditional and original building model but also for the one proposed to improve it.

Finally, two comfort bands for each thermal cycle were defined: from 18 to 26 °C for the refrigeration cycle and from 16 to 24 °C for the heating one.

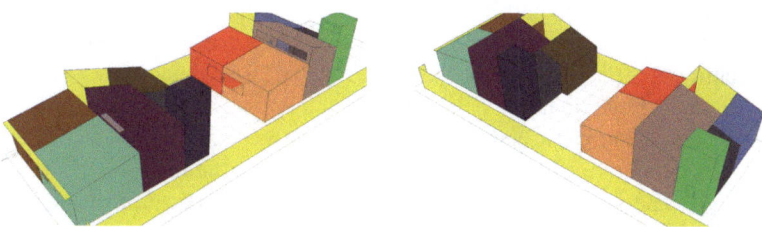

Fig. 3 Thermal zones

6.8 Simulation Results—Diagnosis for Existing Constructive Scenario.

Average results per house	
Energetic demand for thermal load in Kwh/year	
Cooling: 2.07685	Maximum load: 6, 80 Kw at 4 p.m. December 01st. 01
Heating: 4.10008	Maximum load: 9, 53 Kw at 8 a.m. July 14th
Total: 6.17744 Kwh/year	
Energy intensity for thermal load in Kwh/M²/year	
Cooling: 15, 28 Heating: 30, 15 Total: 45, 42 Kwh/M²/year	

6.9 Diagnosis of the Annual Hygrothermic Comfort

For comfort ranges in both thermal cycles (cold: 16–24 °C-heat: 18–26 °C), the audit shows 1440 °C annual hours of discomfort, which are reduced to 1013 °C hours (29, 70% improvement) with the implementation of the improved systems.

7 Conclusions

Environmental performance improvements and the use of resources have a direct impact on the economic component.

The building sustainability audit findings of the dwellings presented conclude that the design strategies used to reflect a good final performance, in general, highlighting the variable primary energy demand, whose final *energy performance index (EPI)* is 92.53 kWh/m2/year (taking as a reference that the national average for this building typology is in between 250 and 300 kWh/m²/year), placing itself on the upper third of the reference classification and presenting, at the same time, a good and considerable improvement potential, easily reachable after the detailed detection of each type of energy demand discriminated for the annual cycle.

49% of the final energetic demand depends on the thermal load for acclimatization, and this one depends on the external casing and the efficiency of the ventilation systems used. The impact of lightning is minimum (7%) because of the strategies designed to cause this.

It is necessary to implement a policy to change the technology of the domestic electrical appliances (especially refrigerators) because they consume 43% (2600 kWh/year).

The improvements in the environmental performance and the use of the resources impact directly on the economic component.

When social housing incorporates the other two components of sustainability (environmental and economic), the possibilities for continuous improvement quickly materialize.

The participation of the users from the initial stage (selection of the land) reduces the distortions produced by the non-appropriation of the built environment, by the users.

We are convinced that it is necessary to reconsider and reflect upon the habitat, the world that we encounter and must leave to future generations.

From the IMV, we commit ourselves responsibly to implement actions that allow to provide housing with greater quality and integrity, caring for the environment and helping create a more environmental friendly place for future generations. With correct and applied policies, we believe that the environmental crises we are suffering nowadays may be reverted in the near future.

Part II

Chapter 4
Future Energy-Related House Renovations in Sweden: One-Stop-Shop as a Shortcut to the Decision-Making Journey

Georgios Pardalis, Krushna Mahapatra, Brijesh Mainali, and Giangiacomo Bravo

Abstract Based on an online survey, this paper analyzes the attitude of detached house owners in Sweden toward future renovations and their perception over a one-stop-shop (OSS) service for deep renovation of these dwellings. With the aid of a house owners' renovation decision-making journey for renovation, personal and contextual variables have been analyzed to identify those house owners having renovation plans in the near future, what they are going to renovate, and which needs to lead them to that decision. Furthermore, we examine if there is an interest in OSS concept and the factors affecting positively or negatively the choice for such a concept. Results suggest that deep renovation is not yet prioritized. The priority for house owners is to change specific components of their dwelling and follow a step-wise approach. Aesthetic renovations are high on the agenda, with some structural and energy-related renovations following them. House owners between 29 and 49 years of age are those mostly interested in more comprehensive renovations. The OSS concept appears to be interesting to a number of house owners capable to verify a business potential. House owners up to the age of 45 years, with dwellings built from 1960 and above and with environmental awareness, are the market segment that can act as early adopters of the OSS concept. When it comes to the decision-making journey for renovations, house owners' future plans, and the factors affecting their

G. Pardalis (✉) · K. Mahapatra · B. Mainali
Department of Built Environment & Energy Technology, Linnaeus University, 35195 Växjö, Sweden
e-mail: georgios.pardalis@lnu.se

K. Mahapatra
e-mail: krushna.mahapatra@lnu.se

B. Mainali
e-mail: brijesh.mainali@lnu.se

G. Bravo
Social Studies/Centre for Data Intensive Studies & Applications, Linnaeus University, 35195 Växjö, Sweden
e-mail: giangiacomo.bravo@lnu.se

© The Author(s), under exclusive license to Springer Nature Singapore Pte Ltd. 2021
R. J. Howlett et al. (eds.), *Emerging Research in Sustainable Energy and Buildings for a Low-Carbon Future*, Advances in Sustainability Science and Technology,
https://doi.org/10.1007/978-981-15-8775-7_4

choice for an OSS provider, we can claim that OSS can act as a guide for house owners from the early stages of their decision-making journey and provide them with a shortcut that will make this journey more secure, while triggering renovation decision of greater extent. In terms of financing, incentives related to energy performance are also suggested as means that could boost greater interest for more comprehensive renovations.

Keywords Future renovations · Energy-related renovations · One-stop-shop · House owners · Renovation journey

1 Introduction

The European Union has a goal of 32.5% energy efficiency in 2030 compared to the levels of 2005 as referred to the Directive 2018/844/EU [1]. Sweden has set an ambitious cross-sectoral target of reducing energy intensity 20% by 2020 compared to 2008 levels and 50% by 2030 compared to the levels of 2005 [2]. For the building sector, Sweden has a national goal to reduce energy consumption by 20% compared to the 1995 level by the year 2020 [3], a goal that seems ambitious considering that several EU member states have already revised their energy efficiency targets to 16.9% by the same year [4]. The Swedish residential sector could be a major contributor to achieve the national goals for reduction of energy consumption, as it is responsible for 22% of the total energy consumption, from which 12% coming from single-family houses [5]. Out of 4.7 million residential dwellings, 51% (2.4 million) are one- or two-family houses (from here onward "detached houses"), and they account for 293 million square meters of floor area, which is larger than that of multifamily houses [6].

According to Statistics Central Bureau (SCB) in Sweden, 86% of the one- and two-family dwellings are about 30 years old. They have poor energy standard and need of renovation. The most common practice in Swedish renovation market is to renovate kitchens and bathrooms (aesthetic type of renovations) as was observed in a recent survey of owners of detached houses in Kronoberg region of Sweden [7, 8].

The adoption of energy-efficient measures or deep renovation for improvement of the energy performance is rather low due to various socioeconomic barriers [9]. Also, house owners in general are not aware of where to seek information regarding deep renovation or which craftsmen offer such a service [10]. The renovation market is dominated by small and locally based craftsmen-owned companies who are interested to offer individual solutions and sell their own product, which leads to lack of trust between them and the house owners [11]. Moreover, the current building code in Sweden (BBR26) obliges the energy performance of a deep renovated house to reach the levels of a newly built one [12]. Regardless of location, the investment cost for such a renovation is relatively high, and this can act as a demotivating factor for many house owners. Additionally, the existing tax incentive does not differentiate between deep renovation and aesthetic renovation [13]. Moreover, the financing mechanisms that could boost deep renovations (green loans) are yet to be developed.

To address those challenges, Mahapatra et al. [11] proposed the one-stop-shop (OSS) concept, in which a single actor could offer a full-service renovation package to house owners. One-stop-shop is a product-service system (PSS) concept, which could address some of the factors that prevent the house owners to renovate their dwellings [14]. In this concept, a single actor coordinates all the involved actors in the renovation process to offer a comprehensive package on energy-efficient reno- vation. In that concept, house owners deal with a unique contact point, and partic- ipating actors work together in a way that redefines their activities and increases their resource efficiency [15]. House owners would receive consulting services for renovation, including energy audit and recommendations for upgrades, facilitation in getting building permits (where it is required), renovation packages with finan- cial schemes based on house owners specific needs and financial situation, supply of quality material and technical re-sources in the value chain, and post-renovation quality checks and guarantees. With OSS, house owners receive a guided journey throughout their renovation and get a renovated dwelling that truly satisfies their needs.

OSS concept has been emerging in several parts of Europe [16] and in Scandi- navian countries, such as Norway [17] and Denmark [18]. In Sweden, the concept is yet to be tested, although the turnkey ("totalentreprenad" in Swedish) concept exists in the construction of all types of new buildings and renovation of multifamily buildings.

In this study, we analyze the future plans of Swedish house owners to renovate their houses. We analyze the underlying reasons leading them to take that decision, the influence of their socioeconomic background, and their preferred type of renovation. Furthermore, we examine to which extent they have knowledge of the renovation services offered and their perceptions over a full-renovation package as envisioned in the OSS concept.

2   Background and Related Work

The low level of adoption of energy-efficient measures by the house owners when they renovate their dwellings is a subject widely examined by literature. A significant number of studies identified factors motivating or preventing house owners toward energy-efficient solutions. They are categorized in related literature as economic [19–22], behavioral [19, 23–27], social [28–30], regulatory [31], and factors related to the physical condition of the dwelling [19, 28, 31–35]. To better understand how all those different factors affect the level of adoption of energy-efficient measures, it is important to understand the "journey" of house owners in deciding to renovate their dwelling.

2.1   The Context of Renovation Decisions

The changing demands of domestic life and the need to adapt to those changes influence the house owners in their decision to renovate and the specific improvements they chose. According to Karvonen [36], "Domestic retrofit is not an activity of changing a house from poor energy performance to exceptional energy performance, but an intervention into the rhythms of domestic habitation". Judson and Maller [28] observed that adoption of energy efficiency measures is usually combined with improvements of other parts of the dwelling. Hence, the decision to adopt energy efficiency measures is not independent or static, but a journey over time in a broader context of renovating the property.

The innovation-decision model of Rogers [37], which has been applied in different contexts like adoption of heating systems [38–40] and solar photovoltaic systems [41, 42], posits that decision to adopt an innovation passes through five stages. The upper part of Fig. 1 represents those stages starting from initial awareness to a final decision. In our conceptual framework, we have adapted those stages to the renovation journey of house owners, which is presented in the lower part of Fig. 1. The decision-making journey of a house owner starts from the point of reflection over the dwelling's situation (step 0). On that step, house owner is not thinking that it is time to renovate. A previous study [41] has found that satisfaction with the physical condition, thermal performance, and aesthetics of the dwelling are the reasons for that decision. When the house owner.

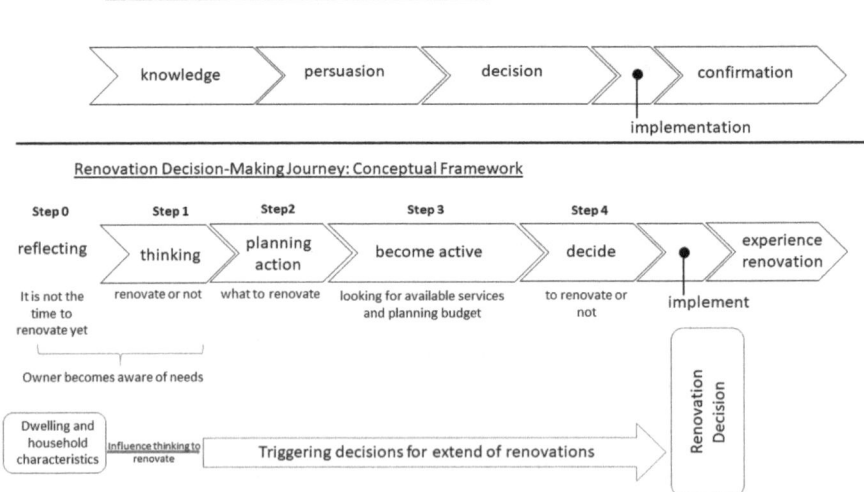

Fig. 1 Decision-making journey for house renovations

becomes aware of a need, like the dwelling itself or a component is old and dysfunctional, energy cost is high, etc., or when he/she becomes more environmentally concerned, proceeds in step 1. In that step, house owner makes an initial decision on whether the dwelling needs renovation or not. If it needs renovation, then step 2 follows where house owner starts to investigate what exactly needs to get renovated. At a later stage, house owner starts becoming active (step 3), looking for available services and planning his/her budget before reaching the "final decision" step (step 4), where after considering all the parameters described on the previous steps proceeds or not in the renovation. There is also the stage of "experience renovation" that relates to how house owners adapt and react to the renovations performed in their dwellings, but this stage is not considered in this paper since, as previously mentioned, we analyze future plans, not actual implementation of renovations. Dwelling's and household's characteristics are influences explaining why house owners start thinking of renovation and typically are not included in decision-making models [43, 44]. Instead, they are used as triggers for personal and contextual influences on renovations.

2.2   One-Stop-Shop Concept for Deep Renovation

One-stop-shop is currently advocated by the energy performance of buildings Directive (EPBD) 2018/844/EU [1], which amends the earlier EPBD 2010/31/EU [45] and Directive 2012/27/EU [46] on energy efficiency (EED). The Article 2A of the 2018 EPBD calls for a long-term renovation strategy, and member states "are required to facilitate access to mechanisms, such as one-stop-shops, which are considered as advisory tools here to inform and assist consumers in relation to energy efficiency renovations and financing instruments" [47]. According to the Article 20(2), "member states shall provide the information through accessible and transparent advisory tools such as renovation advice and one-stop-shops" [45]. Examining the potential interest of Swedish house owners on OSS will allow us to understand the market potential for such a service and to point out which parts of this service need to be reconsidered to reach a broader customer base.

3   Materials and Methods

The analytical framework in this paper is based on a preliminary stage of a larger-scale research project about the renovation-related practices of owners of detached houses in Sweden and the development of an OSS concept offering full-service renovation packages to them. To gain in-depth understanding regarding house "owners'" perception regarding energy consumption in their dwellings and toward renovation, we have designed an online survey in late spring 2017. The questionnaire for that survey was based on literature review and existing theories and pretested with a limited number of house owners, prior to its distribution. Later, the survey was sent to

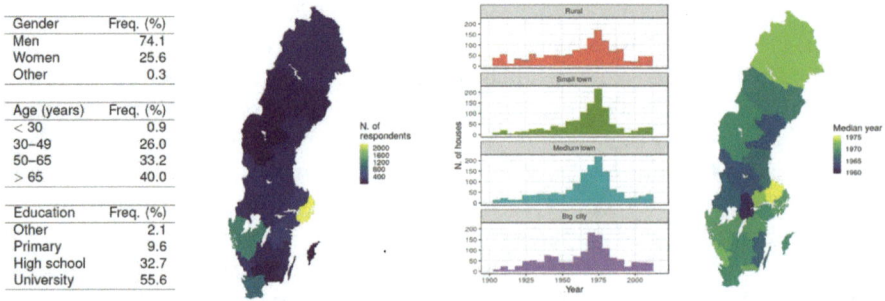

Fig. 2 Basic info on the survey respondents and their dwellings

144,660 members of Villaägarna, which is a non-profit and party-politically unbound consumer and interest organization for residents and owners of single-family houses in Sweden. In total, 12,194 house owners answered, after one reminder, which corresponds to a response rate of 8.43%, which is considered normal for such kind of surveys [48]. In the introductory note of the survey, the participants were informed that their participation was voluntary and that their identity and individual responses would be kept anonymous. Some basic information regarding the respondents and the dwellings can be found in Fig. 2.

The research is built on abductive approach, moving between theory and reality or observation in a systematic way. What will be presented in the following section is a preliminary analysis of the survey findings. The results were analyzed and interpreted in the theoretical context to derive conclusions [49].

4 Preliminary Survey Findings

4.1 Future Renovation Plans (Step 0)

The respondents were asked to share if they had planned to renovate their dwelling in the near future with the possible alternative answers: (a) I have no plans to renovate, (b) Yes, I plan to renovate my whole house at once, (c) Yes, I plan to renovate my whole house but gradually, and (d) Yes, I plan to renovate only a few parts of my house (Fig. 3). We have analyzed the respondents' answers per age and income groups to better understand the influence of socioeconomic attributes on their future renovation plans. Analysis showed that 25% of the respondents had no plan to perform a renovation in the nearest future. The main reasons for not planning a renovation in the future were satisfied with the current state of the dwelling, which confirms the findings of previous studies, and the time lived in the house. House owners living in their dwelling for a few years appear to be less willing to renovate in the close future.

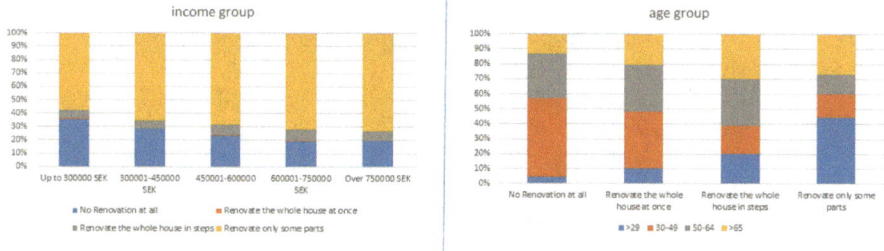

Fig. 3 Percentage of respondents planning to renovate their dwellings per income and age group

From the rest, 7% plan to renovate their whole house in stages, only 0.3% are interested to renovate their house at once, while the remaining are interested in renovation only some components of their dwelling. House owners between 29 and 49 years of age showed the greatest willingness to renovate their dwellings, at least some components, while owners over 50 years of age have shown greater willingness only to renovate parts of their dwelling or rejected the idea of renovation. Medium and high-income groups were more willing to renovate, but this was not the case when looked at the willingness to renovate the whole house at once or in stages. Additionally, houses built before the 1980s were more likely to be renovated.

4.2 Needs Leading to Renovation (Step 1)

The respondents were asked a question about the reasons for them to consider renovating their dwelling. The respondents were given 12 different alternative reasons (e.g., "house is old", "I want to improve indoor environment", etc.), with the possibility to indicate their level of agreement to each alternative on a Likert scale of 1–5, where 1 = disagree, 5 = agree. Their answers are presented in Fig. 4.

The analysis showed that the age of the house and the need to improve its aesthetics are very important reasons for house owners to begin thinking of a renovation project. Furthermore, the perception of increased value of the renovated house and the desire to reduce the energy cost are high on their priority list. On the other hand, parameters like improved indoor environment or influences from the social environment seem not to be reasons leading to the need for renovation.

4.3 Preferred Type of Renovation (Step 2)

Survey participants were asked to specify the type of renovation they were planning to perform. As seen in Fig. 5, they were given a variety of renovation measures to choose, which we have classified into energy, structural, and aesthetic renovations.

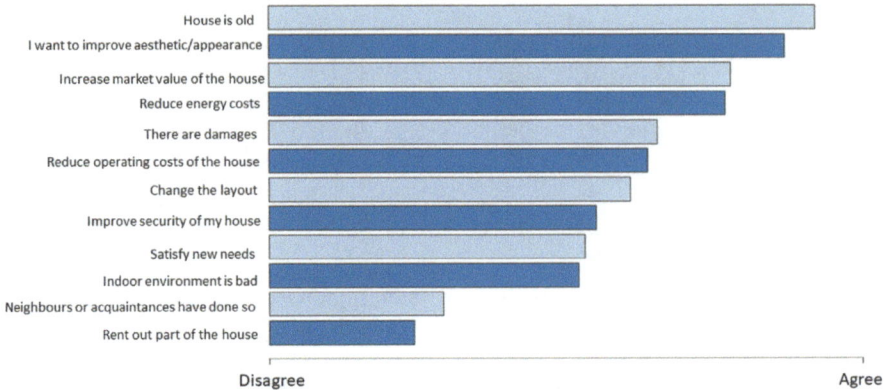

Fig. 4 Trends regarding reasons for considering renovating the dwelling

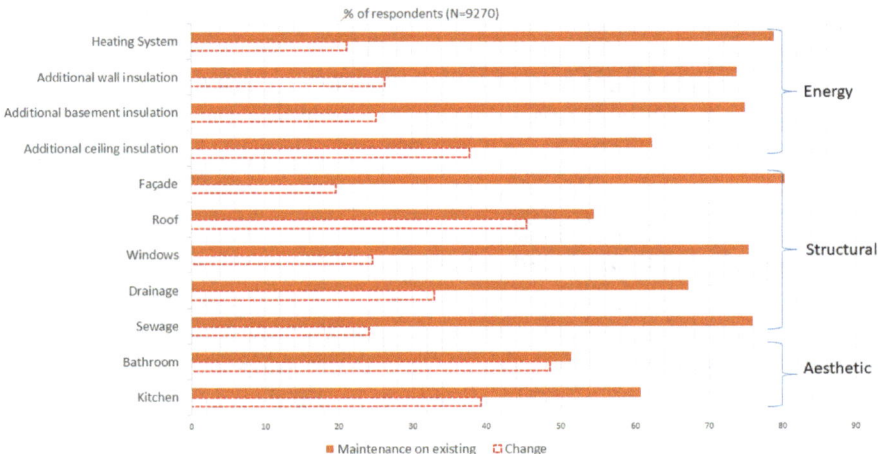

Fig. 5 Depiction of plans to perform maintenance or change house components

Respondents were asked to specify if they planned to perform a maintenance of existing building component or change them completely. Performing a maintenance appears to be the most common plan for the future, especially related to energy or structural renovations. Results regarding complete changes show that a greater number of respondents were willing to perform aesthetic renovations (changing kitchen and bathroom) than energy or structural renovations. Among structural renovation, roof and drainage were more attractive, while ceiling insulation was the most preferred option among energy-related renovations. 20% of the respondents intended to perform only one energy-related renovation (including in combination with structural or aesthetic renovations), while 43% of the respondents planned to implement multiple energy-related measures.

4.4   Awareness of Services and Interest in OSS (Steps 3 and 4)

Without introducing the phrase one-stop-shop, the respondents were asked if they were aware of actors offering full-service renovation packages in their area. Only 18% of them answered that they were aware of such actors in their area. Additionally, the respondents were asked to express their opinion on which actor could offer a full-service renovation package. The majority answered that local craftsmen and small construction/renovation companies could offer such a package, and some of them answered that large construction companies could offer such a package. A considerable number of respondents answered that someone else could be able to offer such a renovation package without specifying who that others could be.

Moreover, house owners were asked to express their interest on a full-service renovation package offered by a single contractor, as one-stop-shop concept suggests if that was offered in their area. More than 1/4 of the respondents answered that are interested or very interested to buy such a package if existing, while almost half of the respondents showed low or no interest at all. Results showed that both men and women owners, up to the age of 45 years and with university or high school education showed greater interest for that type of renovation package (Figs. 6 and 7). Moreover, those house owners own dwellings built from 1960 and above and show greater environmental concern, while they are willing to take action to protect the environment (Fig. 7).

Additionally, house owners were asked about the factors that are important for them to choose and actor offering a full-service renovation package. Respondents were given 16 different factors, and they were asked to evaluate them on a 1–5 Likert scale, where 1 = not important and 5 = important. Guarantees on cost/benefits, guarantees on delivery according to the agreed time schedule, and guarantees on quality of renovation work are the main qualities that house owners look after in order to choose such a package (Fig. 8).

To understand better why almost 50% of the respondents showed low or no interest in the OSS concept, we asked "how important are the following facts for not choosing one-stop-shop". Four different options were given in a 3-point Likert scale, where 1 = not important for me and 3 = very important for me. The main reasons for low or

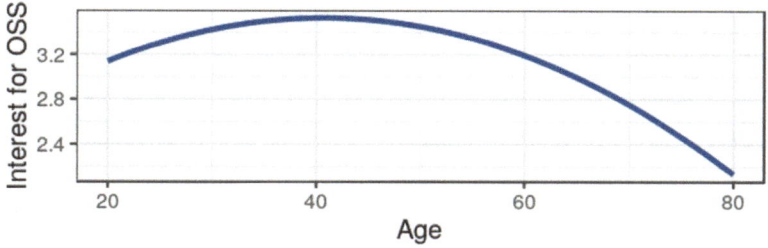

Fig. 6   Correlation of interest for OSS with the age of house owner

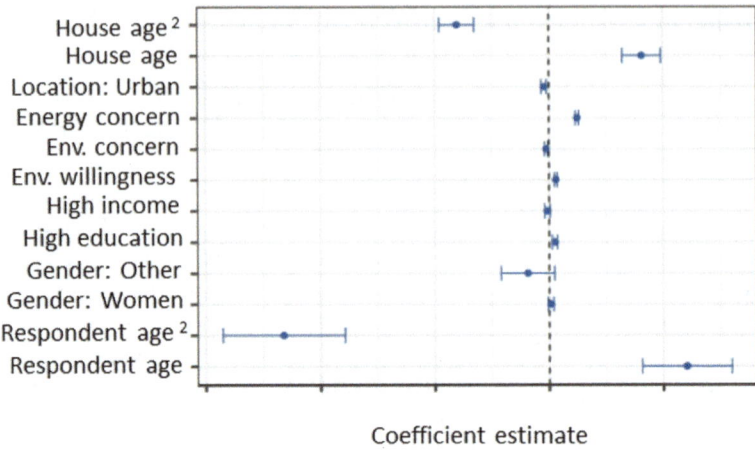

Fig. 7 Variables affecting interest on OSS

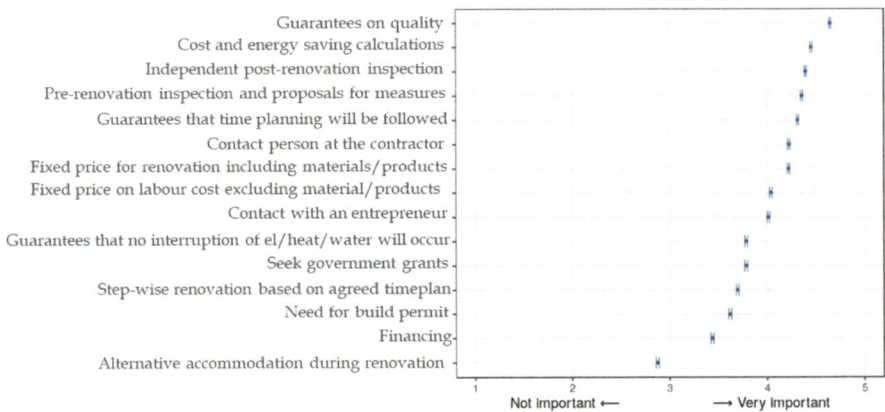

Fig. 8 Factors affecting choice of an OSS provider

no interest in OSS were the perceived high cost of it and the house owners' preference to choose different contractors to perform the different parts of renovation (Fig. 9).

4.5 *Financing the Renovation*

One of the questions asked to house owners was how they were going to finance their planned renovation in the near future. Respondents had to select between three different alternatives, namely own savings, mortgage loan, and private loan. The vast majority.

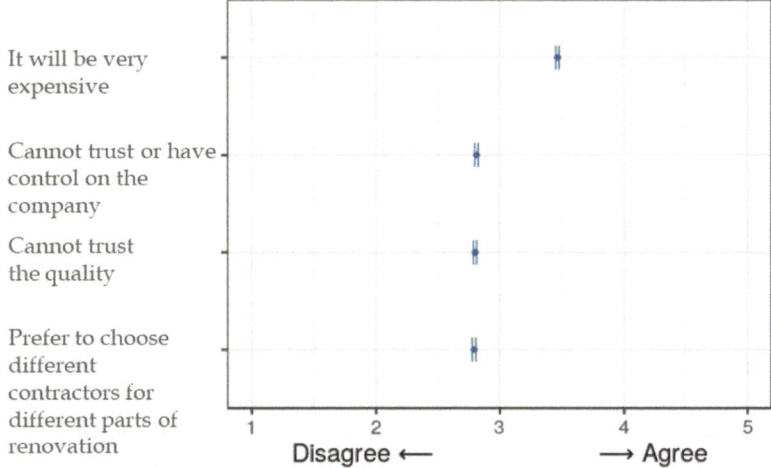

Fig. 9 Factors affecting lack of interest for OSS

of respondents (67.8%) would fund the renovation from their own savings, while very few (3.6%) would choose a personal loan, which usually has a higher interest rate than house mortgage loan. Additionally, house owners were asked to express their opinion on different financial incentives that could affect their interest to perform energy renovation.

Respondents were called to express their positive, neutral, or negative opinion on four preselected financial incentives for energy renovation (Fig. 10). According to them, connection of tax subsidies with energy reduction and the introduction of energy loans could positively incline them toward performing energy renovations.

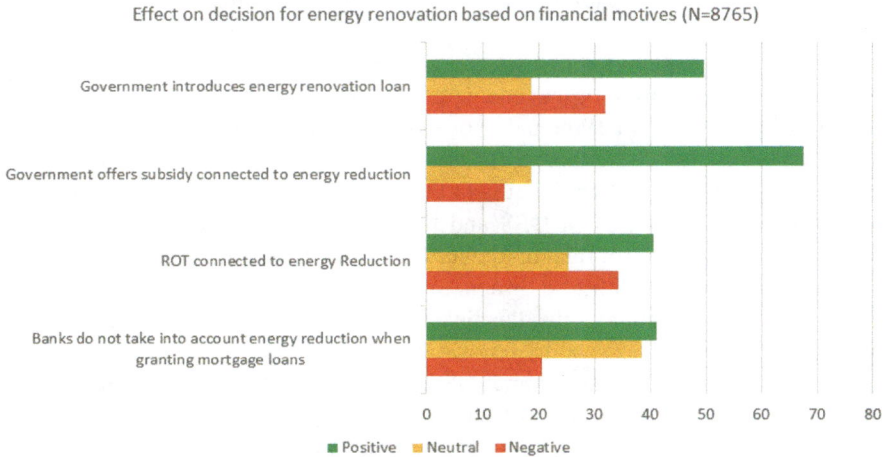

Fig. 10 Opinions on financial incentives that may motivate house owners for energy renovations

On the other hand, return on investment connected to energy deduction after renovation, and mortgage loans which consider the energy reduction post-renovation are financing measures that would lead to less positive inclination among house owners toward energy renovations.

5 Discussion and Conclusions

In this paper, we have analyzed the future plans of Swedish house owners toward renovating their dwellings. We have identified the underlying reasons leading them to take that decision and the influence of their socioeconomic background on it. Furthermore, we have identified the extent of the renovation they would prefer to perform and the components of their dwellings that they consider maintaining or change. Additionally, we have examined to which extent they have knowledge of the renovation services offered and their perceptions over a full-renovation package as envisioned in the OSS concept, as well as the financial incentives that would motivate them to invest in a more comprehensive renovation of their property (adoption of both energy and non-energy-related measures).

Our analysis showed that house owners between ages 29 and 49 years are the most positively inclined to perform renovations, while people aged 50 years and above show lower interest to perform renovations (with those aged 65+ years of age to be the least interested). The most preferable option is to renovate parts of the house. The number of people planning to renovate their whole house at once or in steps is still low. Aesthetic renovations are prioritized, but there is also a significant interest for structural changes (roof) with accompanying energy measures (additional ceiling insulation). Energy-related renovations are part of the house owners' future plans, but not from a holistic perspective (deep renovation). It should also be noted that energy-related renovations are not decided as an only renovation, but as a part of a broader renovation plan.

One-stop-shop appears to be interesting for an amount of house owners capable to verify a business potential for that concept. House owners up to the age of 45 years, with dwellings built from 1960 and above, are the market segment that can act as early adopters of the OSS concept. Additionally, those house owners show greater environmental interest and willingness to take action to protect the environment.

The perceived high cost of OSS and the interest to hold power to select the craftsmen appear to be important hindrances to market for OSS. Own savings are the most preferable option for house owners to finance renovations. Energy loans from financial institutions and tax subsidies linked to energy renovations are suggested to encourage them to perform energy renovations.

It becomes evident that an optimal OSS concept for detached house renovation should allow house owners to have an active participation in the whole process. Having in mind the decision-making journey for house renovations presented in Fig. 1, in the adoption of OSS, house owners are likely to be very engaged in the decision-making process right from step 1. That concept can provide solutions in the direction of making the right decisions on which house components need to be reno-vated, taking under consideration house owners' desires, but also the necessities in each individual dwelling. Moreover, OSS provides house owners with a feasible work and time schedule for the renovation of their house and with a group of craftsmen capable to deliver quality work. That enables to a great extent the decision of house owners to renovate and creates opportunities for the adoption of energy efficiency measures, within the given budget for renovation. Overall, we can claim that one-stop-shop can become the shortcut of the renovation decision-making journey for house owners and the vehicle for a smoother renovation process. Exempting house owners from the pains of renovation can become the trigger for them to further the extent of renovation and orient themselves to the adoption of more energy-efficient measures, despite the expressed financing doubts. Ways to finance a more comprehensive reno-vation need further research. There are though insights from house owners' side that highlight what needs to be addressed. The existing subsidies for house improvements should be linked with improved energy performance of those houses, to create an extra motive to invest in energy-efficient solutions. Own savings are, for nonce, the main financing mechanism of renovations. That makes the need for the development of financing products, like "green loans", imperative, as such loans could create a more favorable environment for investments in comprehensive renovation solutions.

Acknowledgements The authors gratefully acknowledge the financial support from the Kamprad Family Foundation for Entrepreneurship, Research & Charity, Smarthousing Småland, and Euro-pean Union Horizon 2020 project "INNOVATE". They would also like to thank house owners asso-ciation Villaägarna for sharing the questionnaire among its members and the survey respondents for responding to the survey.

References

1. European Parliament (2018) Directive (EU) 2018/844 of the European Parliament and of the Council of 30 May 2018 amending Directive 2010/31/EU on the energy performance of buildings and Directive 2012/27/EU on energy efficiency. Off J Euro Commu 61:75–91
2. Ministry of the Environment and Energy (2019) Sweden's draft integrated national energy and climate plan. Government Offices of Sweden, Stockholm. Published 17 January 2019. Available at: https://www.government.se/reports/2019/01/swedens-draft-integrated-national-energy--and-climate-plan/. Accessed on 18 July 2019
3. SE Agency (Energimyndigheten) (2012) Renovera Energismart Energimyndigheten. Eskil-stuna, , Sweden
4. European Environment Agency (2017) Trends and projections in Europe 2017. Tracking progress towards Europe's climate and energy targets. 10.1016/j.watres.2007.01.052
5. Swedish Energy Agency (2017) Energy in Sweden: facts and figures. Available online: https://www.energimyndigheten.se/en/facts-and-figures/publications/. Accessed on 16 Feb 2019
6. Swedish Energy Agency (2015) Summary of energy statistics for dwellings and non-residential premises for 2014 (1654–7543). Available online: www.energimyndigheten.se. Accessed on 11 April 2019
7. Mahapatra K, Mainali B, Pardalis G (2019) Homeowners' attitude towards one-stop-shop business concept for energy renovation of detached houses in Kronoberg, Sweden. Energy Procedia 158:3702–3708. https://doi.org/10.1016/j.egypro.2019.01.888

8. Bravo G, Pardalis G, Mahapatra K, Mainali B (2019) Physical vs. Aesthetic renovations: learning from Swedish house owners. Buildings 9(1):12. https://doi.org/10.3390/buildings901 0012

9. Friege J, Chappin E (2014) Modelling decisions on energy-efficient renovations: a review. Renew Sustain Energy Rev 39:196–208. https://doi.org/10.1016/j.rser.2014.07.091

10. Klöckner CA, Nayum A (2017) Psychological and structural facilitators and barriers to energy upgrades of the privately owned building stock. Energy 140:1005–1017. https://doi.org/10. 1016/j.energy.2017.09.016

11. Mahapatra K, Gustavsson L, Haavik T, Aabrekk S, Svendsen S, Vanhoutteghem L, Ala-Juusela M (2013) Business models for full service energy renovation of single-family houses in Nordic countries. Appl Energy 112:1558–1565. https://doi.org/10.1016/j.apenergy.2013.01.010

12. Ekström T (2017) Passive house renovation of Swedish single-family houses from the 1960s and 1970s: Evaluation of cost-effective renovation packages.

13. Agency ST (Skatteverket), (2015) Taxes in Sweden (2015) An english summary of tax statistical year-book of Sweden. Swedish Tax Agency, Solna

14. Pardalis G, Mahapatra K, Mainali B (2020) Swedish construction MSEs: simply renovators or renovation service innovators? Build Res Inf 48(1):67–83. https://doi.org/10.1080/09613218. 2019.1662713

15. Mlecnik E, Straub A, Haavik T (2019) Collaborative business model development for home energy renovations. Energ Effi 12(1):123–138. https://doi.org/10.1007/s12053-018-9663-3

16. Innovate Project Report (2018) Inventory of best practices for setting up integrated energy efficiency service package including access to long-term financing to homeowners: extensive analysis of the existing energy efficiency services operators and long-term financing schemes. Available at: https://www.financingbuildingrenovation.eu/wp-content/uploads/2017/ 08/Innovate_Inventory-of-best-practices_public-version.pdf. Accessed on 4th March 2019

17. Risholt B, Berker T (2013) Success for energy efficient renovation of dwellings—learning from private homeowners. Energy Policy 61:1022–1030. https://doi.org/10.1016/j.enpol.2013. 06.011

18. Bjørneboe MG, Svendsen S, Heller A (2018) Initiatives for the energy renovation of single-family houses in Denmark evaluated on the basis of barriers and motivators. Energy Build 167:347–358. https://doi.org/10.1016/j.enbuild.2017.11.065

19. Achtnicht M, Madlener R (2014) Factors influencing German house owners' preferences on energy retrofits. Energy Policy 68:254–263. https://doi.org/10.1016/j.enpol.2014.01.006

20. Murphy LC (2016) Policy instruments to improve energy performance of existing owner occupied dwellings. A+ BE| Arch Built Environ (17):1–242. ISSN 2214–7233

21. März S (2018) Beyond economics—understanding the decision-making of German small private landlords in terms of energy efficiency investment. Energ Effi 11(7):1721–1743. https:// doi.org/10.1007/s12053-017-9567-7

22. Friedman C, Becker N, Erell E (2018) Retrofitting residential building envelopes for energy efficiency: motivations of individual homeowners in Israel. J Environ Planning Manage 61(10):1805–1827. https://doi.org/10.1080/09640568.2017.1372278

23. Ameli N, Brandt N (2015) Determinants of households' investment in energy efficiency and renewables: evidence from the OECD survey on household environmental behaviour and attitudes. Environ Res Lett 10(4):044015

24. Murphy L (2014) The influence of energy audits on the energy efficiency investments of private owner-occupied households in the Netherlands. Energy Policy 65:398–407. https://doi.org/10. 1016/j.enpol.2013.10.016

25. Das R, Richman R, Brown C (2018) Demographic determinants of Canada's households' adoption of energy efficiency measures: observations from the households and environment survey, 2013. Energ Effi 11(2):465–482. https://doi.org/10.1007/s12053-017-9578-4

26. Buser M, Carlsson V (2017) What you see is not what you get: single-family house renovation and energy retrofit seen through the lens of sociomateriality. Const Manage Econ 35(5):276–287. https://doi.org/10.1080/01446193.2016.1250929

27. Nair G (2012) Implementation of energy efficiency measures in Swedish single-family houses (Doctoral dissertation, Mid Sweden University)
28. Judson EP, Maller C (2014) Housing renovations and energy efficiency: insights from homeowners' practices. Build Res Inf 42(4):501–511. https://doi.org/10.1080/09613218.2014. 894808
29. Mlecnik E, Kondratenko I, Cré J, Vrijders J, Degraeve P, van der Have JA, Svendsen S (2012) Collaboration opportunities in advanced housing renovation. Energy Procedia 30:1380–1389. https://doi.org/10.1016/j.egypro.2012.11.152
30. Mahapatra K, Gustavsson L, Haavik T, Aabrekk S, Tommerup HM, Svendsen S, Paiho S, Ala-Juusela M (2011) Possible financing schemes for one-stop-shop service for sustainable renovation of singlefamily houses. Nordic Innovation Centre. Nordic Call on Sustainable Renovation NICe, No. 08191 SR
31. Weiss J, Dunkelberg E, Vogelpohl T (2012) Improving policy instruments to better tap into homeowner refurbishment potential: lessons learned from a case study in Germany. Energy Policy 44:406–415. https://doi.org/10.1016/j.enpol.2012.02.006
32. Guy S, Shove E (2014) The sociology of energy, buildings and the environment: constructing knowledge, designing practice. Routledge, Abingdon, UK
33. Wilson C, Pettifor H, Chryssochoidis G (2018) Quantitative modelling of why and how homeowners decide to renovate energy efficiently. Appl Energy 212:1333–1344. https://doi.org/10. 1016/j.apenergy.2017.11.099
34. Christensen TH, Gram-Hanssen K, de Best-Waldhober M, Adjei A (2014) Energy retrofits of Danish homes: is the energy performance certificate useful? Build Res Inf 42(4):489–500. https://doi.org/10.1080/09613218.2014.908265
35. Baumhof R, Decker T, Röder H, Menrad K (2017) An expectancy theory approach: what motivates and differentiates German house owners in the context of energy efficient refurbishment measures? Energy Buildings 152:483–491. https://doi.org/10.1016/j.enbuild.2017.07.035
36. Karvonen A (2013) Towards systemic domestic retrofit: a social practices approach. Build Res Inf 41(5):563–574. https://doi.org/10.1080/09613218.2013.805298
37. Rogers EM (2005) Diffusion of innovations, 5th edn. Free Press, New York, NY
38. Madlener R (2007) Innovation diffusion, public policy, and local initiative: the case of wood-fuelled district heating systems in Austria. Energy Policy 35(3):1992–2008. https://doi.org/10. 1016/j.enpol.2006.06.010
39. Michelsen CC, Madlener R (2013) Motivational factors influencing the homeowners' decisions between residential heating systems: an empirical analysis for Germany. Energy Policy 57:221–233. https://doi.org/10.1016/j.enpol.2013.01.045
40. Mahapatra K, Gustavsson L (2008) An adopter-centric approach to analyze the diffusion patterns of innovative residential heating systems in Sweden. Energy Policy 36(2):577–590. https://doi.org/10.1016/j.enpol.2007.10.006
41. Nair G, Gustavsson L, Mahapatra K (2010) Owners' perception on the adoption of building envelope energy efficiency measures in Swedish detached houses. Appl Energy 87(7):2411–2419. https://doi.org/10.1016/j.apenergy.2010.02.004
42. Wilson C, Crane L, Chryssochoidis G (2013) The conditions of normal domestic life help explain homeowners' decisions to renovate. ECEEE Summer Study (European Council for an Energy Efficient Economy), Toulon, France
43. Dodds PE (2014) Integrating housing stock and energy system models as a strategy to improve heat decarbonisation assessments. Appl Energy 132:358–369. https://doi.org/10.1016/j.ape nergy.2014.06.079
44. Rommel K, Sagebiel J (2017) Preferences for micro-cogeneration in Germany: policy implications for grid expansion from a discrete choice experiment. Appl Energy 206:612–622. https:// doi.org/10.1016/j.apenergy.2017.08.216
45. Recast European Parliament (2010) Directive 2010/31/EU of the European parliament and of the council of 19 May 2010 on the energy performance of buildings (recast). Off J Euro Union 18(06)

46. European Parliament (2012) Directive 2012/27/EU of the European Parliament and of the Council of 25 October 2012 on energy efficiency, amending Directives 2009/125/EC and 2010/30/EU and repealing Directives 2004/8/EC and 2006/32. Off J L 315:1–56.
47. Boza-Kiss B, Bertoldi P (2018) One-stop-shops for energy renovations of buildings. Ispra: European Commission, JRC113301
48. Baruch Y, Holtom BC (2008) Survey response rate levels and trends in organizational research. Hum Relat 61(8):1139–1160. https://doi.org/10.1177/0018726708094863
49. Voss C, Johnson M, Godsell J (2016) 5 Case research. Res Meth Oper Manag 165

Chapter 5
Crisis of Institutional Change: Improving Restoration and Reconstruction Methods for Estate Cultural Heritage

A. I. Dayneko, D. V. Dayneko⬤, V. V. Dayneko, and S. V. Zykov⬤

Abstract The work presents the problem of institutional changes in Russian Urban planning. Institutional problems in the sphere of cultural heritage are discussed. The necessity to conduct the research is substantiated, and methods and tools for evaluation of the effectiveness of the institutional changes are suggested. The program of the urban development is proposed, and tested methods for estate objects of cultural heritage protection, which are to be implemented further in Russia considering climatic, seismic and ecological peculiarities of the regions, are suggested. The issues of marketing and reforming of institutions in the sphere of cultural heritage protection as well as the ways to manage estate cultural heritage are discussed.

Keywords Institutional change · Cultural heritage · Restoration · Reconstruction · Urban planning · Management · Marketing

JEL Classification B49 · C13 · C21 · D23 · L51 · M13 · O11 · O12 · O17 · O31 · O32

A. I. Dayneko · D. V. Dayneko (✉) · V. V. Dayneko
Irkutsk National Research Technical University, Irkutsk, Russia
e-mail: ddayneko@oresp.irk.ru

A. I. Dayneko
e-mail: aidayneko@mail.ru

V. V. Dayneko
e-mail: vday18@mail.ru

D. V. Dayneko
Irkutsk Scientific Center SB RAS, Irkutsk, Irkutsk, Russia

S. V. Zykov
National Research University Higher School of Economics, Moscow, Russia
e-mail: szykov@hse.ru

1 Introduction

One of the most vital and complicated issues of the modern urban planning is a problem of the cities' growth, of the reconstruction and modernization of the cities' environment. In the context of the global social, economic and ecological changes happening in the post-soviet space, first of all, the important role of institutional changes happening in the country, which influence economic, political, social institutions and the urban planning policy in the whole, is attentively studied. Institutional changes are considered as changes of legal norms and relations of their correlations, as changes of organizational structures and space of their interaction, such as associations or as changes in a common way of thinking. In my work, institutions can be considered as a set of rules and customs, norms and political and economic rules of behavior of society, which are stated in the legislation, decisions and instructions. Alternatively, institutions are established and developed in the country and in the region, mechanisms for coordination of the political and economic processes, such as elections and/or markets. Modern economic science defines "institutions" as "a steady complex of mutual roles and relationships, and behavioral features of social and economic agents". These are the "rules of the game", which are formal (laws, contracts, deals and organizational and legislative structures) and informal (rules of behavior, customs and ethic and ideological norms) [1].

2 The Crisis of Institutional Change in the Sphere of the Cultural Heritage Protection and the Ways for Its Solution

The cultural heritage objects (CHOs) and, first of all, monuments of history and culture of the Russian Federation (RF) are of unique value and are an integral part of the world cultural heritage. The Russian Federation guarantees the preservation of cultural heritage (historical and cultural monuments) in favor of present and future generations.

Today, cultural heritage sites are becoming increasingly important, especially in forming of the urban environment. There are new requirements for the development of the society, corresponding to modern economic, environmental, social and aesthetic norms and settings. To solve the problem of historical and cultural heritage preservation, there is a need to develop new approaches that correspond to the modern conditions of economic, social and political transformations happening in the country. The analysis of various conceptual approaches for the explanation of the ongoing with historical real estate processes in our country allowed the authors to conclude that it is necessary to study the methods and methodology for solving this problem. It is concluded that it is expedient to improve the regulatory and legislative support and development of marketing programs in the field of preservation of cultural heritage. There is a need to improve the attitude to cultural heritage objects too. This is also due to the objective impossibility of ensuring the proper protection and use of all cultural heritage sites solely by the efforts of state and local authorities alone.

The basis for the preservation of the cultural heritage is the relevant legal framework. The existing institutional framework requires improvement, at the federal, regional and municipal levels. For example, the federal law on cultural heritage objects provides for the division of cultural heritage objects into categories of historical and cultural significance (federal, regional and local). This division of cultural heritage, as practice has shown, has no scientifically based criteria, and the system of measures of the state preservation does not depend on the categories of historical and cultural significance, under the modern conditions. In turn, such a division of cultural heritage objects creates unnecessary obstacles to the implementation of the state protection and entails some problems related to differentiation of the state ownership of cultural heritage objects and of their financing at the expense of the state budgets at different levels.

Therefore, as a result of the authors' analysis of the types and methods of management of cultural heritage, it was revealed that the solution to the problem is in the implementation of market mechanisms and of state levers to influence the related decisions. Particularly relevant is the development of marketing programs in the field of real estate cultural heritage preservation of. The urgency and necessity of the work on marketing programs for the preservation of architectural monuments are caused, first of all, by the aggravated situation in the field of cultural heritage protection, by the need to attract investments for the preservation of cultural monuments.

3   The Current Legal Framework in the Field of Estate Cultural Heritage Protection

The most important in the context of the ongoing reforms are the formal institutions, which, first of all, establish the main rules of the business activities. These rules determine the effectiveness of the whole economic system functioning and serve as a basis for the solution of the urban planning issues. Therefore, the role of system and local organizational institutions has to be distinguished.

The major document determining the norms of urban planning and regulating the interaction of the stakeholders in the country is Urban Planning Codex of the Russian Federation (UPC RF), the last edition of which has been ratified on December 29, 2004. The major task of the UPC is to guarantee the following of the legal rights of builders and to protect interests of the local community in the field of planning and building in the territory.

The necessary requirements for urban planning and reconstruction activities are included in the Building Norms and Rules (BNRs). The other normative documents, which regulate urban planning and protect citizens' rights, include federal level documents and documents of the regional and municipal levels. Urban planning for each large city is usually conducted through the development and periodic actualization of the documents about urban planning for the development of the territories. The planning of volumes, pace and turn of activities for capital repairing, modernization, reconstruction and renovation of buildings and structures of the present cities' structures can be conducted as well, based on the territorial scheme of the scheduled developments and implementation of the cities' programs for the complex reconstruction, which is to be conducted with the consideration of the general plan. The general plan of the urban planning development of the large cities is of long-run character.

Back in 1996, Russian Government has adopted the program "Housing: Reconstruction of the Dwelling Blocks of the First Mass Series". The single experiment has been realized in cities of Sankt-Petersburg, Moscow, of Moscow region and in a few other cities of Western Russia. In July 2007, the federal law of the Russian Federation No 185-FZ "About a Fund to Assist the Reforming of the Municipal Housing Economy" was ratified.

The existing legal framework needs improvements and further development, both at the federal and regional levels. The federal law on cultural heritage objects states the division of cultural heritage objects into certain categories of historical and cultural significance (of the federal, regional and local (municipal) significance). This division of cultural heritage practically has shown that it lacks scientifically grounded criteria, and the existing system of preservation measures to protect cultural heritage does not depend on the categories of historical and cultural significance under the modern state. In turn, this division of cultural heritage objects results in unnecessary obstacles in the implementation of the state protection and entails problems of differentiation of state ownership of cultural heritage objects, their financing via the state budget money of different levels.

In the context of the problems of the complex reconstruction of the ongoing urban development, a new project of the Urban Planning Code of the Russian Federation (UPC RF) is of particular interest [2]. In particular, the regulation of issues related to legal zoning. It details the purpose and content of rules of the land use and development, procedure of their preparation, coordination, public discussion, adoption and application, types and characteristics of territorial areas in the rules of land use and development, as well as the scheme of recognition of individual real estate objects that do not comply with urban planning regulations.

The draft of the code defines the procedural issues of preparation of urban planning documentation for the planning of the territory (grounds and procedure for preparation, purpose, types and composition of such documentation). Differences in procedures of preparation, coordination, discussion and acceptance of urban planning documentation for planning of the territory are predetermined by the fact that this documentation can be ordered by various entities of urban planning relations: public authorities of the Russian Federation or of its entities, local governments, individuals and other legal entities. The key point in the preparation of project documentation is to change the regulatory issues of the state examination of project documentation. This procedure is proposed to be implemented only to particularly dangerous, technically complex and unique objects (in the form of verification of project documentation, especially for dangerous and technically complex objects, for compliance with technical regulations, and for unique objects the determination of the feasibility of the project in the absence of technical regulations). Project documentation for other objects in accordance with the draft code can be realized through assessing its compliance with technical regulations, which is carried out by accredited organizations in the manner prescribed by self-regulatory organizations. In addition, the procedure for issuing a construction permit and providing a certificate of acceptance is established for a property built in accordance with the approved design documentation.

The current Russian legislation is lacking the concept of "protected historical and cultural territory", but in some cases, such territories have already been created or are being created in accordance with special legislative acts of regional and local authorities. The strategy for the protection of cultural landscapes requires the inclusion of issues of socioeconomic development of the protected territory. In turn, the program of socioeconomic development of the administrative unit, within which borders the protected territory is located, should be adjusted in accordance with the requirements of the preservation of the cultural landscape [3]. As a rule, unique territories form the framework of regional heritage systems. In most countries of the world, the basis of this framework consists of natural reserves, historical cities and large historical and cultural museums-reserves, but most often regional and national parks, combining elements of natural and cultural heritage [4].

4   The Concept of Estate Cultural Heritage and the Subject of Protection of Cultural Heritage

According to the existing definition of the cultural heritage presented by federal legislature, it is important, in the context of this research, to formulate the following concept of a cultural heritage object. Estate cultural heritage is a special type of real estate, each category of which (building, block and historical center of the city) is unique and has individual features that form its historical and cultural potential. Cultural heritage objects are divided into the following three types, in accordance with the most significant typological features and their functional purpose:

(1) *monuments* are separate constructions, buildings and structures with historically developed territories (including monuments of religious significance: churches, bell towers, chapels, mosques, Buddhist temples, pagodas, synagogues, houses of worship and other objects specially intended for divine services); memorial apartments; mausoleums, separate burials; works of monumental art; objects of science and technology, including military; partially or completely hidden in the ground or under water traces of human existence, including all related to them movable items, the main or one of the main sources of information about which are archaeological excavations or finds (hereinafter: objects of archaeological heritage);

(2) *ensembles* are clearly localized in historically developed territories groups of isolated or united monuments, buildings and structures for fortification, Palace, residential, public, administrative, commercial, industrial, scientific, educational purposes as well as monuments and structures of religious purpose (temple complexes, datsans, monasteries, farmsteads), including fragments of historical planning and settlements constructions, which can be attributed to urban planning ensembles;

(3) *works of "landscape architecture and garden art""*(gardens, parks, squares, boulevards), *necropolis*;

(4) *places of interest* are the ones created by man or joint works of man and nature, including places of folk arts and crafts; historic centers of settlements or fragments of urban planning and construction, memorable places, cultural and natural landscapes associated with the history of people and other ethnic communities on the territory of the Russian Federation, by historical (including military) events, life of outstanding historic figures; cultural layers, remains of buildings of ancient urban, settlements, villages, ancient places of stay; sites of religious rites and worship.

According to the existing legislation, among the categories of cultural heritage objects with historical and architectural, artistic, scientific and memorial value, which are of special importance for history and culture, are distinguished the following:

- objects of cultural heritage of *federal significance*. These are the objects which have historical, architectural, artistic, scientific and memorial value of particular significance for the history and culture of the whole Russian Federation, as well as objects of archaeological heritage;
- objects of cultural heritage of *regional significance*. These are objects having historical, architectural, artistic, scientific and memorial value, having special significance for the history and culture of the Russian Federation entity;
- objects of cultural heritage of *local (municipal) value*. The objects possessing historical and architectural, art, scientific and memorial value having especial significance for history and culture of a municipality.

Such allocation of the immovable monument of history and culture to the corresponding type and category is fixed when drawing up documents of the state account of monuments and entering data to the uniform state register of objects of cultural heritage.

The main objectives of the protection of historical and cultural heritage are declared, as already mentioned earlier, in a number of federal laws and local regulations governing architectural, urban planning and land management activities, as well as property relations. However, the direct legal force of these declarations, according to Academician Slavina T. A., is weakened by the uncertainty of such concepts as "lands of historical and cultural significance", "historical city", "protected areas" and "monument" [5]. According to the federal law of the Russian Federation "about objects of cultural heritage" (monuments of history and culture) of the people of the Russian Federation, objects of cultural heritage are of unique value.

Article 3 of this law defines that the objects of cultural heritage include immovable property with related paintings, sculptures, decorative and applied art objects, science and technology, and other objects of material culture, which appear as a result of historical events, and are significant in terms of history, archaeology, architecture, urban planning, art, science and technology, aesthetics, ethnology or anthropology, social culture, and represent evidence of epochs and civilizations. Hence, they are authentic sources of information about the origin and development of the culture.

As the carrier of culture, heritage contains information potential that can ensure the preservation of the human being as a species, people as a nation, city as a smaller community, the communities itself and families. It is implied that not everything created by man is expedient in this sense of the word and must be preserved at all costs. This thesis is fully applied to the architectural and urban heritage. The criteria for assessing what has been created in the past are historically mobile, and we can rely only on the paradigm of the turn of the century (millennia). However, let us assume that modern society, objectively assessing the path traversed by mankind, is able and obliged to "separate the grain from the chaff".

The subject of protection implies those elements, parameters and characteristics of the architectural and urban development object, which are carriers of its real value. In terms of activity, the subject of protection is the subject of limiting the actions of all participants in the urban planning process, fixed by full-fledged documents from the legal point of view. The subjects of protection are the features of the object of cultural heritage, which served as the basis for its inclusion into the register and mandatory preservation [6].

When insisting on the protection of architectural and urban heritage, we strive to preserve the material evidence of the "memory of the past" and thus provide our contemporaries and descendants with conditions for the production of material and spiritual culture. *Architecture* is a system of buildings and structures that form a spatial environment for the life and activities of people, as well as the art itself to create these buildings and structures in accordance with the laws of beauty.

The existing experience of protection activities for the last decades allowed to reveal numerous objects which require preservation: lands of historical and cultural use, reserved places, historical cities, districts and their planning, compositions, panoramas, etc. Then, for a long time, the necessity was realized to switch from protection of separate objects to protection and maintenance of the whole systems. Architectural heritage is the spatial shell of the past life. It is this life, at its best, that deserves to be extended and, therefore, *protected*.

The mechanism of protection, according to many authors, is *urban regulation* as well as *professional management*, by means of which the relations of all stakeholders in the architectural and urban development process are streamlined. The definition of objects of protection is accompanied by the assessment and ranking of the territory and of its development. Not all the fragments of the city are equal in their architectural quality. When analyzing the environment of the studied zone, the expertise is obliged to accurately formulate the optimal mode of reconstruction according to typological, compositional, functional analysis.

The differentiated approach to monuments, tested on a number of objects in St. Petersburg's historical center, opened up significant reserves for transformation. The search for a compromise between preservation and transformation is a historically established pattern of city life. Today, in order to increase the investment attractiveness of urban areas and economic returns, it is necessary to preserve the historical and architectural potential of urban objects and the territories surrounding these objects.

Stated legally in federal and local laws, in inventories, in zonal regulations and in property documents, the specific definition of the subject of protection allows us to (1) ensure real preservation of historical and cultural heritage; (2) to determine the degree of freedom of the designer, investor and property owner; (3) to improve the quality of the environment where it is required.

Thus, the "subject of protection" is a concept that allows to actualize a large amount of historical and theoretical knowledge, which is not excessive, and allows to understand what the city has and what and how should be preserved in the historical environment.

The monument in its broad sense is an object that forms part of the cultural heritage of the country and society. Protection of historical and cultural monuments is a task of public importance. Literary and historical monuments, works of architecture, fine and decorative arts, archaeological finds and complexes of national and international importance are subject to protection.

5   The Management of Estate Objects of Cultural Heritage

The management of cultural heritage is a system of measures aimed to ensure the preservation of the qualitative state of the object and/or targeted development (restoration) of its qualities. The system of such measures includes research and inventory works (information support, including monitoring); regulation and control of various subjects of activity (legal support and organization of protection); verbal and graphical modeling of the desired states of the object under the management, namely planning actions to preserve or change its qualitative parameters (development of concepts, strategies, projects, programs, plans and schemes); and realization of the planned works, including their logistical, financial and personnel support (implementation of management initiatives and goals achievement) [7]. Some separate objects, property complexes and their territories are distinguished among objects of management of historical buildings and structures.

Most of the identified cultural heritage sites are in state ownership. The existing authorities, responsible for monitoring the proper implementation of obligations for the preservation and use of historical and cultural monuments, are taking a number of appropriate measures. At the same time, the interaction of state bodies for the management of cultural heritage objects also depends on the level of significance of these objects (please, see Fig. 1).

The owner of federal property is the federal agency for state property management (Rosimushchestvo) and its territorial bodies [8]. The owners of cultural heritage

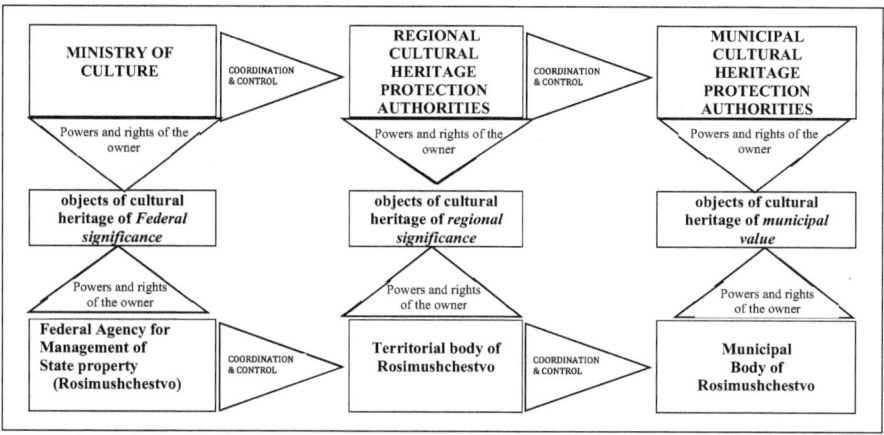

Fig. 1 Interaction of cultural heritage managing bodies

may also be the entities of the Russian Federation and municipalities. In this case, the powers of the owners shall be exercised by the relevant executive authorities authorized to manage such property.

Responsibility for supervision and control over the preservation of federal monuments was assigned to the federal service for supervision of compliance with legislation in the field of protection of cultural heritage (Rosokhrankultura). In August 2011, the decree of the Government of the Russian Federation # 590 of July 20, 2011, was ratified, according to which the functions of Rosokhrankultura were transferred to the Ministry of Culture of the Russian Federation. In accordance with this resolution, the organizational structure of the Ministry of Culture was reformed. The department accepting powers and functions of Rosokhrankultura had been formed. Now the Ministry of Culture of the Russian Federation performs functions on development and implementation of the state policy, normative-legal regulation, control and supervision in the sphere of cultural heritage.

The activities of the Ministry of Culture as regards preservation of heritage sites are realized directly through the territorial authorities in cooperation with other federal executive authorities, executive authorities of the Russian Federation entities, local authorities, public associations and other interested organizations. Among territorial bodies of the Ministry of Culture, we can name, for example, Department of Cultural Heritage, Committee on the State control of use and protection of historical and cultural monuments (KGIOP), Committee on the State protection of objects of cultural heritage, Department of the State protection, preservation, use and popularization of objects of cultural heritage in the Department of Culture and other.

6 State Protection of Historical and Cultural Monuments

Let us study a number of the most important functions of the state protection of cultural heritage. There is state registry kept in respect to all historical and cultural monuments, regardless of ownership or use of. The complex of measures includes identification and inspection of monuments; determination of their historical, scientific, artistic or other cultural value; fixation and study; preparation of record documents; and maintenance of state lists of estate monuments. Documents of the state registration of historical and cultural monuments, including those excluded from the state lists of monuments, are subject to mandatory and permanent storage too.

A registration card containing data on location, dates, character of modern use, degree of safety of a monument and/or again revealed object, about availability of scientific documentation and a place of its storage as well as the short description and illustrative material is developed for each monument representing historical, scientific, art or other cultural values in the corresponding centers of scientific documentation of the state bodies of protection of monuments. Also, a passport is developed for each state monument, containing information describing the history of the monument and its current state; location; assessment of historical, scientific, artistic or other cultural significance; information about its territory, associated structures, gardens, parks, works of art, objects of cultural value, protection zones as well as the main historical, architectural and bibliographic materials. The passport indicates the category of protection and type of the monument with reference to the approving document.

If the object is the one which is newly identified and represents historical, scientific, artistic or other cultural values, it is recognized as a monument of history and culture by including it in the relevant state list of estate monuments of history and culture, depending on the type and category of protection of the monument. State lists of estate monuments of history and culture, which are the main documents of the state registration and protection of monuments, are compiled according to categories and types [9].

The regulations on protection zones for objects of cultural heritage of Russian Federation population, approved by the resolution # 315 of the Russian Federation Government back on April 26, 2008, establish the order of projects development for zones of protection of objects of cultural heritage (monuments of history and culture) of RF people, and requirements to land-use regimes and urban planning regulations within the boundaries of these zones. The project of protection zones of these objects represents the documentation in the plain text format and in the form of maps (schemes), containing the description of borders of the projected zones and borders of territories of the objects of cultural heritage located in the specified zones, projects of modes of use of lands and urban planning regulations in borders of these zones.

Based on the project of protecting zones for the cultural heritage object of federal significance and the positive conclusion by state historical and cultural expertise, the corresponding body of the state power of the Russian Federation entity, in coordination with the Ministry of Culture (except for borders of zones of protection of especially valuable cultural heritage object for people of the Russian Federation or an object of cultural heritage included in the world heritage list), approves the modes of land use and urban planning regulations within borders of these areas. Projects of protecting zones of objects of regional or local (municipal) importance are approved in the order established by the law of the region in which it is located. Data on availability of zones of protection for object of cultural heritage is registered in the uniform state register of objects of cultural heritage (of monuments of history and culture) of the people of Russian Federation and is submitted to the body that is acting as the state cadastre of real estate keeper.

The urban planning regulations define the legal regime of the land lots, as well as everything that is above and below the surface of the land lots and is used in the process of their development and further exploitation of capital construction objects: art. 36 of the Urban Planning Code of the Russian Federation, December 29, 2004 # 190-FZ GSK. According to the fourth paragraph of this article, it does not extend to the land lots within borders of territories of the monuments and ensembles included in the uniform state register of objects of cultural heritage (monuments of history and culture) of the people of the Russian Federation. Only authorized federal executive authorities, authorized executive authorities of the Russian Federation entities and authorized local governments, and in accordance with federal laws, determine the use of these land lots. If the use of these land lots continues and is dangerous for human life or health, for the environment and cultural heritage, it is possible to impose a ban on the use of such land lots and objects.

7 Essence and Principles of Management of Estate Cultural Heritage

The problem of preservation of estate cultural heritage, while reconstructing and if building new facilities in different eras and in different countries, was solved in different ways. However, the exceptional role, played by CHO to ensure historical national continuity and identity, educates the younger generation in the spirit of patriotism, develops tourism and related industries and regional business links, etc. and has rarely been disputed.

In Russia, there are four options for saving CHO: (1) exterior and interior restoration; (2) restoration of external appearance and reconstruction of interiors to the modern level of improvement; (3) complete or partial reconstruction of the exterior and interiors; and (4) transfer of CHO to another location and development of the territory with new capital construction objects.

Today, all these schemes are used to some extent in the large cities of the Russian Federation (such as Moscow, St. Petersburg, Kazan, Nizhny Novgorod, Novosibirsk,

Krasnoyarsk and Irkutsk). This is due to the rapid reconstruction and new construction happening in our country.

The relocation of a monument building is justified only when it is threatened to be completely lost if it remains at the same location. This situation may be the result of climatic, geological or other objective reasons. In fact, the possibilities to relocate an estate heritage object are significantly limited. Basically, this measure is used for the preservation of wooden buildings, whose frame, elements of the roof and truss system and ceilings can be disassembled and mounted on the site where the preservation of the monument is not threatened. For instance, among the striking examples of such a solution in Russia can be named the village of Malye Korely in Arkhangelsk region, where 120 monuments of wooden architecture were collected together on a territory of about 140ga, museum of wooden architecture Taltsy and 47 km away from Irkutsk toward the Lake Baikal (Baikal tract).

The use of other three schemes of the estate objects preservation depends on many factors, including the technical condition of the monument, its cultural and historical value, the possibility of interior recreating, compliance with modern requirements of comfort, reliability and safety. The implementation of these schemes can be realized as part of capital major repairs of the heritage object. However, in the vast majority of cases, the management of cultural heritage objects is carried out as a part of a particular investment project. The reasons for initiating such projects are, for example, non-compliance of the building, which is a monument, with modern sanitary and epidemiological requirements, the discrepancy of utilitarian purpose originally incorporated in the object to its modern use in accordance with the current needs of the society, poor technical or disrepair condition of the object or of its particular structural elements, etc.

The changes carried out at the heritage site in accordance with the market requirements, as a rule, are directed not only to the object's preservation as a building itself, having certain historical and cultural values, but also to increase its cost too, to create with it as with the basis, a real estate product demanded in the active market. This approach to historical objects fits into the framework of the existing system of the real estate transformation of, called the development. This transformation is presented in Fig. 2.

8   Marketing in the Field of Cultural Heritage Preservation

We define *marketing for the preservation of the cultural heritage* as a system of measures to attract new economic agents to the region or city, aimed at the preservation of cultural heritage. It can be realized in the form of marketing of administrative and residential buildings, zones of economic development and adjacent territories, investments, tourism marketing, etc. Specific marketing measures at the regional and municipal level may include publication and distribution of printed materials about cultural heritage sites in the region or city, targeted visits by heads of administration, meetings with chiefs of organizations willing to invest in cultural heritage objects

Fig. 2 Historical estate development activities

in the region or city, campaigns conducted mutually with the chamber of commerce and industry, etc. In fact, marketing in the field of preservation of cultural heritage can be an effective tool for socioeconomic development of the region and the city.

Therefore, marketing in the field of preservation of estate cultural heritage involves actions to increase the attractiveness of the object in the view of investors, tenants, renters, buyers or another target group for whom the object is intended after its transfer into operation. With the professional approach, marketing should be applied from the very beginning of the works for the preparation of the estate cultural heritage object and continues throughout the life cycle of the object.

In the course of the life cycle of an estate cultural heritage object, the following directions of marketing in the field of its preservation can be distinguished:

- marketing of restored/reconstructed objects (business idea—completion of construction);
- marketing of completed objects and land lots (object circulation);
- marketing of services.

Development and implementation of the marketing programs in the field of preservation of cultural heritage are a complex and urgent problem of economic development of all entities of the Russian Federation:

- firstly, because not always they understand in the regions the importance and necessity of systematic promotion of positive information about cultural heritage in order to create a favorable attitude to the region and monuments of architecture as well as to provide information about the benefits of doing business and/or living

in monuments of architecture, i.e., information marketing for the preservation of cultural heritage;
- secondly, because organizations which are performing functions of information marketing in many entities of the Russian Federation, as a rule, perform these functions only partially and haphazardly.

The relevance and necessity of the marketing programs development for the preservation of monuments of architecture are caused, first of all, by the aggravated situation in the field of protection of cultural heritage, the need to attract investment for the preservation of cultural monuments. The problem is also complicated by the fact that institutions of history and culture do not have the right to spend budget money on commercial advertising, and social advertising is not available for them.

The need to develop marketing programs is also associated with a weak focus on marketing. Often the thinking of workers, whose work is related to history and culture, to objects of historical and cultural heritage, is focused on the product/service that they provide or directly on the object itself, rather than on the consumers or investors with their desires. Therefore, most often, they go the way of modifying traditional services or search for new forms with old content.

To present the marketing of an object of cultural heritage as a complete system, it is necessary to distinguish the principles of marketing of an object, subjects, components, goals, elements, methods and composition of the marketing program.

As a concept of market strategic management of an object, marketing of CHO requires the following of these principles:

- purposefulness, which is marketing goals definition;
- development of the strategy of the object development based on the analysis of market opportunities, identification of strengths and weaknesses of the object, determination of its competitiveness;
- selection of target markets and of positioning method;
- coordination of interests of the parties (complexity of marketing of CHO consists that such an object can serve different functions: a residence, a place of rest, a place of business and/or production, representation of part of history and culture of the city, region, the country or the world). There may be conflicts of interest among groups representing these functions. Private interests of potential investors may conflict with the interests of the centers of preservation of history and culture of the city, region or the world as a whole. The resolution of these contradictions can be accomplished on the basis of the principle of striving for the satisfaction of the parties involved in the conflict;
- development of organizational structure of the object marketing;
- development of tactics of realization and audit of object marketing.

Marketing of the object's components is the most important element of marketing of the whole object, since components are its bearing skeleton and the basis at the same time. In addition, such advantages as reliable communications, good insolation and aeration, cleanliness and safety are signs, which are quite attractive to potential investors. Their availability is not a guarantee of demand growth, but their absence or poor condition makes this growth impossible.

For the permanent rehabilitation and renovation of the object, it is not enough only to repair it. Resource constraints, requirements of heritage agencies and interdepartmental relations create the need for the development of a special type of management activity which is management of cultural and historical heritage, associated with a wide range of works, from monitoring the condition of objects to their systematic improvement with the help of modern methods and technologies.

Providing support from citizens, politicians, organizations, services and agencies in marketing in the field of cultural heritage protection requires the development of interaction marketing aimed at the development of cultural and historical capital of the region. This strategic direction of cultural heritage marketing may include support of public organizations.

Stimulation of development of cultural and historical capital and support of public initiatives can be carried out by local administrations: in the form of regular competitions for creation of projects of protection and preservation of cultural and historical heritage, projects of development of objects of cultural heritage and ongoing historical architectural research projects.

9   Reforming of the Ownership Institution for Historical and Cultural Heritage

One of the most important directions of the reforms realized in the country is the reform of the property institution. Ownership is the right on the basis of which an entity (legal or an individual) manages a property at its own discretion. The concept of ownership includes to the full and indivisible extent the right to use: the right to use the fruits resulting from the property, which is the subject of the ownership, and the right to use this property and to manage it at own discretion. The modern economic interpretation defines property as a set of rights of the subject to manage the conditions of economic activity and its results.

Some authors believe that property is realized only through the appropriation by the owner of income generated by their means of production [10]. Others, appealing to the fact that property is a complex system and its implementation should reflect this complexity, argue that it is realized through the entire system of production processes [11], i.e., " the content of property can be realized only in the concrete historical reproduction process" [12]. Along with the monistic points of view, there are also dualistic ones. So, Afanasenko I. D. believes that at the level of society, property is realized through production, exchange, distribution and consumption, i.e., through the entire system of production relations and, in general, at all levels through appropriation [13].

The functional structure of ownership, which includes the property ownership itself, and the way it is used and/or disposed, has to be implemented through the allocation of appropriating and managing entities according to corresponding functions. If applying this theoretical approach, it can be concluded that the realization of state property involves the distribution of property functions among all participants in these relations. The state form of ownership in accordance with its functional structure, reflected in the levels of its entities (state, collectivity and individual), for each entity in the production phase is implemented in management as a function of ownership, i.e., specific historical forms of labor organization (organizational and economic relations), and at the distribution phase: the appropriation of the relevant part of income.

The institution of private property is a guarantee of inviolability of economic freedom, and it minimizes the possibility of forced redistribution of the wealth generated. Transformation of relations and of the ownership structure is one of the most important problems. The role of the property itself, as the basis of the economic system, determines the systemic nature of both: the reforms in this area and the reforms in general in the context of the transition economy.

It is the owner of the object of cultural heritage who is responsible for the maintenance of the object belonging to him, which is included in the register, or the identified one, and considering the requirements of the federal law. Therefore, the privatization of monuments of architecture and history by private individuals is considered as a panacea by many cultural figures. In their opinion, only this way we can save the cultural heritage of Russia.

They note that the transfer of historical and cultural monuments to private owners abroad is quite a common phenomenon. For example, in the Czech Republic, someone who owned a piece of property (say, a castle) before 1945 can get it back now and then sell it off or rent it out. The estate that remains in the state ownership is offered at special auctions. In France, the owners of monuments have to maintain them according to the historical era and pay quite large taxes if they do not use them for excursions. In Russia, many officials and businessmen also acquire cultural monuments into the ownership not for the first year. The officials believe that the privatization of cultural monuments is quite timely in Russia too, since they still can be sold out to private owners in order to save them.

The private and collective (joint or shared) ownership of the property arises as a result of the privatization. Some researchers also include in this definition the process of modifying the management model of the state enterprise without alienation of property rights: that is, via an agreement, a lease, contracts, full or particular changes in the legal or financial status of the state enterprise. Synonymous with this interpretation is largely the concept of "denationalization". The latter, a broader interpretation, is usually not accepted (as a method of privatization) by Russian ideologues. At the same time, it is possible to speak with a certain degree of conditionality about the beginning of such denationalization in Russia back in 1987–1988 already and having in mind spontaneous privatization too as a certain prepared stage of transformation in this sphere [14].

In the context of this work, a special theoretical and practical interest may be the study of the interdependence of privatization and deprivation as the possible methods of state influence on the formation of the institute of ownership of CHOs, the relationship of owners and renters with the state.

Therefore, the task is "not to give away property, state property, but to give it into the hands of a really better owner. The prerequisite for this is the formation of a genuine private entrepreneurial motivation" [15]. "The main role that institutions play in society is to reduce uncertainty by establishing a stable (though not necessarily effective) structure of interaction among people" [1]. One of the tasks of the institutions is to remove the uncertainty arising from the interactive activities of enterprises and processes of economic exchange. For example, the long-term development of the industry requires reliable property rights and of property management, guaranteeing its inviolability and protection by law.

10 The Justification of the Choice of the Way to Manage an Object of Cultural Heritage

Taking into account the existing options for the preservation of cultural heritage, a decision-making model was developed for choosing the OCH management option (Fig. 3). This model takes into account a number of parameters of an object, such as the state of structures, state of the internal infrastructure, the relation of the architectural and planning solution of an object to the heritage to be protected.

The choice of the optimal control option is done depending on the required characteristics of the object and the acceptability of changes. Based on this, the final version of the project is determined. At the same time, regardless of the chosen option of object management, the procedure and conditions for ensuring public accessibility of estate cultural heritage are defined in special sections of heritage rent agreements

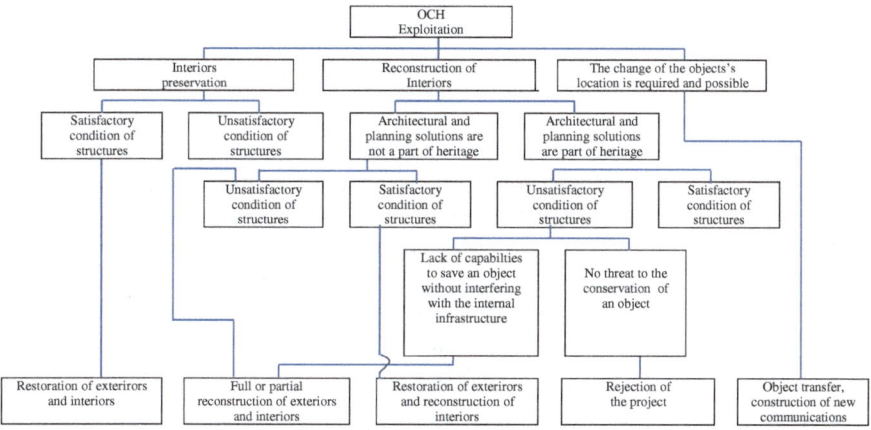

Fig. 3 Adapted model of decision-making process to choose the management option for the cultural heritage object

and in other documents, regardless of the type, form of ownership, departmental affiliation, nature and mode of use.

Owners and users of estate cultural heritage and of related land lots are obliged to observe the established procedure and terms of accessibility by public and to ensure the compliance with the requirements of state protection, preservation, use and promotion of estate cultural heritage during its realization.

When changing the functions of the building and its space-planning solutions to realize these functions, the major task is preserving the exteriors of the building, within its historical environment, its structure, plastics and silhouette. It is possible to accomplish this task through partial dismantling of a historical building, while saving only its facade and plastic of details. A special permission has to be granted by the state heritage protecting body for this design solution. There is a need for a special project to strengthen the facade at the stage of its connection with the new object, and during disassembly, it is necessary to very carefully protect the facade from accidental breakdowns, shocks, unexpected deformations and destruction.

11 Conclusion

It is important to create terms and conditions in Russian cities, to develop an effective and adequate policy of historical heritage management, aimed primarily to restore and reconstruct monuments of architecture and cultural heritage and attract investment, for elaboration and implementation of plans for the territories development, taking into account cultural attractions and historical features of the objects. In this regard, it is necessary to develop and implement a long-term concept of preservation and the integrated management of the cultural and historical environment in the region.

When marketing cultural heritage objects, four important issues or components should be addressed: (1) place or geographical location of the object; (2) product or inherent advantages and disadvantages; (3) price and/or investors and philanthropists, buyers/renters of the object expenses, related to their stay or business on the site; (4) promotion or information marketing of the object.

Professional disclosure and promotion of information about cultural heritage objects, in standards that are understandable both for professional circles and for ordinary citizens, is one of the most important strategic directions. In Russia, there are still very few instances when regional and municipal authorities are seriously engaged in the problem of information promoting and maintaining of favorable image of the site and objects of cultural and historical heritage in particular. Today's situation is such that without information, there is no trust. Without trust, there will be no investment and preservation of cultural and historical heritage. Without the management of heritage sites, there will be no history and attractions, originality and individuality of Russian cities.

Currently, many Russian cities get the understanding about the need for marketing in the sphere of historical and cultural heritage preservation and for information marketing. However, the efforts of cities in this direction are usually very small, with few exceptions. The surveys of the heads of departments of executive bodies and specialized NGO "center of preservation of cultural heritage" in Irkutsk city had revealed that marketing functions in the field of conservation of cultural heritage objects were realized only partially. The institution of private ownership of estate objects of cultural heritage experiences the formation stage. There is literally no evaluation of the effectiveness of marketing and of institutional changes happening.

The specificity of the ongoing systemic changes in the country is such that the problems of urban development should be considered within the framework of the institutional structure. This approach relates to the issues of territories development and landscape planning, building up, the role of education and cultural and historical heritage and problems of environmental and economic development. Three types of institutional changes can be distinguished by their origin in the context of the problem of preservation of cultural heritage objects: changes carried out by federal authorities, changes carried out by regional administrations and changes initiated by municipalities. In addition, there are endogenous and exogenous transformations.[1]

When we talk about institutional changes to address the problems of urban planning and the preservation of cultural heritage, we mean changes that need to be made in the relevant institutions (please, see Fig. 4). Namely, these are property and investment and financial institutions; in the regulatory and legislature framework; institutions providing the development, implementation and realization of regulatory documents; institutions providing professional training and education of personnel whose work is related to the solution of existing and potential problems of urban planning; as well as institutions responsible for information support.

As a result of the analysis, it was determined that all activities in the sphere of protection of objects of historical and cultural heritage may include the implementation of scientific researches aimed at the study, classification, cataloguing and publication of monuments, as well as the preparation and publication of legislative acts on the recognition of objects as monuments of history and culture, ratification of laws prohibiting damage, destruction or reconstruction of monuments, their export abroad, etc. Therefore, instructions are developed on the procedure for the registration, storage and restoration of artistic and historical valuables, and the works for *conservation, restoration* and *reconstruction* of estate objects of cultural heritage are conducted regularly.

Now, it is possible to create the basis for the use of cultural heritage as an instrument of urban policy, in determining the approaches and models of transformation of society, both social and economic components of this process. In our case, cultural heritage and its associated range of concepts are considered not only as a value in itself, but also in the context of the current social, economic, environmental and other problems of the society's existence, as well as prospects for its development.

To solve the problem of reconstruction of Russian cities, we propose a consistent set of activities presented in the program, which includes three consecutive stages.

---

[1]Endogenous transformations are transformations of the institutional structure, which are carried out by evolutionary changes in the existing rules and norms that form the basis of institutions. Exogenous institutional changes are inherently more radical.

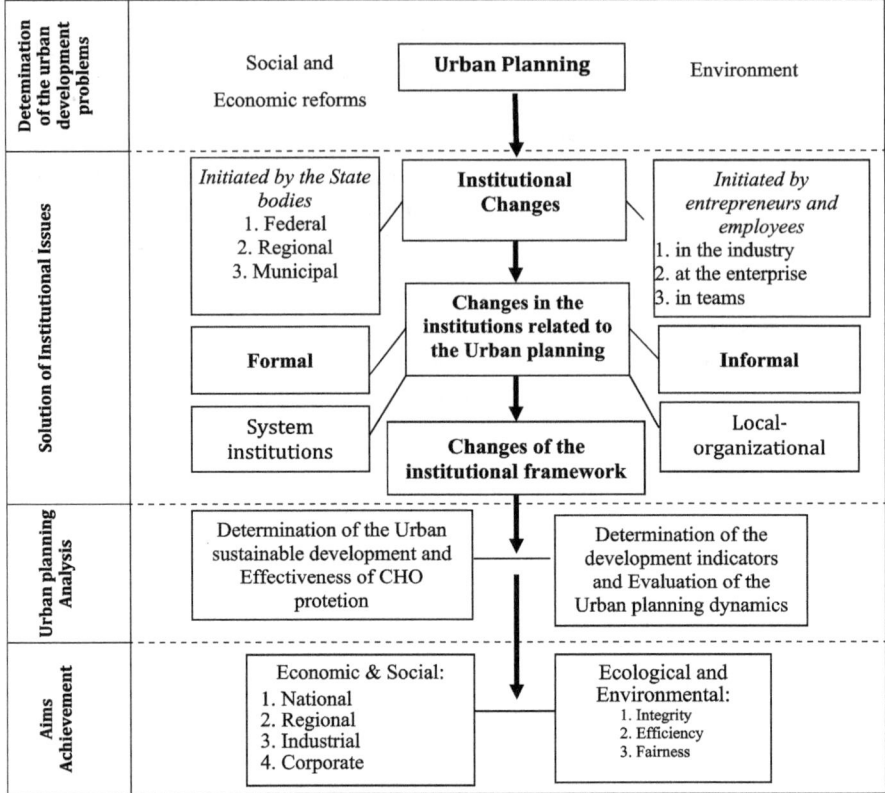

Fig. 4 Relationship between urban planning and institutional changes

The urban planning problems of reconstruction solved in the proposed program are related to preservation of the buildings having historical and architectural value, increase in density of housing stock by means of construction on free sites, ensuring hygienic conditions of building up (insolation and aeration modes, level of gas pollution and noise), optimization of structure and capacities of social, engineering and transport infrastructure. The program proposed includes the following steps [16]:

1. Reconstruction and modernization of old and dilapidated buildings and structures,
2. Complex per block reconstruction,
3. Construction of residential complexes.

At least, three institutional "dimensions" can be considered in the context of sustainable territorial development: formal rules, informal constraints and the effective enforcement of these rules and constraints. The creation of a favorable environment for living is based on the development of a favorable regulatory and legislative framework. In this regard, the need to improve the institutional structure should be noted, which in Russia and in regions still remains complicated and confusing.

Economic reforms and new principles of public administration resulted in the abolition of the old system of planning and development. Complex urban problems, including the prioritization of development and reconstruction of the city in accordance with social, environmental and economic goals and criteria, the need to coordinate the development and implementation of long-term and current programs for the development of urban sectors, its infrastructure, plans for development of administrative districts and other territorial units, can be solved effectively only within a united system of planning for development of the city.

For the further research of institutional changes, it is necessary to define tools and methods of evaluation of institutional changes effectiveness and to define levels of effectiveness of institutional transformations too. To assess the effectiveness of institutional changes in urban planning, we suggest to use Q-methodology and IADF method. Another modern approach proposed for the study of institutional problems of urban planning is the method of modeling, and to evaluate the effectiveness of institutional changes, a cross-sectional regression model is proposed.

The economic compound in the management and development of the cultural heritage objects should take into account its individual value and sociocultural significance. It is difficult to overestimate the role of cultural heritage in the process of social and economic development and preservation of the environment, as an instrument of urban policy, in determining the approaches and models of transformation of society, and prospects for its development.

References

1. North D (1997) Institutions, institutional changes and economic performance (in Russian). Fond economicheskoyknigi "Nachala", Moscow
2. The Urban planning Code of the Russian Federation issued on December 29, 2004 # 190-FZ (ed. August 2, 2019) (with ed. and extra, effective since November 01, 2019) (in Russian)
3. Comparative analysis of cultural landscape management practices. Heritage Institute, Moscow, 1999. pp 38, 39. (in Russian)
4. Mazurov YL (1994) (in Russian) Мазуров Ю.Л. Unique territories: the conceptual approach to identification, protection and use. Unique territories in the cultural and natural heritage of the regions. Moscow
5. Slavina TA (1997) Subject of protection. Monuments of history and culture of St Petersburg. 4:10–13
6. Methods of economic evaluation of cultural heritage objects (historical and architectural monuments of the people of the Russian Federation. ANO, Center for Independent Evaluation, Moscow, 2012. (in Russian)
7. Kuleshova ME (2002) Management of cultural landscapes and other objects of historical and cultural heritage in national parks (in Russian). Publishing House of the Center for wildlife protection, Moscow
8. Gryaznova AG et al (2010) Real estate valuation: a textbook for students of higher school, studying specialty Finance and credit (in Russian) In: Gryaznova AG, Fedotova MA (ed) Financial Acad. under the Government of Russian Federation, Institute of professional evaluation. Ed. 2nd, revised and expanded. Finance and statistics, Moscow
9. Afonina AB (2009) Architectural Monuments as the object of civil rights: a practical guide. Information and legal system (in Russian). ConsultantPlus, Moscow

10. Yagodkin VK (1984) On the issue of economic realization of public property on the means of production (in Russian). Economic Sciences. No. 7; Socialist property and improvement of forms of organization of production. Kazan
11. Cherkovets VN, Pokrytan AK (1970) Industrial relations and economic laws of socialism. Moscow
12. Morovaya-Minsk AP (1992) Nature, structure and factors of formation of economic relations (in Russian)
13. Afanasenko ID (1989) Restructuring and ownership. The Soviet Economist (in Russian)
14. Gaidar E et al (1998) Economies in transition. Essays on the economic policy of post-Communist Russia 1991–1997 (in Russian)
15. Kornai J (1990) The way to a free economy (in Russian). Ekonomika, Moscow
16. Dayneko DV, Dayneko AI (2017) Institutional problems in urban planning and modern methods of reconstruction for Siberian cities. Matec web of conferences iop conference series: materials science and engineering (MSE) 262 012065. 10.1088/1757-899X/262/1/012065

Chapter 6
Greening Existing Garment Buildings: A Case of Sri Lanka

Thanuja Ramachandra and Achini Shanika Weerasinghe

Abstract Green retrofitting is the justifiable solution for contemporary issues such as global warming, resource depletion, and greenhouse gas emissions which have arisen due to the existing conventionally built environment. Nevertheless, the building owners are less willing to invest in green retrofits due to the contradictory views associated with the first cost and payback period implications of the green retrofit. In that context, this chapter presents an assessment of the first costs and life-cycle saving implications of fourteen (14) energy and water-efficient retrofits incorporated into four (04) garment buildings in Sri Lanka to find the retrofit options which are financially sound. The green retrofits such as skylights, LED lights, steam line insulation, compressed airline modification, biomass boiler, evaporative cooler, energy-efficient chiller, and VSDs were implemented in the garment buildings in Sri Lanka as energy and IEQ measures. Additionally, the selected buildings were upgraded using green retrofits such as subsystem-level water meters and low water flow push taps. All the selected green retrofits are financially practical with positive NPVs and SPB periods of less than 5 years. This information would provide some cost-based considerations for green investors in the selection of retrofits for other industrial buildings and thereby contribute to promoting sustainable developments.

Keywords Green retrofits · Industrial · Costs · Savings · Sustainability · Sri Lanka

1 Introduction

Globally, buildings are responsible for increased energy consumption and greenhouse gas (GHG) emissions. For example, buildings construction and operations as end-use sectors consume about 36% of global final energy and produce 39% of energy and process-related CO_2 emissions in 2018 [20]. The impact of the building and construction industry on the GHG emissions and depletion of natural resources

T. Ramachandra (✉) · A. S. Weerasinghe
Department of Building Economics, University of Moratuwa, Moratuwa, Sri Lanka
e-mail: thanujar@uom.lk

© The Author(s), under exclusive license to Springer Nature Singapore Pte Ltd. 2021
R. J. Howlett et al. (eds.), *Emerging Research in Sustainable Energy and Buildings for a Low-Carbon Future*, Advances in Sustainability Science and Technology,
https://doi.org/10.1007/978-981-15-8775-7_6

is staggering. Therefore, the issues of greenhouse gas (GHG) emissions, energy consumption, and depletion of natural resources challenged the sustainable development goals (SDGs) [45]. Consequently, building owners have pledged to reduce emissions to slow down the rate of warming, besides ensuring environmental, social, and economic sustainability [35].

As of today, green buildings are an important concept, which addresses the issue of resource consumption. Green buildings have now become the flagship of sustainable development (SD) that takes the responsibility for balancing the three pillars of the SD: social, environment, and economic or the triple bottom line approach [41]. However, to demolish an existing building and to build a new green building in its place is counter-productive to the idea of energy conservation. By some estimates, it would take more than 65 years to regain the energy savings of demolishing an existing building and replacing it with a new green building [51]. However, most existing buildings will still be in use for the next 50–100 years due to its long lifespan nature [29]. Thus, green retrofits have forefronted in countries.

The challenge of climate change could be effectively addressed by retrofitting the vast stock of existing conventional buildings [36]. Similarly, the green retrofitting of existing buildings would reduce energy consumption and GHG emission [27]. The built environment will have less responsibility to deal with global warming unless the rate of green retrofits amplified [47]. However, this is on account of the number of buildings built every year in developed nations that only corresponds to 1.5–2% of the existing building stock [8]. Although there is increasing recognition that green buildings outperform conventional buildings in terms of a variety of environmental, economic, and social indicators, much less is known about how green building initiatives might be incorporated into existing buildings [36].

Unfortunately, retrofitting existing buildings is significantly more difficult than creating a new green building from scratch. For example, in existing multi-tenant commercial buildings, any sustainability retrofit or technology upgrade needs the cooperation and participation of a wide range of stakeholders [23, 25]. Moreover, due to issues such as long payback periods [22, 34], excessive cost, and limited access to capital [12, 22, 60], the industries are unenthusiastic about green retrofit. Other barriers are lack of retrofit experience and lack of understanding of the available retrofit technologies [12, 36]. The main challenge met is the unwillingness of building owners to pay for retrofits [30].

Although there is a wide range of retrofit technologies readily available, finding the most cost-effective retrofit measures for projects is still a major technical challenge. If there is no financial support from the government, the cost of the retrofit falls onto building owners while the benefit often flows primarily to the tenants. Therefore, the significance of the subject is the identification of the most appropriate retrofit options based on the potential expenses and effects involved [3].

In the context of Sri Lanka, the existing building stock is substantially large and represents one of the biggest opportunities to reduce energy, waste, air pollution, and global warming. However, a limited number of buildings have been certified for the incorporation of green features in Sri Lanka [56]. For instance, 51 buildings have been certified to date, while there are only seven existing buildings which were converted as green buildings [52]. Further, the green building investors fail to appreciate the subsequent benefits received by those buildings during the operational phase [55].

Therefore, a detailed analysis of the costs and benefits of green retrofits is essential at the pre-retrofit stage. This research aims to assess the green retrofits towards identifying the potential retrofit technologies which can be integrated with full awareness of their cost commitments and saving potentials and thereby enhance the rate of sustainability transition of the existing building stock.

Accordingly, this piece of work enables the identification of the rightest retrofit options based on the analysis of costs and benefits of green retrofits implemented in green buildings in Sri Lanka.

The following three objectives were set towards achieving the aim of this research:

- Objective 1: To review the concept of green retrofitting, economic drivers and challenges, and cost implications and potential savings of green retrofitting,
- Objective 2: To identify the green retrofits technologies incorporated in the selected garment manufacturing buildings,
- Objective 3: To assess the costs and benefits of green retrofit technologies implemented in the green-certified garment manufacturing buildings in Sri Lanka.

The literature review section of this chapter addresses the first objective while the second and third objectives are achieved through data collection and analysis presented in the results section of this chapter.

2 Literature Review

2.1 Green Retrofitting

In the past, Sanvido and Riggs as cited in [21] explained retrofit projects as the modification or conversion of an existing facility, structure, and process. Unlike complete replacement, retrofits involve additions, rearrangements, deletions, and replacements of one or more parts of the facility. Later, Egbu as cited in [2] more elaborately defined the retrofit as 'refurbishment to encompass renovation, rehabilitation, extension, improvement, conversion, modernization, fitting out, and repair which is undertaken on an existing building to permit its reuse for various specified purposes' (pp.684). Besides, Latham as cited in [21] viewed retrofitting as 'a process that reaps the benefits of the embodied energy and quality of the original building dynamically and sustainably' (pp. 1361). Subsequently, USGBC [53] defined the retrofit as,

'any type of upgrade at an existing building that is wholly or partially occupied to improve energy and environmental performance, reduce water use, improve comfort and quality of space in terms of natural lighting, air quality, and noise, all done in a way that it is financially beneficial to the owner'.

Similarly, green retrofitting is an effective strategy to enhance the sustainability of existing facilities [29]. Particularly, green retrofit can be defined as the 'incremental improvement of the fabric and systems of a building with the primary intention of improving energy efficiency and reducing carbon emissions' [27]. Furthermore, the USGBC is continuously stressing that the buildings should sustain over time. For instance, USGBC [52] noted that green retrofits range from minor work that involves installing new heating, ventilating, and air-conditioning (HVAC) components, rooftop solar panels, and placing a bike rack outside the building to major work that involves multiple complex renovations on building facades and interior spaces [52]. Similarly, green retrofitting includes an envelope, structural upgrades, and spatial layout changes as well as improvements to the building's environmental systems; mechanical, electrical, plumbing, etc. and systems influencing thermal comfort; exterior insulation, roof insulation, etc. [18].

Furthermore, the existing buildings need retrofitting rather than demolishing and rebuilding existing buildings to meet new energy standards at an excessive cost [6]. The same authors argued that even though the reduction in operational energy is not always the primary reason, any retrofitting should include full consideration and renewal of the energy systems for the building. In particular, the term 'deep energy retrofitting' is used where the entire fabric and conventional systems of the building must be evaluated, redesigned, and reconstructed in an integrated way [16]. Similarly, the deep energy-efficient retrofits such as any aesthetic or functional upgrades reduce the demand for non-renewable energy resources and help the society and environment sustainability [18].

On the other hand, the implementation of green retrofits follows the specific standard rating system of LEED for existing buildings, which evaluates operations improvements and maintenance [28]. As per the green building rating system, the greening of the existing building incorporates different strategies and technologies. However, the green retrofitting is only a part of greening existing buildings, which only addresses the physical improvement to the building and site, whereas the greening of existing buildings also includes planning, monitoring, controlling, managing certain aspects, etc.

2.2 Economic Drivers and Challenges

2.3 Economic Drivers

The owners or occupiers of existing buildings may have varying economic drivers, should a building be retrofitted and the conflictions about time and the way of implementing the retrofit. For example, Fuerst and McAllister [15] explained that high rent, occupancy rate, and tax reduction will motivate the owners to implement retrofit projects. On the other hand, Gucyeter and Gunaydin [17] pointed out that occupiers are interested in energy cost savings. Even though the owners invest in energy-efficient retrofits, it is the occupiers who receive the most direct benefits of energy cost savings [30, 38, 42]. Similarly, low rent [35] and productivity [50, 60] are other drivers that motivate the occupiers to retrofit their existing facilities.

Moreover, for both owners and occupiers, the main motivations for green retrofitting are low operation costs, followed by a high return on the green investment, higher asset values, improved tenant satisfaction, and competitive advantage [33]. As earlier studies highlighted, the main driver for green retrofitting is the company's corporate responsibility for sustainably leading the market to increase environmental awareness in society and employee satisfaction [1]. Similarly, increasing environmental awareness would eventually improve the company's corporate image. Furthermore, these authors observed that the client-driven approach and employee satisfaction are other reasons for greening existing buildings [39, 19].

From a stakeholder's perspective, a drastic reduction in GHG emissions is vital to mitigate global warming and climate change. Therefore, it is a duty to look at methods of reducing emissions from the vast stock of existing buildings that drive green retrofitting [59]. Further, green retrofitting of existing buildings reduces the energy use worldwide [8]. Indeed, policymakers have acknowledged the need for more retrofitting projects in the company's vision, and the environmental policy of the client is a way of reaching sustainability in the built environment [59].

Reviewing the earlier studies, it was identified that the reduced whole life-cycle costs are the major economic drivers which motivate the owners and occupiers to invest in green retrofits [1, 7, 61], which is followed by rising energy costs [1, 61]. Energy-efficient retrofits reduce the cost of energy and the operational costs, which later contribute to the reduction of the whole life cost of the building [30, 33, 38]. Moreover, high return on major investments within a short payback period such as renewable energy projects also motivates the investors to go for retrofitting [61]. Green retrofitting enhances building value and thus increases the property value, and the owners will be able to earn a high rental [7]. Additionally, the green retrofitted existing buildings have high resale value, less construction cost compared to new construction, less depreciation in rent, and reduced-price in insurance cost and have a greater chance of lease renewal [61]. All these together drive the owners and occupiers to invest in green retrofitting.

2.4 Economic Challenges

The stakeholders of existing building stock contribute to drastic reductions in GHG emanations globally [59]. Even with the growing concerns of the stakeholders over sustainability aspects, green retrofitting is still not winning its place at the forefront due to the challenges that exist [40]. According to Wilkinson [58], the earlier researches show that building stakeholders are less likely to agree with green retrofitting due to the challenges that exist when deciding whether to execute a retrofitting project.

The major barriers affecting the green building implementations are higher costs for green design and energy-saving material, technical difficulties during the construction process, lack of knowledge and awareness of green technologies, and conflict of interest between various stakeholders [62]. An interesting finding is that the owners do not perceive the cost as a barrier for greening their existing buildings. However, in the case of new construction, this is one of the main reasons for the companies to be reluctant [13, 19, 33, 62]. The reason why the cost is not considered as a barrier in these cases is primarily the fact that the converted buildings are mostly commercial, and the owners see the certification as an opportunity to enhance their corporate image, so they are more flexible with their budget.

The high initial cost is the main challenge in designing and retrofitting existing buildings towards achieving sustainability. Therefore, the initial cost for green space may be acceptable for new construction, but any improvements to existing space need capital expenditure [32].

A study by Pedini and Ashuri [40] revealed that the lack of knowledge and experience leads to deciding not to implement green retrofit. For instance, company financial plans, which are usually unstructured results in difficulties to track life cycle cost (LCC) for a project and longer-term gains, which are hard to record. Hence, the initial capital cost of green buildings, lack of life-cycle costing knowledge, insufficient funding, and the fluctuation of the price of green materials are the factors affecting green retrofit projects financially.

Shift in government priorities, such as supporting regulations, but removing tax incentives and subsidies are other barriers in implementing sustainability in existing buildings. Removal of tax incentives and subsidies contributes to the barriers in sustainability implementation [5]. As a result, lack of financial incentives, loss of financial incentives, unevenness and the difficulty to obtain tax and regulatory incentives, uncertain end dates of incentives, low investment, and the absence of government and private sectors' involvement in the green building development affects the implementation of the green retrofit projects by stakeholders in existing buildings.

According to Reza et al. [43], the information of green structures and items should be available to significant parties in the building industries as well as to the overall population to get more attention for green buildings. For instance, owners and investors who do not have access to enough information failing to realize that green building is the best course of action to pursue, unaware of the benefits of green building and also the lack of knowledge about financial institutions have primarily affected stakeholders from implementing green retrofit projects.

2.5   Cost Implications and Saving Potentials of Green Retrofits

In this section, the major green retrofit studies carried out in different contexts are reviewed, and the outcome is in terms of potential savings: energy saving, CO_2

emission reduction, cooling load reduction, and water savings. The cost implications such as cost saving, payback period, additional cost, and investment cost of individual green retrofits are summarized in Table 1.

As shown in Table 1, individual green retrofits reviewed are categorized into the five (05) sustainable features. Considering the SS feature, an experimental study on energy and environmental performance of the green roof system by Santamouris et al. [44] found that the energy saving due to the reduction of the cooling load of the green roof system was between 15 and 49%. Another study, [1] using actual site measurement, highlighted that giving space for bicycle parks as an alternative transportation facility saves 26% of energy.

In terms of WE feature, the same authors, Aktas and Ozorhon [1], showed that implementing subsystem-level water meters and sensor faucets with low flow rates equally saves 40% of water, while gray water recycling saves another 43% of potable water use for non-portable purposes by re-directing the treated gray water. Another study conducted on an industrial factory in China by Li et al. [26] using simple energy calculations and site readings indicated that rainwater harvesting through the permeable pavement, garden space, roof greening, landscape pool, and other runoff control measures absorbs 5716.3 kg carbon dioxide each year and 20% more rainwater is absorbed to the ground through the runoff water infiltration.

With respect to EA feature, Stefano [48] reported that the potential to save electricity and reduce electricity-related carbon dioxide emissions were achieved by modelling the installation of four energy-efficient lighting technology alternatives to replace 1.2 m fluorescent lighting fixtures.

The four energy-efficient lighting technology alternatives include electronic ballasts, T8 magnetic ballasts, T8 electronic ballasts, and T5 electronic ballasts that would result in energy savings of 13.9, 20.5, 24.4, and 64.9%, respectively. In another study, Mahlia et al. [31] highlighted the significant savings of energy and cost of $37 to $111 million through retrofitting incandescent lamps with compact fluorescent lamps. Recently, Si [46] calculated energy saving, investment cost, and payback of lighting retrofits of a university building in the UK. The findings of this study reveal that 30w halogen lamps supply substantial saving of energy by 1800 kWh for an investment cost of £298, and the investment cost could be recovered within 0.8 years compared to other energy-efficient lighting retrofits such as T5 lamps and lighting timers. However, implementing T5 lamps gives a little saving (378 kWh) for a high investment cost of £440 which needs 8 years to pay back.

Further, Si [46] indicated that integrating a building management system (BMS) provides an annual energy saving of 18,413 kWh for an investment of £3,000 which could be paid back within 2.4 years, and installing secondary glazing on all single-glazed windows provides an annual energy saving of 20,160 kWh which would pay back the investment cost of £11,200 within 16 years.

The energy savings and cost-effectiveness of individual retrofit options in single-family buildings were studied by Cohen et al. [11] based on analyzing metered energy consumption and actual installation costs. The results showed that the replacement of windows is not a good retrofit option since it has a very small normalized annual energy saving of 2–5% with an average cost of conserved energy greater than $15/GJ,

Table 1 Summary of cost and saving effects of green retrofits

Sustainable features	Green retrofits/technologies	Potential savings				Cost implications			
		Energy	CO_2 emission	Cooling load	Water	Cost saving	PB (years)	Additional cost	Investment cost
Sustainable sites (SS)	Provide bicycle racks	26%	–	–	–	–	–	–	–
	Green/vegetated roof	–	–	15-49%	–	–	–	–	–
	High albedo and vegetated roof	–	–	6-33%	–	–	–	–	–
Water efficiency (WE)	Subsystem-level water meters	–	–	–	40%	–	–	–	–
	Sensor faucets with low flow rates	–	–	–	40%	–	–	–	–
	Gray water recycling	43%	–	–	–	–	–	–	–
	Rainwater recycling	–	5716.3 kg	–	20%	–	–	–	–
Energy and atmosphere (EA)	Electronic ballasts over 1.2 m fluorescent lamps	13.9%	13.9%	–	–	–	–	–	–
	T8 magnetic ballasts over 1.2 m fluorescent lamps	20.5%	20.5%	–	–	–	–	–	–
	T8 electronic ballasts over 1.2 m fluorescent lamps	24.4%	24.4%	–	–	–	–	–	–
	T5 electronic ballasts over 1.2 m fluorescent lamps	64.9%	64.9%	–	–	–	–	–	–

(continued)

Table 1 (continued)

| Sustainable features | Green retrofits/technologies | Potential savings | | | | Cost implications | | | | |
		Energy	CO$_2$ emission	Cooling load	Water	Cost saving	PB (years)	Additional cost	Investment cost
	Replacement of incandescent bulbs by 25%, 50%, and 75% of compact fluorescent lamp (CFL)	–	–	–	–	$37, $74 & $111 million	–	–	–
	T8 lamps replacement with T5 lamps	378 kWh	–	–	–	–	8	–	£440
	Replacing 50 W halogen spotlights with 30 W halogen lamps	1800 kWh	–	–	–	–	0.8	–	£298
	Use of time-scheduled control of lighting	1620 kWh	–	–	–	–	1.2	–	£240
	Wall insulation	2%	–	–	–	–	–	–	18,600 EUR
	Window replacement and upgrading	2–5%	–	–	–	>$15/GJ	–	–	–
	Cladding replacing and insulations	–	–	–	–	10%	–	–	–
	Improvement of heating, preheat upgrade	19–34 GJ	–	–	–	<$2.70/GJ.	–	–	–
	Floor insulation	–	–	–	–	–	–	30–100 Yuan/m2	–

(continued)

Table 1 (continued)

Sustainable features	Green retrofits/technologies	Potential savings				Cost implications			
		Energy	CO_2 emission	Cooling load	Water	Cost saving	PB (years)	Additional cost	Investment cost
	Heat recovery	5%	–	–	–	–	–	–	17,000 EUR
	Heating system retrofits								
	Ground source heat pump compared to electrical heating	–	11.3 tons	–	–	–	0.25	–	–
	Boiler efficiency improvement	64.3%	–	–	–	74%	0.64	–	–
	Low-E double glazing	12%	–	–	–	–	–	–	76,000 EUR
		20,160 kWh	–	–	–	–	16	–	£11,200
	Building management system	18,413 kWh	–	–	–	–	2.4	–	£3000
	Solar collectors and PV cells	25%	–	–	–	–	–	75% of NPV	–
		2–6%	16.8 tons	–	–	–	–	–	–
Indoor environmental quality (IEQ)	CO_2 sensors	60%	–	–	–	–	–	–	–
	Air filtration, air sealing of ventilation system	11%	11%	–	–	–	–	–	7600 EUR
	system	10%	–	–	–	–	–	–	–

while the heating system retrofit produced significant savings of 19–34 G J/year with an average cost of conserved energy less than \$2.70/GJ. Stovall et al. [49] performed a series of experiments that were performed by another study to examine wall retrofit options such as replacing the cladding, adding insulation under the cladding, and multiple sealing methods that can be used when installing replacement windows in well-built or loosely built rough openings. The results from the experimental tests were later applied to an energy model to estimate whole-house energy impacts, and it was found that the annual utility cost savings are equal to 10% for most of the wall retrofit options. Another study in the EA category, Zhang et al. [62], proved the incremental costs of applying floor insulation ranges between 30 and 100 Yuan/m^2.

Similarly, Ascione et al. [4] studied the energy-efficient retrofit of historical buildings and proposed a multi-criteria approach for the energy refurbishment of historical buildings and employed a numerical energy model to simulate the effectiveness of energy performance and economic feasibility of several retrofit actions. The results showed that wall insulation, heat recovery, and a double-glazing system with low-emissive coating involve an investment cost of around 18,600 EUR, 17,000 EUR, and 76,000 EUR and save 2, 5, and 12% of annual energy, respectively. Doherty et al. [14] conducted an experimental investigation on implementing a ground source heat pump in a university building concluded that the predicted CO_2 savings for the system are 11.3 tons per annum and the payback period for the system is 0.25 years compared to electrical heating. Another simulation study was done by Ciampi et al. [10] on a historical building in Italy highlighted that the use of a natural gas-fired condensing boiler saves 64.3% of energy, reduces the operating cost by 74.0%, and the SPB equals 0.64 years.

In terms of renewable energy, Verbeeck and Hens [54] discussed the economic viability of different retrofit measures through the use of the NPV method concluded that solar collectors and PV cells reduce the total primary energy consumption by 25%; however, this retrofit potion is not financially optimal due to the increase of the total net present value by 75%. Similarly, Li et al. [26] estimated that the solar thermal system and the PV together can produce 2 ~ 6% of the total energy consumption and the CO_2 emissions will be reduced by 16.8 tons per year.

Considering the IEQ feature, Li et al. [26] estimated the energy saving due to CO_2 sensors was 25.99 kWh/m^2, 60% of the predicted usage. Moreover, Ascione et al. [4] and Nabinger and Persily [37] investigated the impacts of air tightening retrofits such as installing house wrap over the exterior walls; sealing leakage sites in the living space floor; tightening the insulated belly layer; and sealing leaks in the air distribution system showed that for the two studies, the energy reduction of heating and cooling was 11% and 10% respectively.

As per the foregoing review, most green retrofits were in the EA category and contribute to the reduction of energy consumption and CO_2 emission, while green retrofits related to WE, SS, and IEQ are comparatively less. Among the selected green retrofits, energy-efficient lighting retrofits are important retrofit measures for any kind of buildings, while solar collectors and PV cells, Low-E double glazing, heat recovery, wall insulation, HVAC systems, and air filtration are not financially best due to the incremental costs and long payback periods. Further, a lack of integration of retrofits of the material and resources (MR) category is visible from the earlier studies. As evidenced by the review and summarized in the table, retrofit technologies vary widely in terms of their implications on initial cost, operational costs and savings, and finally their contributions to overall sustainability. However, in most of the previous studies, retrofits are assessed based on a single parameter, either potential savings or cost implications, and the trade-off between initial investment commitments and saving potentials is seemed to have given less priority in integrating the retrofit technologies. Moreover, most of these reviewed studies were based on numerical simulations instead of reporting actual cost implications and potential savings in implementing desired green retrofits. Only a few studies have conducted an analysis that accounts for the NPV and payback period of retrofits based on actual cost. Thus, it is essential to assess the overall cost-effectiveness of these retrofits and integrate the most effective retrofits and thereby achieve sustainability. Detailed analysis of the actual cost implications would increase the level of confidence of building owners to retrofit their buildings for better sustainable performance.

3 Methodology

Quantitative researches are more suited to finding out the extent of variation and diversity in any aspect of social life and involve quantitative data [24]. This research was approached using quantitative methods involving quantitative data collected through document analysis. Four (04) garment buildings certified under the LEED O + M Existing Building category were selected for this study as those buildings were in the similar business category and were certified under the same rating system. Relevant data, the costs, and savings due to green retrofits were collected from those buildings by referring to documents related to green retrofit projects. The collected data on costs and economic savings of green retrofits implemented were analyzed using net present value (NPV) and simple payback (SPB) to determine the lifetime gain and time to recover the initial capital cost of the retrofits, respectively.

Assumptions made in performing NPVs and SPBs are as follow:

- Energy savings from retrofitting are the difference between the energy consumptions before and after retrofitting. Annual cost saving of retrofit is a function of energy savings (kWh) and the price of a unit of electricity (LKR 14 per kWh).
- Water savings from retrofitting are the difference in water consumption before and after retrofitting. Cost of water equals to LKR 0.60 per one liter.
- The cost savings achieved due to the reduction of energy and water consumption through the implemented green retrofits are the cash inflows, and the first investment cost is cash outflow of the projects.
- The market interest rate (r) is assumed as 9%.
- The discount rate (i) equals to 4.26% at the average inflation rate (e) of 4.5% for the period of 2012–2016, obtained from the Central Bank Annual Report 2016 [9] which was used in calculating the life-cycle saving of each retrofit (Table 2).

Table 2 Profile of LEED-certified existing buildings

Building	Rating system (Version)	Rating level	Green space	Business
GB1	LEED O + M: Existing buildings (v2009)	Platinum	Industrial manufacturing	Garment
GB2		Gold		Garment
GB3		Gold		Garment
GB4		Silver		Garment
GB5		Gold		Spirits and wines
GB6	LEED O + M: Existing buildings (v2.0)	Platinum		Garment
GB7	LEED O + M: Existing buildings v4.0	Gold	Warehouse	Logistics

$$i = \frac{1+r}{1+e} - 1$$

- In the selected buildings, the retrofitting was carried out in different period (2013–2015) as indicated in Table 3. However, the national consumer price index (2013 = 100) for the said years does not seem to change significantly [9].
- The retrofits will give an equal annual monetary saving throughout the lifetime of the project and no scrap value at the end of the project.
- The year 2015 was considered as the base year for the calculation of NPV, using the formula below where FV-future value, discount rate (i), and n-number of years.

$$NPV = \sum_{i=0}^{n} FV \frac{(1+i)^n - 1}{i(1+i)^n}$$

- SPB of each retrofit equals.

Table 3 Profile of the selected green buildings

Building	Gross floor area (ft2)	Age of buildings	Number of employees	Year of green retrofit	Life cycle (Years)	Type of function
GB1	168,000	10	2600	2015	50	Garment
GB2	155,200	19	2200	2015	50	Garment
GB3	181,048	19	2400	2014	50	Garment
GB4	124,000	25	1800	2013	50	Garment

$$\text{SPB} = \frac{\text{First Cost}}{\text{Annual Saving}}$$

• The lifetime of individual retrofits was assumed based on the market information. As the lifetime of retrofits varies, an equivalent annuity cash flow was calculated to compare the retrofits for net savings.

4 Results

4.1 A Case of Garment Buildings in Sri Lanka

According to the profile of green-certified buildings under the LEED O + M Existing Building category, in Sri Lanka, six (06) industrial manufacturing buildings and one (01) warehouse building were transformed to green with the integration of green retrofits. Among, most of the buildings (5 out of 7) were certified under LEED O + M: Existing Buildings (v2009) category. Those five (05) buildings include four (04) garments manufacturing buildings and one (01) building used for the production of spirits and wines, as shown in Table 2. Therefore, four (04) buildings of similar business, garment, and certified LEED O + M: Existing Buildings (v2009) were considered for the study. However, two of these buildings, GB2 and GB3, are 'Gold' certified while GB1 and GB4 are certified as 'Platinum' and 'Silver', respectively. Although these buildings are certified with different certification levels, there are no differences in the implementation of green retrofits by the certification. However, the cost implications of each retrofit vary due to the physical condition of retrofit technologies such as capacity, type, model, and scale. Therefore, the implications of costs and savings due to retrofits implemented in each of these buildings were analyzed separately, and comparisons were made between the technologies toward recommending the most economical retrofit technologies.

Table 3 presents the profile of four (04) garment manufacturing buildings certified under LEED O + M: Existing Buildings (v2009) and achieved platinum, gold, and silver rating levels.

According to Table 3, selected buildings are of approximately similar range, in terms of gross floor area, land area, number of employees, and life-cycle years of 50. All buildings are large scale, leading garment factories in Sri Lanka with the age of above 10 years. The next sections present the green technologies implemented and the cost implications of those technologies.

4.2 Green Retrofits Implemented in Garment Buildings

Firstly, the green retrofits/technologies implemented in the selected four garment buildings were identified. As seen from Table 4, irrespective of green certification

Table 4 Summary of green retrofits implemented in selected green buildings

No.	Retrofits	Sustainability criteria	GB1	GB2	GB3	GB4
[1]	Replace existing chillers with the evaporative cooler	EA/IEQ	✓	✓	–	✓
[2]	Replace oil fired steam boiler with the biomass boiler	EA	✓	✓	✓	✓
[3]	Replace existing chillers with energy-efficient chillers	EA/IEQ	✓	✓	✓	–
[4]	Replace clutch motors with servo motors	EA	✓	✓	✓	–
[5]	LED lights	EA/IEQ	✓	✓	✓	✓
[6]	Insulate steam lines	EA	✓	✓	✓	✓
[7]	Fluorescent lamps with skylights	EA/IEQ	✓	✓	✓	✓
[8]	Compressed air line modification	EA	✓	✓	✓	✓
[9]	Biogas project	EA	✓	–	✓	✓
[10]	Install variable speed driver (VSD) for chiller	EA/IEQ	✓	✓	–	–
[11]	Install variable speed driver (VSD) for compressor	EA/IEQ	✓	–	✓	✓
[12]	Recovery of flash steam for water heating	EA	✓	–	–	✓
[13]	Install low water flow push taps	WE	✓	–	–	–
[14]	Install subsystem-level water meters	WE	✓	–	–	–

grading, i.e., 'Gold' or 'Platinum', almost all buildings have incorporated a similar set of technologies except few technologies. Altogether 14 technologies were found and of which 11 technologies were implemented in all four buildings except GB4 where the building continues to operate with existing chillers and clutch motors.

From the sustainability perspective, most of these retrofits implemented in selected buildings belong to EA and IEQ categories, while some retrofits related to WE and retrofits related to SS and MR categories are not at all implemented in these buildings. SS offers several green retrofit technologies that could be incorporated when converting an existing building to a green. However, the selected buildings have already been incorporated with features of parking spaces, light-colored roofing and paving surfaces, and low reflectance surfaces. This facilitated the transformation of the existing building into a green building with the least cost and higher probability of achieving green certification. In terms of green retrofits related to WE, in most of the selected buildings, technologies such as water meters, automatic controls, dry fixtures and fittings, and gray water recycling helped to convert the existing buildings into green buildings without going for any retrofits. However, GB1 has installed low water flow push taps and subsystem-level water meters as retrofit measures. Considering MR, the green rating system, LEED O + M: Existing Buildings (v2009) promotes sustainable purchasing of consumables and solid waste management. Therefore, retrofit technologies in this category were given the least priority in the selected buildings.

Energy retrofits have been given the topmost priority over other retrofits due to the economic savings. The focus on energy efficiency was less at the first construction of those buildings; therefore, a considerable amount of improvements was done when converting those existing buildings into green buildings. The respective buildings include the green retrofits such as skylights, LED lights, steam line insulation, a biomass boiler, and a compressed air line modification to improve energy-efficient performance. Further, few of the energy retrofits show both energy efficiency and IEQ, and those retrofits ensure the ventilation and lighting aspects of the buildings. For example, the installation of skylights enables the daylight into the building, LED is used as task lighting for the sewing machines, and use of the evaporative cooler, energy-efficient chiller, and VSDs ensure the demand control and air infiltration of the ventilation system. The main reasons for the limited adoption of green retrofits are those features were already in the selected buildings and the building owners focused on achieving sustainability via sustainable strategies (without involving any additions, rearrangements, deletions, and replacements of one or more parts of the facility).

4.3 Cost Implications of Green Retrofits Implemented in Garment Buildings

As discussed above, the selected buildings have incorporated most of the retrofit technologies in common. Besides, the retrofitting was carried out in all four buildings during the year 2013–2015, where there were no significant changes in costs and prices of retrofit technologies. Therefore, a comparison was made in terms of first (initial) cost, annual savings, life-cycle savings, NPVs, and SPBs of each retrofit technology among the selected buildings, and finally, an average of NPVs and SPBs of all buildings was calculated for each retrofit technology. Table 5 and Table 6 in annexure present the initial cost, annual and life-cycle cost savings, NPVs, and SPBs of all four buildings.

As shown in Table 7, in considering the first cost of retrofit technologies, the use of energy-efficient chillers and biomass boilers is expensive to retrofit technologies, involved the very high cost of over 10 million Sri Lankan rupees for the initial installations. This could be one of the reasons, and the building, GB4, has not incorporated this technology and also received the certification of 'Silver' rating level. There are other technologies such as florescent lamps with skylights, use of the evaporative cooler, LED lights, servo motors, and installation of subsystem-level water meters which are in the range model level cost, varying from 1 to 5 million Sri Lankan

Table 5 Cost implications of green retrofits of the garment buildings—GB1 and GB2

Retrofits	GB1				GB2				Life cycle (Years)
	First cost (LKR)	Annual savings (LKR)	NPV (LKR)	SPB (Years)	First cost (LKR)	Annual savings (LKR)	NPV (LKR)	SPB (Years)	
[1]	4,098,510	13,100,735	138,947,770	0.31	5,250,000	13,520,000	142,374,214	0.39	15
[2]	10,745,700	7,086,423	83,381,330	1.52	10,745,700	8,105,000	96,910,812	1.33	20
[3]	20,475,000	4,782,960	38,814,197	4.28	25,700,000	4,856,000	34,494,595	5.29	18
[4]	1,797,600	3,160,690	27,423,193	0.57	1,580,000	2,856,000	24,823,914	0.55	12
[5]	2,241,366	1,902,169	10,429,069	1.18	3,250,000	2,033,066	10,292,346	1.60	8
[6]	183,000	2,499,420	8,834,314	0.07	245,000	2,350,000	8,233,242	0.10	4
[7]	4,680,000	1,124,312	5,714,341	4.16	3,452,500	1,062,307	6,368,600	3.25	12
[8]	795,450	1,613,178	2,235,859	0.49	765,600	1,695,200	2,419,836	0.45	2
[9]	1,265,540	505,008	2,098,341	2.51	–	–	–	–	8
[10]	475,000	169,638	654,967	2.8	556,000	213,600	866,799	2.60	8
[11]	150,600	53,396	205,073	2.82	–	–	–	–	8
[12]	80,000	54,992	118,398	1.45	–	–	–	–	4
[13]	50,450	12,100,608	53,428,094	0.004	–	–	–	–	5
[14]	1,003,507	7,987,200	34,295,862	0.13	–	–	–	–	5

Table 6 Cost implications of green retrofits of garment buildings—GB3 and GB4

Retrofits	GB3				GB4				Life cycle (Years)
	First cost (LKR)	Annual savings (LKR)	NPV (LKR)	SPB (Years)	First cost (LKR)	Annual savings (LKR)	NPV (LKR)	SPB (Years)	
[1]	–	–	–	–	7,856,000	15,200,500	158,117,511	0.52	15
[2]	9,362,000	7,356,800	88,356,374	1.27	12,500,000	8,555,000	101,133,739	1.46	20
[3]	15,600,500	3,657,000	29,731,385	4.27	–	–	–	–	18
[4]	1,350,000	2,565,000	22,363,599	0.53	–	–	–	–	12
[5]	2,855,000	1,565,000	7,569,537	1.82	1,853,000	1,235,500	6,376,722	1.5	8
[6]	195,000	2150,500	7,563,494	0.09	225,000	1,850,000	6,449,361	0.12	4
[7]	4,450,000	927,000	4,120,178	4.80	3,653,000	820,890	3,936,184	4.45	12
[8]	526,500	1,125,080	1,587,628	0.47	850,000	1,320,050	1,630,494	0.64	2
[9]	–	–	–	–	856,500	435,400	2,043,719	1.97	8
[10]	–	–	–	–	–	–	–	–	8
[11]	1,050,000	53,400	2,067,369	2.24	275,000	122,500	540,978	2.24	8
[12]	–	–	–	–	105,000	82,000	190,837	1.28	4
[13]	–	–	–	–	–	–	–	–	5
[14]	–	–	–	–	–	–	–	–	5

Table 7 Comparison of the first cost of green retrofits (LKR Million)

Retrofits	GB1	GB2	GB3	GB4
[3]	20,475,000	25,700,000	15,600,500	–
[2]	10,745,700	10,745,700	9,362,000	12,500,000
[7]	4,680,000	3,452,500	4,450,000	3,653,000
[1]	4,098,510	5,250,000	–	7,856,000
[5]	2,241,366	3,250,000	2,855,000	1,853,000
[4]	1,797,600	1,580,000	1,350,000	–
[9]	1,265,540	–	1,050,000	856,500
[14]	1,003,507			
[8]	795,450	765,600	526,500	850,000
[10]	475,000	556,000	–	–
[6]	183,000	245,000	195,000	225,000
[11]	150,600	–	150,600	275,000
[12]	80,000	–	–	105,000
[13]	50,450			

rupees. On the other hand, the technologies such as compressed air line modification, installations of VSDs for chiller and compressor, streamline insulation, recovery of flash steam for water heating, and installing low water flow push taps are most economical and involved less than one (01) million Sri Lankan rupees. It is worth noting that some of the buildings have not even integrated these technologies.

Table 8 presents the comparison of annual savings of green retrofits implemented in the selected buildings. As seen from Table 8, the use of evaporative cooler is placed at the first rank in contributing to the operational saving of all buildings considered in the study. The water efficiency technologies implemented only in one building, GB1, receive the next places in terms of annual saving. The technologies, biogas project, installations of VSDs for chiller and compressor, and recovery of flash steam for water heating are the least contributory technologies for energy savings and indoor air quality improvements.

Compiling the initial costs and annual savings presented in Tables 7 and 8, respectively, the net effect, NPV of each technology implemented in each building was calculated based on the assumed lifetime of each technology as stated in Table 6. Then, the average value of the calculated NPVs of each technology of each building was calculated. Besides, the equivalent annuity cash flow and the average SPBs of all four buildings were calculated using the data presented in Tables 5 and 6. Table 9 presents the average NPVs, equivalent annuity cash flow, and SPBs.

As seen from Table 9, all the technologies have positive NPVs, which indicate the potential of having substantial savings compared to the first cost incurred in implementing the retrofits. Among, the top four technologies such as the use of evaporative coolers, biomass boilers, energy-efficient chillers, and servo motors give substantial annual savings of over rupees 2.0 million with the payback periods of

Table 8 Comparison of annual savings of green retrofits (LKR Million)

Retrofits	GB1	GB2	GB3	GB4
[1]	13,100,735	13,520,000	–	15,200,500
[13]	12,100,608	–	–	–
[14]	7,987,200	–	–	–
[2]	7,086,423	8,105,000	7,356,800	8,555,000
[3]	4,782,960	4,856,000	3,657,000	–
[4]	3,160,690	2,856,000	2,565,000	–
[6]	2,499,420	2,350,000	2,150,500	1,850,000
[5]	1,902,169	2,033,066	1,565,000	1,235,500
[8]	1,613,178	1,695,200	1,125,080	1,320,050
[7]	1,124,312	1,062,307	927,000	820,890
[9]	505,008		468,000	435,400
[10]	169,638	213,600	–	–
[12]	54,992	–	–	82,000
[11]	53,396	–	53,400	122,500

Table 9 Comparison of NPVs and SPBs of green retrofits

Retrofits	Average of NPVs—all buildings (Rs. Million)	Equivalent annuity cash flow (Rs. Million)	Average of SPBs—all buildings (Years)
[1]	146,479,832	13,415,193	0.41
[2]	92,445,564	6,959,833	1.23
[3]	34,346,726	2,770,809	4.61
[4]	24,870,235	2,690,108	0.55
[5]	8,666,919	1,301,135	1.18
[6]	7,770,103	2,153,718	0.07
[7]	2,717,321	293,921	3.11
[8]	1,613,922	858,884	0.37
[9]	1,116,855	167,670	1.60
[10]	280,721	42,144	2.70
[11]	216,293	32,471	1.89
[12]	82,918	22,983	0.73
[13]	53,428,094	12,090,540	0.00
[14]	34,295,862	7,761,001	0.13

mostly less than one year and six months. It is worth noting that all the technologies require less than five years on average to repay the initial investments.

5 Discussion and Conclusions

The application of green retrofits in the garment buildings in Sri Lanka was examined through an in-depth analysis of costs and savings of implemented retrofits. Overall, the findings confirmed that the respective buildings have implemented energy, IEQ, and water-related retrofits, while in terms of sustainable sites and materials, green features have been incorporated at the initial stage of the building and there were no features added during retrofitting, sustainability transition stage. The current study highlighted that the selected industrial buildings in Sri Lanka have incorporated green retrofit technologies which include evaporative coolers, biomass boilers, energy-efficient chillers, servo motors, skylights, LED lights, steam line insulation, compressed air line modification, and VSDs as EA and IEQ strategies, in common. However, in terms of WE strategies, only one building has been upgraded with the use of subsystem-level water meters and low water flow push taps. Further, some of the selected buildings have not incorporated some of the above retrofits but still have achieved the green certification. For example, the use of evaporative cooler is the topmost significant technology, which contributes to substantial energy saving with less than one year of payback period, but the building GB3 has not incorporated it. This could be due to its initial cost which was not affordable for the investor. However, this particular building investor was able to integrate some other technologies (i.e., biomass boiler and energy-efficient chillers) which are expensive and less contributor to energy saving than the use of the evaporative cooler, as evidenced from the analysis. Therefore, the possible reason in this context may be due to the client's unawareness about the potential savings and the initial investment payback period.

Similarly, the building GB4 has not integrated the two (02) of the top four (04) technologies (use of efficient chillers and servo motors) in terms of contributing to savings. As evidenced by the study, this decision of the client could have been highly influenced by the initial cost and absence of awareness about operational savings and time taken to recover the first cost. Therefore, the current study confirms the general perceptions that the retrofit decisions of investors are influenced by the initial costs, which has been the barrier of implementing retrofits. In this context, the current study concludes that although some of the technologies are expensive to implement, all of them result in life-cycle savings, with positive NPVs, whereas contradictory views were presented by earlier studies on the initial cost of green retrofits [32, 61]. Further, all technologies except the use of efficient chillers and florescent lamps with skylights require less than 3 years to recover the initial cost of implementing the technologies while the literature findings showed that the green retrofits involved long payback periods [22]. Therefore, the findings of the current study give confidence to the investor in decision making concerning the economic selection of retrofit technologies. Moreover, earlier studies have considered building envelope retrofitting as a key to improve the energy performance of buildings [18, 49]. However, the selected green buildings have not implemented any upgrades to the building envelope and the existing building conditions remain. The current study further revealed that energy retrofits such as the use of the evaporative cooler, biomass boiler, energy-efficient chillers, and servo motors give higher economical savings over the life cycle. Similar findings were indicated in the study of Bond [7] for the renewable energy projects that provide a high return on major investments within a short payback period.

The study concludes that the most economical/appropriate green retrofit technologies such as evaporative cooler, biomass boiler, energy-efficient chillers, low water flow push taps, water meters, servo motors, LED lights, steam lines insulation, skylights, compressed air line modification, and biogas project can be implemented to an industrial manufacturing building as they provide a higher return within a short payback time.

The lack of knowledge on life-cycle cost and long-term return of green retrofits leads to decisions of not implementing green retrofits. Further, the investors who do not have access to enough information will not realize the contribution of green retrofits toward energy, environmental and water performance, comfort, and quality of space, etc. Lack of knowledge of financial institutions and unawareness of the benefits of green retrofits has primarily affected building owners from implementing green retrofit projects. To this end, the findings of the current study highlight the financial viability of the implemented retrofit projects of underwater efficiency, energy, and IEQ with positive NPV values and fewer SPB periods. Moreover, considering the lifetime financial returns of those retrofits, each shows significant benefits compared to the initial cost. Therefore, the study recommends building investors and owners to apply those retrofit technologies in existing buildings that can maximize the sustainability of their buildings while minimizing the required cost.

The current study was limited to four (4) selected garment buildings that are certified under the LEED O + M Existing Building category and retrofit technologies implemented in those buildings. The analysis of changes in initial costs and savings failed to consider the effects of capacity, type, model, scale, etc. of the retrofit technologies used. Further, in the calculation of NPV and SPB, the study assumed a single discount rate of 4.26% and the set of life span for individual retrofits. However, the changes in these values could influence the economic status of technologies. Therefore, it is believed that further study would address these limitations of the current study. Although the calculation was performed based on 2015 cost data, the conclusion is free from this effect as the aim of the study is to recommend the most economical retrofit technologies based on the detailed analysis of cost and savings of implemented retrofits.

Acknowledgements Authors greatly acknowledge the financial support provided by the Senate Research Committee of University of Moratuwa under the Grant SRC/LT/2018/11.

Annexure

See Tables 5 and 6.

References

1. Aktas B, Ozorhon B (2015) The green building certification process of existing buildings in developing countries: cases from Turkey. J Manage Eng 31:050150021–9
2. Al-Kodmany K (2014) Green retrofitting skyscrapers: a review. Buildings 4:683–710
3. Asadi E, Silva MG, Antunes CH, Dias L (2012) Multi-objective optimization for building retrofit strategies: a model and an application. Energy Build 41:81–87
4. Ascione F, Rossi F, Vanoli GP (2011) Energy retrofit of historical buildings: theoretical and experimental investigations for the modelling of reliable performance scenarios. Energy Build 43:1925–1936
5. Azizi NSM, Fassman E, Wilkinson S (2010) Risks associated in implementation of green buildings. Beyond Today's Infrastructure
6. Baldwin AN, Loveday DL, Li B, Murray M, Yu W (2018) A research agenda for the retrofitting of residential buildings in China—a case study. Energy Policy 113:41–51
7. Bond S (2010) Lessons from the leaders of green designed commercial buildings in Australia. Pac Rim Prop Res J 16:314–38
8. Bullen PA (2007) Adaptive reuse and sustainability of commercial buildings. Facility 25:20–31
9. Central bank of Sri Lanka (2017) Annual report 2016, Colombo, Central Bank of Sri Lanka
10. Ciampi G, Rosato A, Scorpio M, Sibilio S (2015) Energy and economic evaluation of retrofit actions on an existing historical building in the south of italy by using a dynamic simulation software. Energy Procedia 78:741–746
11. Cohen S, Goldman C, Harris J (1991) Energy savings and economics of retrofitting single-family buildings. Energy Build 17:297–311
12. Davies P, Osmani M (2011) Low carbon housing refurbishment challenges and incentives: architects' perspectives. Build Environ 46:1691–1698
13. Dewick P, Miozzo M (2004) Networks and innovation: sustainable technologies in Scottish social housing. R&D Manage 34:323–333
14. Doherty P, Al-Huthaili S, Riffat S, Abodahab N (2004) Ground source heat pump description and preliminary results of the eco house system. Appl Therm Eng 24:2627–2641
15. Fuerst F, McAllister P (2011) Green noise or green value? measuring the effects of environmental certification on office values. Real Estate Econ 39:45–69
16. Gillott M, Loveday DL, Vadodaria K (2013) Airtightness improvements and ventilation systems in domestic refurbishment. In: Loveday DLL, Vadodaria K (eds) Project CALEBRE (Consumer appealing low energy technologies for building retrofitting) a summary report. Loughborough, Loughborough University
17. Gucyeter B, Gunaydin HM (2012) Optimization of an envelope retrofit strategy for an existing office building. Energy Build 55:647–659
18. Gultekin P, Anumba CJ, Leicht RM (2013) Case study of integrated decision-making for deep energy-efficient retrofits. Int J Energy Sect Manage 8:434–455
19. Häkkinen T, Belloni K (2011) Barriers and drivers for sustainable building. Build Res Inf 39:239–255
20. IEA and the UNEP (2019) Towards a zero-emission, efficient and resilient buildings and construction sector. In: 2019 Global status report for building and construction. https://www.iea.org/reports/global-status-report-for-buildings-and-construction-2019. Accessed 23 Nov 2019
21. Jagarajan R, Asmoni M, Mohammed A, Jaafar M, Mei J, Baba M (2017) Green retrofitting—a review of current status, implementations and challenges. Renew Sustain Energy Rev 67:1360–1368
22. Kasivisvanathan H, Ng RTL, Tay DHS, Ng DKS (2012) Fuzzy optimisation for retrofitting a palm oil mill into a sustainable palm oil-based integrated bio refinery. Chem Eng J 200:694–709
23. Korkmaz S, Messner JI, Riley DR, Magent C (2010) High-performance green building design process modeling and integrated use of visualization tools. J Archit Eng 16:37–45
24. Kumar R (2018) Research methodology: a step-by-step guide for beginners, 5th edn. SAGE Publications Ltd., London

25. Lapinski AR, Horman MJ, Riley DR (2006) Lean processes for sustainable project delivery. J Constr Eng M ASCE 132:1083–1091
26. Li Y, Ren J, Jing Z, Jianping L, Ye Q, Lv Z (2017) The existing building sustainable retrofit in China-a review and case study. Procedia Eng 205:3638–3645
27. Liang X, Peng Y, Shen GQ (2016) A game theory based analysis of decision making for green retrofit under different occupancy types. J Cleaner Prod 137:1300–1312
28. Lockwood C (2008) The dollars and sense of green retrofits. Deloitte, Washington DC
29. Love P, Bullen PA (2009) Towards the sustainable adaptation of existing facilities. Facility 27:357–367
30. Ma Z, Cooper P, Daly D, Ledo L (2012) Existing building retrofits: methodology and state-of-the-art. Energy Build 55:889–902
31. Mahlia TMI, Said MFM, Masjuki HH, Tamjis MR (2005) Cost-benefit analysis and emission reduction of lighting retrofits in residential sector. Energy Build 37:573–578
32. McDonald C, Ivery S, Gagne CM, Scheuer K (2008) Greening leased spaces: opportunities and challenges. In: ACEEE summer study on energy efficiency in buildings. ACEEE, pp 147–56
33. McGraw-Hill Construction (2009) Green building retrofit and renovation: rapidly expanding market opportunities through existing buildings. McGraw-Hill Construction, New York
34. Menassa CC (2011) Evaluating sustainable retrofits in existing buildings under uncertainty. Energy Build 43:3576–3583
35. Menassa CC, Baer B (2014) A framework to assess the role of stakeholders in sustainable building retrofit decisions. Sustain Cities Soc 10:207–221
36. Miller E, Buys L (2008) Retrofitting commercial office buildings for sustainability: tenants' perspectives. J Prop Invest Finan 26:552–561
37. Nabinger S, Persily A (2011) Impacts of air tightening retrofits on ventilation rates and energy consumption in a manufactured home. Energy Build 43:3059–3067
38. Newsham G, Mancini S, Birt B (2009) Do LEED-certified buildings save energy? yes, but…. Energy Build 41:897–905
39. Osmani M, O'Reilly A (2009) Feasibility of zero carbon homes in England by 2016: a housebuilder's perspective. Build Environ 44:1917–1924
40. Pedini AD, Ashuri B (2010) An overview of the benefits and risk factors of going green in existing buildings. Int J Facil Manage 1:1–15
41. Du Plessis C (2007) A strategic framework for sustainable construction developing countries. Constr Manag Econ 25:67–76
42. Rey E (2004) Office building retrofitting strategies: multicriteria approach of an architectural and technical issue. Energy Build 36:367–372
43. Reza EM, Marhani MA, Yaman R, Hassan AA, Rashid NHN, Adnan H (2011) Obstacles in implementing green building projects in Malaysia. Aust J Basic Appl Sci 5:1806–1812
44. Santamouris M, Pavlou C, Doukas P, Mihalakakou G, Synnefa A, Hatzibiros A, Patargias P (2007) Investigating performance and analysing the energy and environmental of an experimental school green roof system installed in a nursery building in Athens Greece. Energy 32:1781–8
45. Seneviratne SI, Donat MG, Pitman AJ, Knutti R, Wilby RL (2016) Allowable CO2 emissions based on regional and impact-related climate targets. Nature 529:477–483
46. Si J (2017) Green retrofit of existing non-domestic buildings as a multi criteria decision mak-ing process. University College London, London
47. Steemer K, Towards A (2003) Research agenda for adapting to climate change. Build Res Inf 31:291–301
48. Stefano JD (2000) Energy efficiency and the environment: the potential for energy efficient lighting to save energy and reduce carbon dioxide emissions at Melbourne University. Australia Energy 25:823–839
49. Stovall T, Petrie T, Kosny J, Childs P, Atchley J, Sissom K (2007) An exploration of wall retrofit best practices: thermal performance of the exterior envelopes of buildings X. In: Proceedings of ASHRAE THERM X. clearwater

50. Thomas LE (2010) Evaluating design strategies, performance and occupant satisfaction: a low carbon office refurbishment. Build Res Inf 38:610–624
51. Township's Boards of Historical and Architecture Review (2008) Historic preservation and sustainability. CHRS Inc. of North Wales, Pennsylvania
52. USGBC: United States Green Building Council (2019). https://new.usgbc.org/. Accessed 17 May 2019
53. United States Green Building Council (2003) Building momentum: national trends and prospects for high performance green buildings. Washington, DC: Author
54. Verbeeck G, Hens H (2005) Energy savings viable, in retrofitted dwellings: economically. Energy Build 37:747–754
55. Waidyasekara KGAS, Fernando WN (2012) Benefits of adopting green concept for construction of buildings in Sri Lanka. In: Sustainable built environment 2012 proceedings of 2nd international conference ICSBE, Kandy
56. Weerasinghe AS, Ramachandra T (2019) Costs and benefits of green retrofits: A case of industrial manufacturing buildings in Sri Lanka. In: The 10th Int Conf Eng, Project, and Prod Manage (EPPM2019), Berlin
57. Weerasinghe AS, Ramachandra T, Nawarathna A (2019) Effects of Green Retrofits: A Case of Industrial Manufacturing Buildings in Sri Lanka. In: Gorse C & Neilson CJ (Eds) Proceedings of the 35th Annual ARCOM Conference, Leeds
58. Wilkinson S (2012) Analysing sustainable retrofit potential in premium office buildings. Struct Surv 30:398–410
59. Wilkinson SJ, James K, Reed R (2009) Using building adaptation to deliver sustainability in Australia. Struct Sur 27:46–61
60. Xu P, Chan EH, Qian QK (2011) Success factors of energy performance contracting (EPC) for sustainable building energy efficiency retrofit (BEER) of hotel buildings in China. Energy Policy 39:7389–7398
61. Zhai X, Reed R, Mills A (2014) Addressing sustainable challenges in China: the contribution of off-site industrialisation. SASBE 3:261–274
62. Zhang X, Platten A, Shen L (2011) Green property development practice in China: costs and barriers. Build Environ 46:2153–2160

Chapter 7
Sustainable Cultural Wagon

Pablo González, Romina Sangoy, Lucia Rodríguez, Soledad Cormick, Lucas Daher, Edgardo Suarez, Hercilia Brusasca, Antonella Caballero, and Danae Conti

Abstract Within the actions developed by the **Municipal Institute of Housing and Municipal Housing Infrastructure—I.M.V. and I.H.**, we present the cultural wagon as an example of urban transformation based on sustainability; not only as conscious awareness but also as induced, given that traditional energy services are impossible to come by. The cultural wagon is part of a technical and transferable reality showing concrete, hands-on labour built on the basis of understanding that there is a growing room. The knowledge and experience acquired through the cultural wagon is part of a constant learning process, which may be used in further projects.

Keywords Urban sustainability · Social transformation · Solar energy · Technology · Learning

1 Introduction

In the last few decades, the world has gone through an accelerated outburst process towards unsustainability. For cities to be sustainable, it is necessary to change from linear city models—where consumption of energy and levels of contamination are high—to circular models where cities are able to find a balance between intake and output within themselves, consuming less resources, prioritizing recycling processes, thus taking contamination levels to a bare minimum. Citizens must be empowered to anticipate and face constant social changes and environmental economic challenges. Life experiences are a way of doing this. In this particular case, **The**

P. González (✉) · R. Sangoy · L. Rodríguez · S. Cormick · L. Daher · H. Brusasca · A. Caballero · D. Conti
Municipal, Municipal Institute of Housing and Municipal Housing Structure (I.M.V.e I.H.) from Villa Maria, 5900 Villa Maria, Cordoba, Argentina
e-mail: estropablo@gmail.com

E. Suarez
Institute of Building Sustainability–Architects' Association (ISE), 5000, Córdoba, Cordoba, Argentina

Municipal Institute of Housing and Municipal Housing Infrastructure offers citizens opportunities to develop abilities, skills and attitudes which lead to sustainable development.

The City of Villa María is located 146 km away. southeast of the province of Córdoba and in the geographical center of the Argentine Republic. It is the head of the Department General San Martín. Third city in importance of the province. It is located in the humid pampas on the banks of the Ctalamochita River. The population of Villa María is 80,006 inhabitants, becoming the most developed city in the General San Martín department, consisting of 36 neighbourhoods.

Las Playas neighbourhood, name with which this sector is identified, is closely linked to the history and operation of the railway. In the sector were located the manoeuvring beaches of locomotives and freight trains, not only for its proximity to National routes but also for being Villa María a strategic point of confluence of numerous train branches.

Given the importance that agricultural production has had and still has in Villa María, the railroad is extremely important in terms of means of transport. These manoeuvring beaches were considered one of the most modern in South America and the only one in the country destined to carry out gravitation manoeuvres.

The confluence of the branches in Villa María, and the magnitude of the activity, made these spaces necessary in order to better manage the constant movements of wagons, which caused traffic jams in the urban radius with its consequent problem in transit in the passages at city levels.

This construction had a great impact on the economic activity of Villa María. It began in 1925 when the first families settled in these places called by the possibility of work.

Within the framework of this project, it led to the emergence of the most remote neighbourhood in the urban area of Villa María where 14 homes were built. The railway company acquired several hectares around Las Playas and, by then, its subdivision was projected into small lots for the construction of houses. In this way, the seat of workers linked to the railway activity was produced, starting the sector called "Los Chaleses" in reference to the buildings built with the chalet typology. These residences are located near the cultural Wagon and this is where the building interventions were carried out, which we will mention next.

This educational space takes place in the old train yard in a neighbourhood called Las Playas, in Villa María, province of Córdoba. This area is characterized by representing marginalization, abandonment, and conflicts. The presence of the State is crucial. The intervention with habitat improvement projects, revaluation of the community area, which today is a square/park and the recovery of a wagon transformed into a cultural hub show the accessibility problems suffered by parts of the society to reach the formal city and places the issue on the public agenda. The work shared herein is part of this commitment, guaranteeing rights to that area of the city and introducing sustainability to the public sphere. This intervention was recognized and distinguished by UNESCO—United Nations Educational, Scientific and Cultural Organization, which granted the Cordobese City of Villa María the prize *City of Learning 2017.*

1.1   Objectives

- To design and execute infrastructure and equipment which promote sustainable transformations in an integral way, providing quality to the urban and social habitat through continuous improvement.
- To promote sustainable and efficient solutions to the use of resources.
- To foster a lifelong Culture of Learning.
- To diagnose the final performance in the use of resources, mainly the demand for required energy and to promote the improvement in its use, using design tools first and technological ones later.

2   Background Information and Related Works

The area interventions took place after systematic and periodic diagramming meetings with neighbours, which provided results of the citizens' quality of life based on quantitative and qualitative studies. People are considered subjects of rights and social, cultural, and symbolic bearers that will allow themselves to change their current situation.

The actions consisted of the recovery of the property and successively improving the sustainable performances of the spaces already recovered.

Some tasks performed were levelling and compacting the land in flood sectors and access roads to the neighbourhood, the execution of gutter cord, general cleaning and weeding, fencing parallel to the FFCC tracks and placement of LED lighting throughout the area.

The main square was recovered by installing playground equipment, planting tree species, income zoning and perimeters ensuring the correct division with the restricted access street to vehicles.

The environmental quality of the area analysed in its first diagnostic stage was studied under an integral vision of the environment, which implies conferring both public space and private spaces of the RAIL ROAD cars and later to the neighbourhood, some features of cultural, social and recreational expression; that allow motorizing new perspectives in a space developed for collective activities.

Once the cultural wagon project was finished a building audit was performed. Such an audit contemplates the evaluation of the 33 variables (18 calculation engines) which are part of the **eSe protocol/Etiquetación de Sustentabilidad Edilicia (Labelling of Building Sustainability)**, developed by the iSe from the Architects 'association in Córdoba Province. The eSe protocol from the iSe is an orderly ensemble of objectives, requirements, and strategies that, as an operational instrument, allows the improvement and integral efficiency of the design, construction and effective use of the buildings. The environmental factor and the economic and social aspects are especially considered in this system to find a balance within the sustainable development framework.

3 Cultural Wagon

All interventions in the sector were carried out from systematic and periodic meetings that were held with the neighbours, obtaining results that allowed the quantitative and qualitative growth of the quality of life of the inhabitants; from a perspective that raises man as a subject of rights and bearer of social, cultural and symbolic capital that will allow him to reverse his current situation.

The cultural wagon was built as an answer to neighbours' requests for cultural and creative spaces in nearby areas, thus the proposal for restoring one of the wagons in disuse was widely accepted. The wagon was built with the following features: ramps for easy access, thermal-acoustic insulation, a photovoltaic panel system, indoor LED lighting, cross ventilation, DVH openings, and a community accessible hot-water pump station powered by solar panels. We believe it is important to make every public project sustainable, encouraging energy production through renewable sources, contributing to the reduction of energy consumption and the generation of clean energy with endless economic and environmental benefits (Figs. 1 and 2).

3.1 Learning Process

Learning transforms lives, and culture as a space of expression is a tool to activate inclusion and paths for social transformation. Artistic expressions are closely related to social life and favour the creation of emotional bonds and trust, allowing people to transcend barriers and difficulties. Learning transforms lives; It is the foundation of sustainability.

Fig. 1 Before the intervention

Fig. 2 Art workshop for children

These spaces seek to generate sustainable awareness by learning new ways of inhabiting, and then take actions that provide added value, achieving a better quality of life.

4   Sustainable Economy of the Project

There is no sustainable prosperity without the transformation of the economic system. The crisis has affected many people living in peripheral sectors of cities and this is related to inequality in access to education and culture, health care, income-generating opportunities and the right to property.

In this sense, the restoration of the Wagon and the public space of "Los Chalets" used recycled elements and labour of the families within the community, thus reducing unnecessary cost and at the same time investing in wasted natural capital.

The aim is to accelerate the processes of transition towards a sustainable economy, demanding less resources and generating the use of renewable energies, while possible.

Power is provided through solar panels conveniently located on the upper deck of the car. A sanitary water heating device was installed for public use, located meters from the main structure.

The IMV's objective is to implement a strategy called "green for everyone", with new approaches in energy supply, transport, housing and waste management, which combine technical, economic and structural change with social empowerment. Since 2018, IMV has worked together with ISE in developing Management Manuals and Strategies for sustainable intervention at different scales, among other things.

One of the main characteristics of the IMV and IH economy is to offer different opportunities for economic development, eradicate poverty, and include families in these public areas without wasting the city's natural resources. The wagon was restored and the area public space was recovered, costing a total of € 66.813,57. Money was invested to accelerate the transition to a sustainable economy and renewable energy was used as a rentable strategy to protect the energy sources of our planet.

5   Results

The purpose of this analysis is to diagnose the final performance of the use of resources, particularly of the primary energy demanded and to promote the improvement of their use, first by the use of handheld tools and then the implementation of technologically designed tools.

Figure 3 shows the Designated order of strategies according to their impact.

The audit of building performances is carried out, which consists of a complete diagnostic scan, of the 33 variables (18 calculation engines), which make up the eSe/Edilicia Sustainability Labelling protocol, developed by the iSe of the College of Architects of the Province of Córdoba, of the building (cultural wagon).

The eSe protocol of iSe, is an ordered set of objectives, requirements and strategies that, as an operational instrument, allows the improvement and sustainable integral efficiency of the design, manufacture and effective use of buildings. The environmental factor and economic and social aspects are especially taken into account in this system, in order to find a better balance in the framework of sustainable development.

Within this context, occupying the wagon for a total of 2100 h annually, over a base of 48 m^2 and the reference of 15 m^2/person (classrooms-academies), the modified building offers resources for 35 people in simultaneous use, in which demands are as follows.

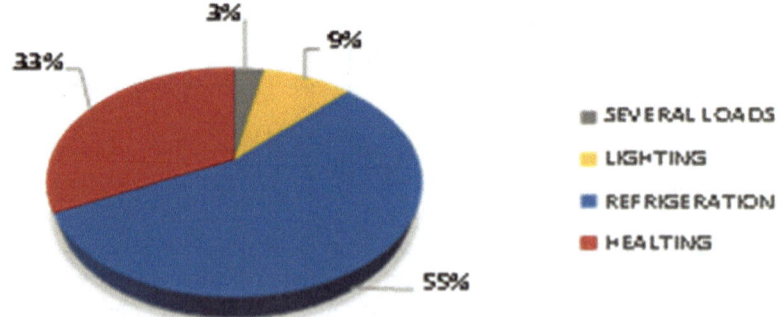

Fig. 3 Annual energetic demand

5.1 Safe Drinking Water

Audit: 445 lt/day or 14 lt/person/day. A demand reduction of 205% is obtained with improvement strategies of the replacement of devices (hydro-mechanical download/flush, with manual pressure and self-closing, for a temporalized low-pressure discharge of 6 s): 355 lt/day or 11 lt/person/day, below the sustainable reference base (12 lt/person/day).

5.2 Effluent Septic Systems

Audit: 355 lt/day impulsion of sewage effluents. Considering the potential of reducing the use of safe drinking water, the amount of sewage liquid poured into the septic system can be reduced, on average, to 75 lt/day.

14 lt/person/day Audited safe drinking water demand 445/lt/day 355, 75 lt/day 12 lt/person/day Sustainable reference 381/lt/day 283, 23 lt/day 11 lt/person/day safe drinking water strategy 354/lt/day Strategies of daily effluents.

5.3 Rainwater Management System/Sigall

The wagon has a usable recollection capacity of 3000 L per month. It is suggested to install a 3000-liter tank system under the wagon walkway. This tank will allow the resource provision for efficient irrigation (trickle irrigation system) proposed by the System of Equivalent Vegetal Cover.

5.4 Equivalent Vegetal Cover (Cobertura Vegetal Equivalente/CVE)

The audit detected that the equivalent environmental correction generated by the project is 454 m^2, which in turn is equal to 22 trees with an average treetop size of 6 meters in diameter. This optimal tree size is reached, on average, after two years of the plants being planted.

5.5 Superficial Reflectance Index/IRS

The final objective of this index is to quantify, to then reduce the overheating impact on the constructed environment associated with the heat island effect (ICU). The IRS index of the existing cover is REGULAR 4577. With IRS improvement strategies, like designing "cold" covers which involve the use of "cold" materials or of high Index of Reflectance, an improvement of 32%: 66 GOOD is obtained.

5.6 Enclosure Bio-climatic Design

The evaluation accurately determined the need to improve the solar protection redesign for the southeast side, which could reduce 40% of overheating in summer. On the northeast side, it was suggested that eaves should be installed, completing 100% of the necessary summer protection, and thus not generate a higher energetic demand for air-conditioning.

5.7 Enclosures Heat Insulation

Audit: Enclosure: 073 W/m^2K, Floor: 208 W/m^2 K With systems improved. Enclosure: 028 W/m^2 K, Floor: 027 W/m^2 K

5.8 Energetic Demand of Primary Energy/IPE

The total IPE is 108, 48 kWh/m^2/year, which corresponds to the necessary energy for climate control 9497 kWh/m2/year (88%). Implementing a series of proposed passive climate control strategies, (earth warming tubes—pozo canadiense—and Provençal well-pozo provenzal) the thermic demand is reduced to 4101 kWh/m^2/year (57% saving).

5.9 Diagnosis of the Annual Hydrothermal Comfort

For comfort ranges in both thermal cycling (cold: 16–24 °C-heat: 18–26 °C), the audit shows 1440 °C annual hours of discomfort. By implementing the improved systems this is reduced to 1013 oc hours (2970% improvement).

5.10 Efficiency Diagnosis of the Photovoltaic System

W^*6 panels $= 969$ Wh/day.IPE : 5.237 kWh/m^2/year.

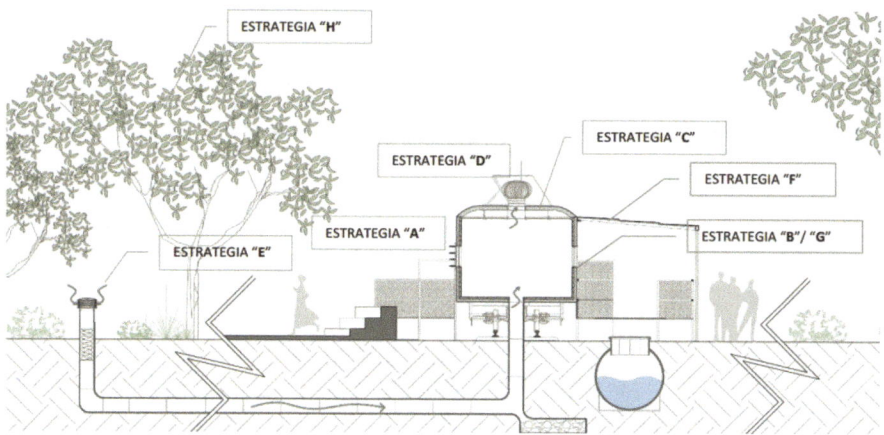

Fig. 4 Intervention strategies

5.11   *Greenhouse Gas Emissions/GEI*

Audit: 3977 kg co_2/m^2 year, the integration with renewables represents a saving of 235 kg co_2/m^2 (5%) (Fig. 4).

6   Conclusions

By performing the audit, the execution of the improvement strategies of energy performance can be started. These strategies will allow us to stabilize the indoor temperatures and save 37% of the final demand of primary energy.

The urban intervention called "Cultural Wagon" was recognized and distinguished by UNESCO, which awarded the City of Villa María with "the City of Learning 2017" award. This prize recognizes the effort and decision of the local governments in terms of participatory and sustainable teaching policies.

This area was characterized by being a space of separation, abandonment, and conflict. By intervening it, space which generates positive exchanges is transformed and, at the same time, it tends to reduce many social conflicts offering its citizens a public space, enhancing and valorizing the natural environment and eco-form of the territory as well as reinforcing the sense of belonging to the community.

Urban sustainability is fundamental in the development of present and future generations, creating inclusive and environment-friendly spaces that preserve and respect the natural environment. We pursue social and environmental goals creating mixed territories, which contain culture, housing, and work. More than urbanizing, we construct a city. The crisis suffered are not two separate ones, one relating to the environment and the other to social instability, rather it is a complex socio-environmental crisis [1].

References

1. Laudatio Si' (2015) Pope Francis' encyclical. UNESCO http://www.uil.unesco.org/es/apr endizaje-lo-largo-vida/ciudades-aprendizaje/dieciseis-ciudades-recibiran-premio-unesco-ciu dad-del

Part III

Chapter 8
Unearthing the Factors Impeding Sustainable Construction in Developing Countries—A PLS-SEM Approach

Douglas Aghimien⊙, **Clinton Aigbavboa**⊙, **Lerato Aghimien**⊙, **Ayodeji Oke**⊙, and **Wellington Thwala**⊙

Abstract This study presents the result of the findings on the factors impeding the sustainability of construction projects in developing countries using Nigeria as a case study. The study sought responses from construction managers, project managers, and quantity surveyors from the six different regions of the country. Data gathered were analysed using factor analysis and structural equation modelling. The findings revealed that issues surrounding regulation and policy, information and management, sustainability knowledge and sustainable materials and technology availability have a significant relationship with the poor sustainable construction in the country. It is believed that the findings of the study will help construction participants in the country and other developing countries particularly in Africa where the construction practice is similar, in understanding the core issues to tackle in the quest for sustainable construction in the country.

Keywords Sustainable construction · Sustainability · Structural equation modelling

1 Introduction

There is no gainsaying that construction industries all over the world contribute significantly to a nation's development, especially in the area of infrastructure delivery and socio-economic growth [1, 2]. However, in doing this, the industry relies mostly on the use of human and natural resources from within the environment. As a result, a significant strain is placed on these resources. Rapid depletion of natural resources cum pollution of the natural habitat has become a common aftermath of the activities of the industry [3–5]. If there is any hope for the future generations to meet their

D. Aghimien (✉) · C. Aigbavboa · L. Aghimien · A. Oke · W. Thwala
SARChi in Sustainable Construction Management and Leadership in the Built Environment, Faculty of Engineering and the Built Environment, University of Johannesburg, Johannesburg, South Africa
e-mail: aghimiendouglas@yahoo.com

own needs that require the use of natural resources, then there must be significant consciousness in terms of inter-generational equity in the use of natural resources now [6, 7]. This begs the need to preserve and sustain these resources. According to Oke and Aigbavboa [8], across the globe, sustainability has been a subject of concern for people from different countries, professions and disciplines. Even agencies such as the United Nations (UN) have set their goals and objectives towards attaining a more sustainable environment. Similarly, in construction, the negative impact of the industry has led to the demand for sustainable construction (SC) which involves delivering more economical, social and environmentally responsive construction projects [9, 10].

Albeit this call for a more sustainable form of construction, one that strives to preserve the natural environment, while promoting the socio-cultural lifestyle of the people and in the process providing economic gain for the investors [11], the issue of attaining SC in developing countries like Nigeria is still a mirage than a reality [12–14]. This scenario is not peculiar to Nigeria alone as similar observations have been made in other African countries such as Ethiopia [15], Ghana [16], Zambia [17] and South Africa [18, 19]. Over time, studies geared towards ascertaining the reason for this failure to attain SC has emanated. Alabi [12] noted that the major culprit of this problem is the poor education and lack of experience of construction professionals in sustainability-related issues. Similarly, Davies and Davies [14] noted issues surrounding lack of expertise and awareness of SC related issues. Dalibi et al. [20] mentioned the issue of perception of the high cost of introducing the concept of SC into projects as a significant hindrance. Aigbavboa et al. [21] and Lowe and Zhou [22] have earlier made the same observation and stated that this assumption is made without a thorough evaluation of the actual cost and whole-life cycle cost of adopting this concept. Aigbavboa et al. [21] further described this assumption as a "lazy view" of construction participants, and this according to Lowe and Zhou [22] is a serious problem to the proper adoption of SC in most countries around the world.

It is imperative at this stage to note that despite the existence of these studies emanating from Nigeria, most of them have attempted to rank the different factors affecting SC in the country. While this approach is commendable, the findings do not reveal in a real sense the relationship between these factors and poor SC delivery in the country. Hence understanding the major issues to tackle becomes unclear. It is based on this knowledge that this study revisits the issues surrounding the attainment of SC with a view towards revealing clearly the areas wherein the improvement is needed for better SC delivery in the country.

2 Factors Impeding Sustainable Construction

Evidence of the poor sustainability of construction projects in most developing countries exists in the body of literature. Most projects in these countries have failed to conform with social, environmental, economic and overall project performance [12, 13, 16, 17]. Furthermore, several factors have been held accountable for this poor

SC, and it is only by understanding these factors, that effective measures can be put in place to overcome them. Kibert [23] identified the perceived higher first costs as one of the issues affecting the adoption of SC. In agreement [24] and [25] pointed out that there is a fear of higher investment costs for SC when compared to traditional buildings. Dalibi et al. [20] also noted the same issue that the perception of the high cost of introducing the concept of SC into projects as a major hindrance in Nigeria. To aid the promotion of SC, [26] suggested that life cycle costs should be incorporated during the assessment of the various costs and their implications.

Aside from fear of high cost, [25] noted that the absence of building codes and regulations would hinder the adoption of SC. Powmya and Abidin [27] further stated that government plays a crucial role in the enforcement of regulation, revision of existing legislation and policies, the introduction of building codes, incentives and other financial instruments to spear-head SC adoption. Similarly, Oke et al. [28] noted the need for government involvement in enforcing regulations and revising existing building legislations to accommodate the concept of SC. William and Dair [29] identified the lack of knowledge, information, and understanding as to the significant barriers to the delivery of sustainable structures. Besides, Opoku and Ahmed [30] recognised the importance of public awareness and proper knowledge and understanding of sustainability as being essential to the successful promotion of SC practices. Alabi [12] observed that there is a low level of awareness of the concept of sustainability among construction participants in Nigeria. In a similar vein, Aghimien et al. [31] noted that sustainability awareness and knowledge related factors are the second most crucial barrier to SC in the country. In Kuwait, Al-Sanad [32] also discovered that SC implementation is low, and this can be as a result of a lack of awareness of the concept within the country. Baron and Donath [15] on the other hand, observed although there is considerable awareness of the concept of SC in Ethiopia, the major barrier SC face is incorrect implementation. It was observed that SC in the country in most cases is either neglected due to; budget constraints, lack of alternative building materials, or knowledge, or it is reduced to the issue of sustainable resource management. Aghimien et al. [31] therefore conclude that the poor understanding of the concept of SC in its holistic form can be a major barrier towards achieving SC.

Davies and Davies [14] submitted that the implementation of SC in Nigeria is also deterred by the construction industry's inability to move away from the traditional methods of construction. This is aside from the issue of lack of expertise and awareness that is already hampering SC in the country. Aghimien et al. [31] also observed construction-related issues affecting the country's SC practices. The study noted that if sustainability in construction is to be achieved within the country, then the construction industry must be ready to abandon the traditional method of construction for more innovative sustainability-oriented methods. Pitt et al. [33] and Powmya and Abidin [27] found that increased involvement and constructive interaction from the demand side which includes the clients, buyers and users would see an improvement in the number of structures completed using SC practices. Aside from demand issues, [5] and [30] found that there is a shortage of skilled employees involved in the implementation of sustainable practices. Shi et al. [26], on the other

Table 1 Factors impeding SC

Factors	Authors
Clients fear of high investment cost	[21, 22, 25, 35–37]
Inadequate technology and technological process	[23]
Limited access to relevant historical data	[38]
Lack of adequate exemplar 'demonstration project	[24]
Industry's resistance to change	[31, 39, 40]
Low demand for sustainable products	[16, 29, 30, 33]
Client worries about profitability	[24]
Clients preference	[34, 41]
Lack of commitment of top management	[24, 42]
Inadequate government policies and support	[24, 27, 30–32, 40]
Inadequate building codes and regulations	[25, 41, 43]
Inadequate sustainable measurement tools	[40, 44]
Method of selecting SC material	[40]
Unstable prices of SC materials	[45]
Inadequate knowledge and understanding of the concept of sustainability	[10, 16; 21, 31, 39]
Lack of expert opinions on sustainable construction	[21, 46, 47]
Limited availability of sustainable materials	[15, 46, 41]
Unreliability of Suppliers	[46]
The perception that SC materials are of low status	[39]
Level of integration of life cycle cost	[24, 26]

hand, observed the issue surrounding green materials availability. One other factor observed by Mousa [34] was that the client-driven nature of the industry leaves little room for sustainable products. This is because clients with insufficient knowledge prematurely eliminate any sustainable alternative that is not commonly used. This tends to result in fewer sustainable solutions offered at high prices due to a lack of interest from consumers and lack of healthy competition. It is based on this wealth of knowledge on the different issues serving as drawbacks to SC, that this assessed the restrains of SC in Nigeria.

3   Research Methodology

To critically identify the restrains of SC in the Nigerian construction industry, the conventional research approach of understanding the key area of the study, conducting a review of existing literature, quantitative data gathering, data analysis and drawing of inference from the result of the data analysed, was adopted. Since the study was conducted across the country, a questionnaire survey was deemed necessary due to its ability to cover a broader range of respondents within a short period and at the same time allow for quantifiability and objectiveness in the research [48, 49]. The questionnaire adopted was designed in sections with the first section geared towards gathering background information of the respondents to ascertain whether the respondents were fit to participate in the survey or not. The second section sought answers to the factors that are affecting the adoption of SC practices in the country. Respondents were provided with a total of 20 variables, as earlier indicated in Table 1. These respondents were asked to rate them based on their level of severity using a 5-point Likert scale, with 5 being very severe, 4 severe, 3 moderate, 2 less severe, and 1 not severe. The last section sought answers to the level of conformance of construction projects to economic, social, environmental sustainability as well as the overall performance of construction projects in the country. The data used were gathered from construction managers, project managers, and quantity surveyors in the six regions of the country. These set of construction professionals were selected due to their role in the use of construction materials, management of construction projects and estimation of these projects. Since the study cut across the entire country, an electronic questionnaire was adopted for easy sending and collection of feedback from respondents. As a result of the difficulty encountered in getting the details of these professionals from their professional bodies, a snowball approach was adopted. Heckathorn [50] described snowball sampling as a technique that assumes that a link exists between the initial sample and others in the same targeted population, thus, allowing a series of referrals to be made within a circle of acquaintance. Atkinson and Flint [51] also noted that this approach could be beneficial when there is a need to increase the sample size, as in the case of this current study. A similar approach was adopted in the study of Rahman [52]. Based on the snowball approach adopted, the exact number of distributions cannot be determined, hence calculating the total response rate is almost impossible [53]. At the end of the survey, 70 responses were collected.

In terms of data analysis, the background information of the respondents was analysed using percentages. Since a total of 20 factors are considered within the study, there is the likelihood of some factors leading to similar underlying effects. Based on this understanding, Exploratory Factor Analysis (EFA) was conducted to reduce the large group of factors into a smaller number of underlying grouped factors. Confirmatory Factor Analysis (CFA) was then conducted using SmartPLS 3.2 to confirm the reliability and validity of the grouped variables from EFA. Partial Least Square Structural Equation Modelling (PLS-SEM) was further used to ascertain the significance of each factor to the attainment of SC within the country. This became necessary as EFA only categorizes these factors without showing clearly the relationships between variables and their latent constructs. Moreover, Hedaya et al. [54] have earlier noted that the "PLS-SEM function better than other multivariate techniques including multiple regression, path analysis and factor analysis when analysing the cause-effect relationships that exist among latent constructs". The constructs identified from the EFA conducted were measured against four specific SC outcomes which are conformance to economic sustainability (SC1), conformance to environmental sustainability (SC2), conformance to social sustainability (SC3), and conformance to overall project performance (SC4).

3.1 Partial Least Square Structural Equation Modelling (PLS-SEM)

PLS-SEM is an extension of standardized regression modelling, which is useful in determining the causal relationships that exist between different factors [54]. This analytical tool does not place emphasis on the distribution of the data under review and has gained significant recognition among researchers in recent times due to its ability to accommodate small samples in a study [55–57]. PLS-SEM can also be adopted when the data gathered is skewed, the correctness of the model cannot be ensured due to one or more shortcomings in the variables selected or their linkage, but predictive accuracy still needs to be ensured [56, 57]. Since the sample size of this study (70) is relatively small, PLS-SEM became the most preferred compared to the co-variance base SEM approach which requires a large sample size, a normally distributed data and accurately specified model [56, 58]. In determining the sample size suitable for PLS-SEM to be conducted, Marcoulides and Saunders [59] gave certain guidelines in terms of the maximum number of arrows pointing to a latent variable in a model in relation to the minimum sample size needed. For example, for a latent variable with 7 arrows pointing to it, a minimum of 80 samples is required. Looking at the initial model designed for this study, it is evident that the construct with the highest arrows is the knowledge and awareness factors with 7 variables. This implies that going by Marcoulides and Saunders [59] a minimum of 80 respondents was needed for the study. However, in their '10-times rule,' Hair et al. [58] suggested that the sample size should be 10 times greater than the highest number of arrows pointing to a latent variable. This implies that for 7 arrows, at least 70 samples are required, thus, justifying the sample size adopted for this current study. Moreover, PLS-SEM studies with a small sample size have emanated over the years [60–63]. PLS-SEM has become popular within the construction industry. Hedaya et al. [54] assessed the issues surrounding cost overrun in Bahrain using this method, while Adeleke et al. [60] adopted it in a preliminary study to assess some factors for effective risk management in Nigeria. Similarly, in India, Shanmugapriya and Subramanian [64] investigated that factors that could influence construction productivity using PLS-SEM, while Memon and Rahman [65] assessed inhibiting factors of cost in large projects in Malaysia using this same method. The use of PLS-SEM is also evident in sustainability studies [61, 66, 67]. Smart PLS 3.0 software was used in conducting the PLS-SEM analysis for this current study.

4   Results

4.1   Background Information of Respondents

Based on the analysis of the data gathered on the background information of the respondents, more response was gotten from respondents in the south-west (48.6%) and the north-central (24.3%) regions, respectively. The high response rate in both regions can be attributed to the presence of Lagos state which is described as the country's commercial city with lots of construction companies, and construction activities taking place daily in the south-west, and the Federal Capital Territory, Abuja, in the north-central which is the country's administrative city and houses the head offices of most construction organisations and professional bodies. The south-south, north-west, south-east, and north-east had a response rate of 11.4%, 10%, 4.3% and 1.4% respectively. A total of 64.3% of the respondent work with private organisations, with only 35.7% working within government establishments. The average years of working experience of the respondents are 8.6 years, while the highest academic qualification of the respondents is a master's degree (47.1%). This is followed by a bachelor's degree and postgraduate diploma with 31.4% and 11.4% respectively. The least academic qualifications are higher national diploma, ordinary national diploma and PhD with 5.7%, 2.9% and 1.4%, respectively. This result shows that the target professionals were adequately represented, and they have reasonable years of experience and considerable academic background to understand the questions of the study and give insightful answers to the questions asked.

4.2   Exploratory Factor Analysis

The factorability of the 20 identified factors was first ascertained by assessing the sample size of the study. Unfortunately, there has been no agreement regarding the idle sample for EFA to be conducted. Overtime most studies have advocated the use of large sample size (between 150 and 200) while others believe that as long as the communalities derived is reasonably high (from 0.5 above), and the expected number of extraction is low, then significant emphasis should not be placed on the sample size of the study [68–71]. Communalities of between 0.529 and 0.859 were generated for this study, as seen in Table 2. This result confirmed that the use of FA was appropriate for the study. Furthermore, Kaiser–Meyer–Olkin (KMO) value of 0.880, which is higher than the 0.6 thresholds required for EFA was derived [71]. Bartlett test also gave a chi-square value of 982.37 and a significant p-value of 0.000, which follows Tabachnick and Fidell [70] submission that the p-value of a Bartlett test must be significant for factorable data. EFA was conducted using PCA with varimax rotation.

Table 2 Exploratory factor analysis result

Factors	Component				Comm Extract
	1	2	3	4	
Sustainability knowledge related factors					
Clients fear of high investment cost (SK1)	0.778				0.697
Inadequate awareness and knowledge (SK2)	0.714				0.713
Low demand for sustainable products (SK3)	0.703				0.694
Industry's resistance to change (SK4)	0.701				0.708
Method of selecting SC material (SK5)	0.693				0.529
Lack of expert opinions on SC (SK6)	0.636				0.610
Clients preference (SK7)	0.595				0.621
Regulation and policy-related factors					
Inadequate sustainable measurement tools (RP1)		0.793			0.752
Inadequate building codes and regulations (RP2)		0.781			0.728
Inadequate government policies/support (RP3)		0.768			0.859
Client worries about profitability (RP4)		0.666			0.613
Unstable prices of SC materials (RP5)		0.498			0.604
Sustainable materials and technology-related factors					
The perception that SC materials are of low status (SMT1)			0.729		0.743
Unreliability of suppliers (SMT2)			0.712		0.788
Inadequate technology and technological process (SMT3)			0.677		0.766
Limited availability of sustainable materials (SMT4)			0.588		0.666
Information and management related factor					
Limited access to relevant historical data (IM1)				0.792	0.614
Lack of commitment of top management (IM2)				0.688	0.747
Inadequate exemplar 'demonstration project (IM3)				0.631	0.651
Level of integration of life cycle cost (IM4)				0.548	0.778
Kaiser–Meyer–Olkin Measure		0.880			
Bartlett's Test of Sphericity	X^2	982.37			
	Df	190			

(continued)

Table 2 (continued)

Factors	Component				Comm
	1	2	3	4	Extract
	Sig	0.000			

Note: X^2 = Chi-square, Df = Degree of freedom, Comm. = Communalities

A total of 4 extractions with eigenvalue of 1 and above were generated, and their combination accounts for approximately 69.4% of the total cumulative variance. The first principal component extracted accounts for 50.5% of the total variance explained and has the highest number of variables loading on it. This component accounts for the most issue facing the SC practices in the country. The variables loading on this component are, clients fear high investment cost, inadequate knowledge and understanding of SC, low demand for sustainable products, construction industry's resistance to change, method of selecting SC materials, lack of expert opinions on SC, and clients preference. This component is subsequently named 'sustainability knowledge related factor' (SK). The second principal component extracted accounts for 8.1% of the total variance explained and has variables such as inadequate sustainable measurement tools, inadequate building codes and regulations on sustainability, inadequate government policies/support, and client worries in profitability loading on it. Subsequently, this component was named 'regulation and policy-related factor' (RP). Component three accounts for only 5.5% of the total variance explained with the perception that sustainable construction materials being of low status, the unreliability of suppliers, inadequate technology and technological process, and limited availability of sustainable materials loading on it. This component was subsequently named 'sustainable materials and technology-related factor' (SMT). The last extracted component has limited access to relevant information and historical data, interest and commitment of top management, inadequate exemplary demonstration project, and level of integration of life cycle cost loading n it and accounts for 5.2% of the total variance explained. This component is seen as 'information and management related factor' (IM).

4.3 Structural Equation Modelling

4.3.1 Confirmatory Factor Analysis

In the model development, two main features exist; the inner model and the outer model. The inner model is the main constructs represented in blue circles linked by arrows to the expected outcomes of SC (as seen in Fig. 1). The outer model includes the measurement variables of the different constructs linked to their respective constructs with an arrow. Confirmatory factor analysis (CFA) was conducted on the outer model for factor reliability and validity. This ensures that the variables

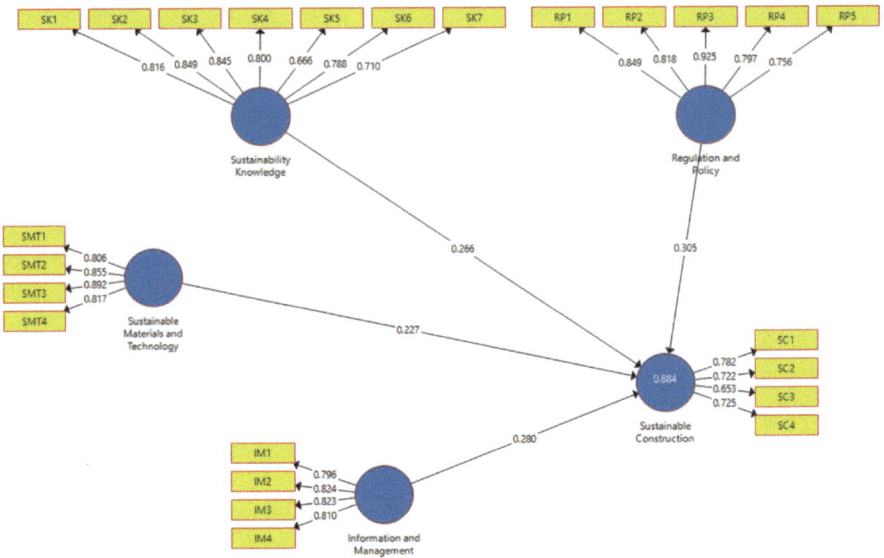

Fig. 1 First iteration of the hypothesised model

selected to measure each construct measures them. To achieve this, internal consistency using Cronbach alpha (α), roh (ρA) coefficient, and composite reliability (CR), as well as the convergent validity using Average variance extracted (AVE), were assessed. Based on existing works, the cut-off for Cronbach α, ρA, CR, and AVE were 0.7, 0.7, 0.7, and 0.5 [57, 72–74]. Discriminant validity was also assessed using heterotrait–monotrait (HTMT) with a cut-off of below 0.85.

The first step was to assess the factor loadings for each variable in the outer model. The threshold for an acceptable factor loading is 0.7 [55, 57]. Looking at the loading of the first iteration in Table 3 and Fig. 1, it is evident that only SK5 and SC3 had a factor loading of less than 0.7, thus, creating the need for a second analysis with this variable eliminated as suggested by Hedaya et al. [54]. However, in deleting these variables, Hulland [74] suggested caution. Careful assessment of the importance of the variable being deleted and the influence of the deletion on the reliability and validity of the model was assessed. Since SC3 is a construction project meeting social sustainability which is considered a crucial element of sustainability in construction, this variable was retained. However, SK5, which is the method of selecting SC materials for construction delivery, was eliminated, and the whole CFA process was repeated.

Looking at the result from the second iteration, the internal consistency of the outer model revealed that Cronbach's α gave a value range of 0.830–0.893, and this is above the cut-off of 0.7 sets. CR gave a value of 0.887–0.918, which is also higher than the set cut-off. Since it has been observed that the true reflection of the reliability of a measurement variable can be sought in the ρA [75, 76] an assessment of the same was conducted. This test gave a value range of 0.832–0.899, which is equally above

Table 3 Summary of internal consistency and convergent validity

Analysis 1							
Construct	Variables	Loading	α	ρA	CR	AVE	VIF
Sustainability knowledge	SK1	0.816	0.895	0.905	0.918	0.616	2.453
	SK2	0.849					3.380
	SK3	0.845					3.193
	SK4	0.800					2.462
	SK5	0.666					1.592
	SK6	0.788					2.435
	SK7	0.710					2.140
Regulation and policy	RP1	0.849	0.887	0.893	0.918	0.691	2.709
	RP2	0.818					2.861
	RP3	0.925					4.251
	RP4	0.797					2.209
	RP5	0.756					1.932
Sustainable materials and technology	SMT1	0.806	0.864	0.865	0.907	0.711	1.902
	SMT2	0.855					2.130
	SMT3	0.892					2.691
	SMT4	0.817					1.963
Information and management	IM1	0.796	0.830	0.832	0.887	0.662	1.583
	IM2	0.824					1.916
	IM3	0.823					1.829
	IM4	0.810					1.874
Analysis 2							
Construct	Variables	Loading	α	ρA	CR	AVE	VIF
Sustainability knowledge	SK1	0.826	0.893	0.899	0.918	0.653	2.453
	SK2	0.857					3.369
	SK3	0.848					3.132
	SK4	0.792					2.256
	SK5	-					-
	SK6	0.800					2.433
	SK7	0.716					2.123
Regulation and policy	RP1	0.849	0.887	0.893	0.918	0.691	2.709
	RP2	0.818					2.861
	RP3	0.925					4.251
	RP4	0.797					2.209
	RP5	0.756					1.932
Sustainable materials and technology	SMT1	0.806	0.864	0.865	0.907	0.711	1.902

(continued)

Table 3 (continued)

Analysis 1

Construct	Variables	Loading	α	ρA	CR	AVE	VIF
	SMT2	0.855					2.130
	SMT3	0.892					2.691
	SMT4	0.817					1.963
Information and management	IM1	0.796	0.830	0.832	0.887	0.662	1.583
	IM2	0.825					1.916
	IM3	0.823					1.829
	IM4	0.810					1.874

Table 4 Heterotrait-monotrait ratio

	Information and management	Regulation and policy	Sustainability knowledge
Information and management			
Regulation and policy	0.790		
Sustainability knowledge	0.751	0.748	
Sustainable materials and technology	0.801	0.833	0.774

the set threshold of 0.7. These results confirmed that the measurement variables have a reliable internal consistency to measure their respective constructs. Convergent validity ascertained using AVE revealed that the measurement variables are valid as a value range of between 0.653 and 0.711 was derived for the four constructs. This is above the 0.5 thresholds set for a valid convergence. This result, coupled with the results from the internal consistency analyses done using CR, Cronbach α and ρA implies that the constructs possess acceptable reliability and convergent validity.

Next was to assess the discriminant validity of the factors under assessment. Hulland [74] noted that the function of this test is to substantiate the extent of the difference of a given construct from others. While most studies have employed the square root of the AVE in each construct to determine discriminant validity, the HTMT has, however, been the preferred choice in most recent PLS-SEM studies [75, 76]. The assumption here is that for clear discrimination between two factors to exist, an HTMT value of 0.85 and below must be derived [76, 77]. The result in Table 4 revealed that all the assessed constructs meet this criterion as they all have a ratio of below 0.85.

4.3.2 Structural Model Assessment

In order to ensure the structural model is accurate and robust in nature, it is important to assess its collinearity, its path coefficient (β), determination coefficient (R^2), predictive relevance (Q^2) and its Goodness-of-Fit (GoF). For this aspect, the result from the second iteration is presented. The collinearity of the model was assessed using the variance inflation factor (VIF) values of all constructs. According to Hair et al. [78], a VIF below 3.0 is good for collinearity issues not to exist, however, VIF of < 5.0 is also acceptable. A look at the last column in Table 3 reveals that the VIF for all variables are all below 5.0 Thus implying that there is no multicollinearity issue with these variables. A look at Fig. 2 revealed an R^2 of 0.886. The ideal value of R^2 varies among different fields of studies [75]. However, the threshold for R^2 according to Hair et al. [78] is 0.75, 0.50, and 0.25 for substantial, moderate and weak. Findings from this current study reveal that the four constructs and their variables account for about 89% (0.886) of the issue of poor SC in the country. Furthermore, using SmartPLS blindfold procedure, the predictive relevance of this inner model Q^2 is derived as 0.396. This figure reveals high predictive relevance when assessed in line with Chin's (1998) threshold for Q^2, which stated that 0.02 is weak, 0.15 is medium, while 0.35 is regarded as high predictive relevance. To ascertain the approximate fit of the hypothesised model, the standardized root means square residual (SRMR), as well as the Bentler and Bonett normed fit index (NFI), were assessed as suggested by Hair et al.. [75] and Henseler et al. [76]. The threshold for these model fits criteria are 1.0 or below for SRMR [79, 80] and between 0.6 and 1.0 for NIF [81]. For this current study, and SRMR of 0.10 was derived with a NIF of 0.722, which are

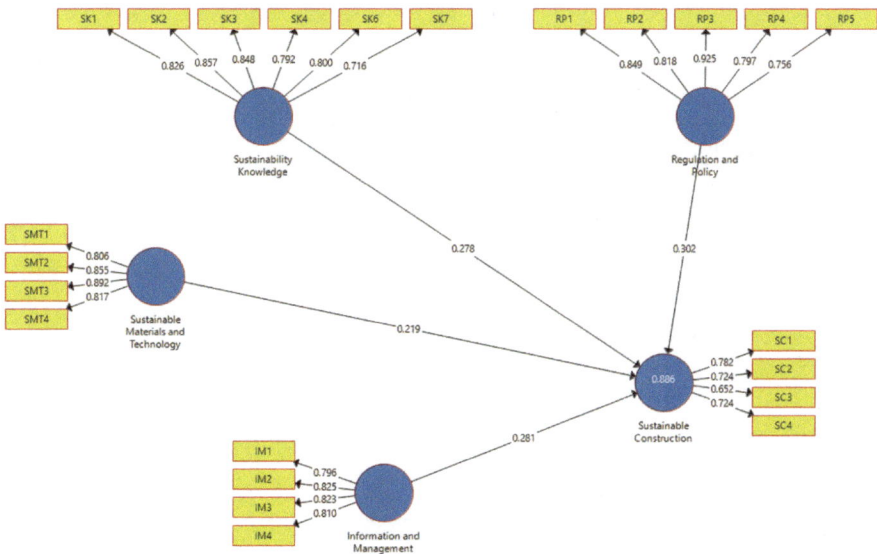

Fig. 2 Final iteration of the hypothesised model

both with the given thresholds, thus, confirming a reasonable level of approximate model fit for the hypothesised model. The absolute fit of the model was also evaluated using the Goodness-of-fit (GoF) approach, which is the geometric mean of the average communality and the average of R^2. The calculated GoF gave a value of 0.775. According to Akter et al. [82], a GoF of 0.1 and below is considered as small, 0.25 is considered as moderate while 0.36 and above is good. Thus, it can be said that the hypothesised model has a significant GoF.

$$GoF = \sqrt{\overline{AVE} \times \overline{R}^2}$$
$$AVE = 0.678$$
$$R^2 = 0.886$$
$$GoF = \sqrt{0.678 \times 0.835} = 0.775$$

The relationship between the constructs and the SC outcome was assessed. In doing this, the non-parametric bootstrapping approach was adopted since PLS-SEM does not require the data under assessment to be parametric in nature. The bootstrapping was conducted using 5000 subsamples as encouraged by Wong [57]. This analysis gives the path coefficient, the sample mean, standard deviation, t-statistics and a significant p-value which shows the significance of the path coefficient. From Table 5, it is evident that the path coefficient for all the constructs is significant. This is major because at two-tailed t-test, with a significance level of 1% (99% confidence interval) the t-statistics of these constructs are greater than 2.58 table value [57, 64]. For example, the relationship between regulation and policy-related factor and SC in the country is significant as a β value of 0.302, and a t-statistics of 3.762 were derived.

Table 5 Summary of path coefficient and significance levels

	β	M	STDEV	t-statistics	p-values	Remark
Regulation and policy -> SC	0.302	0.290	0.080	3.762	***	Supported
Information and management -> SC	0.281	0.272	0.078	3.598	***	Supported
Sustainability knowledge -> SC	0.278	0.290	0.092	3.009	***	Supported
Sustainable materials and technology -> SC	0.219	0.225	0.083	2.619	***	Supported
R^2		0.886				Substantial
Q^2		0.396				Acceptable
SRMR		0.10				Acceptable
NFI		0.722				Acceptable
GoF		0.775				Good

Note = *** significant at 99% confidence interval, M = Sample mean, STDEV = Standard Deviation

The implication of this result is that while all four factors are significant in the poor SC delivery in the country, regulation and policy has the strongest influence with 30%. This is followed by information and management issues (28%) and sustainability knowledge (28%) with least being sustainable materials and technology with 22% influence.

5 Discussion

Based on the findings from the study, it is evident that the factors impeding the attainment of SC in Nigeria are majorly related to sustainability regulations and policies, information and management related issues, sustainability knowledge, and sustainable materials and technology-related issues. It is imperative to note that although EFA revealed issues surrounding sustainability knowledge to have the highest percentage of total variance explained, PLS-SEM revealed that its path coefficient is lower than that of issues on regulations and policies, and information and management. This, therefore, insinuate the importance of having regulations and policies in place as well as adequate information from past projects and support from management within construction organisations in the quests for projects that are economically, socially, and environmentally sustainable as well as projects that can attain overall project performance. Thus, contrary to the findings of Alabi [12] and Davies and Davies [14] submissions that lack of awareness, poor education, lack of experience of construction professionals in sustainability-related issues as the major culprit of poor SC in the country, the finding of this current study is geared towards regulations and policy issue as the bane of this problem.

The role of government in the creation of legislation that will support the adoption of SC practices and the amendments of existing building regulations to accommodate SC concepts have been noted in past studies [28, 30, 32]. There is no doubt that without the right building codes and regulations to promotes SC, as well as government supporting SC through creating and enforcing SC policies, the actualisation of SC in the country will not come to reality. In the same vein, without adequate sustainability measuring tools, attaining SC will be impossible as there will be no benchmark to use in determining the sustainable nature of projects. The finding of this study further corroborates the submission of Häkkinenand Belloni [25] that building codes and regulations are hindering the adoption of SC. It is also in tandem with Al-Sanad [32], Ametepey et al. [24] and Osaily [40] submission that government support is an important factor affecting SC in similar developing countries like Kuwait, Ghana, and Palestine.

Furthermore, the fact that information is a key component of the adoption of any new concept cannot be ignored. Thus, adequate information on the success and failure of past projects wherein SC concepts have been adopted would go a long way in promoting the use of the concept in project delivery. This is because, through modelling SC projects that have been achieved successfully in the past, a roadmap for the attainment of more sustainable projects can be created. Similarly, construction participants can learn from the failure of past projects wherein in the adoption of SC concepts was not successful. This finding further corroborates the submission of Ametepey et al. [24] that lack of information and exemplary demonstration projects can serve as hindrances to SC. In the same vein, there it is almost unlikely for a new concept to be adopted within an organisation without the support of top management [83]. This is a crucial factor affecting SC in the country, as observed from the findings. This further affirms Abisuga and Oyekanmi [42] submission that the support of management within an organisation places a significant role in the adoption or non-adoption of SC practices within organisations in Nigeria.

Past studies have placed emphasis on the need for improved sustainability knowledge among construction experts as this significantly affects the understanding as well as the adoption of the concept. William and Dair [29] have earlier noted that the concept of sustainability is still vague to most construction experts and stakeholders in developing countries. The symptoms persist in this modern-day as observed from the findings of this current study. This has led to the resistance from construction participants and the fear of an increase in cost on the part of the clients, which invariably affect their demand for the use of SC materials on their projects. This finding is in tandem with the submissions of Al-Sanad [32], Baron and Donath [15], and Nguyen et al. [84] who noted that sustainability awareness and knowledge related factors are key hindrances to the attainment of sustainable construction in developing countries around the world. If this is to change, there is the need for more public awareness and proper enlightening of construction stakeholders as to the concept and inherent benefits of SC as observed by Opoku and Ahmed [30] and Pitt et al. [33].

The importance of SC materials in the delivery of SC projects has been reiterated in previous studies [26, 39]. The perception of SC materials being of low status has also been identified by Hwang and Tan [85]. Findings of this study in tandem with these submissions as the fact that SC materials such as the use of earth blocks for construction which is believed to be highly sustainable [11] are perceived to have low status. This is adversely affecting the attainment of economic, social and environmental SC as well as overall project performance in Nigeria. The finding of this study also supports Abidin et al. [41] submission that the use of technology to improve project process and construction methods can go a long way in attaining SC.

Thus, the implication of the findings of the study is that the UN sustainable development goal number nine (industry, innovation and infrastructure) which is aimed at delivering basic infrastructure [86] is most likely unattainable in Nigeria. This is because the construction industry that is saddled with the responsibility of delivering these sustainable infrastructures is being deterred from doing the same by the identified impediments. This is rather disheartening because these infrastructures are not only supposed to meet the need of the present generation but also fulfil the needs of generations to come. Therefore, if Nigeria is to contribute its own quota to the UN dream of attaining a poverty-free and protected planet by 2030, construction industry practitioners and stakeholders need to ensure sustainability regulations and policies are in place, information and management of SC and sustainability knowledge are optimised, and that sustainable materials and technologies are available.

6 Conclusion and Recommendations

This study assessed the factors that could affect the attainment of SC in Nigeria. Using a questionnaire survey administered to construction managers, project managers, and quantity surveyors from within the six different regions of the country, the study has been able to reveal the principal culprits of poor economic, social and environmental SC as well as overall poor project performance in the country. Based on the findings, the study concludes that the poor SC being experienced in the country is as a result of issues relating to regulation and policies, information and management support, sustainability knowledge among construction participants, and sustainable materials and technology availability. Therefore, if SC construction is to be attained in the country, the government must be ready to champion the course of SC attainment through the creation and enforcing of SC policies in the country. Also, there is the need for the development of appropriate sustainability measurement tools and building codes and regulations that will help check the construction processes and ensure they are sustainable. There is a need for support from top management and proper documentation of information on SC projects being conducted in order to have a means of reference when handling new SC projects. Similarly, enlightening construction clients and other participants through seminars and workshops organised by construction professional bodies, on the concept of SC and the overall benefit of implementing SC concepts are important. This will help eliminate the fear of the possible high cost of construction when SC concept is adopted. Similarly, through this enlightenment, the understanding of the SC concept will increase among construction experts, particularly in the aspect of available SC materials and technologies needed.

Base on the findings of the study, it is clear that there is a void in the aspect of regulation and policy implementation in the country, particularly with respect to SC. In the same vein, information storage and retrieval are crucial issues for the construction industry in the country. If the industry is to move past its issue of poor SC project delivery, then the government and other professional bodies such as the Council for Registered Builders of Nigeria responsible for establishing and overseeing the country's construction industry development, have a duty to create legislation and enforce regulations that will promote SC in the country. Top management must also play their part in supporting SC within their organisation and ensuring proper information storage and transfer from one project to another. Thus, this study contributes to the body of knowledge as it reveals the different factors affecting SC and their relationship to SC delivery in the country. Its findings can also prove useful to other developing countries, particularly in Africa, where construction practices are similar. However, while this study contributes significantly to the body of existing knowledge, care must be taken in generalising its result as the study was conducted among only three sets of construction professionals in the country. Further study can be done with a wider range of construction stakeholders to get a much broader view of the topic and compare results.

References

1. Anaman KA, Osei-Amponsah C (2007) Analysis of causality links between the growth of the construction industry and the growth of the macro economy in Ghana. Constr Manage Econ 25(9):951–961
2. Oke AE, Aghimien DO, Adedoyin AA (2018) SWOT analysis of indigenous and foreign contractors in a developing economy. Int J Qual Reliab Manage 3(6):1289–1304
3. Ametepey O, Aigbavboa C (2014) Practitioners perspectives for the implementation of sustainable construction in Ghana. In: Proceedings of the DII-2014 conference on infrastructure investments in Africa, Livingstone, Zambia, 25–26 September, pp 114–124
4. Baloi D (2003) Sustainable construction: challenges and opportunities. In: Proceeding of association of researchers in construction management (ARCOM), held at the school of the environment at the university of Brighton, 3–5 September
5. Du Plessis C (2002) Agenda 21 for: sustainable construction in developing countries—a discussion document. In: International council for research and innovation in building and construction, CIB, United Nations Environment Programme International Environmental Technology Centre, UNEP-IETC. Available at https://researchspace.csir.co.za/dspace/handle/10204/3511
6. Brundtland Report (1987) Our common future, united nations assembly, report of the world commission on environment and development. Annex to General Assembly Document A/42/427
7. Pearce D (1988) economics equity and sustainbel development. Futures 20(6):598–605
8. Oke AE, Aigbavboa CO (2017) In: Sustainable value management for construction projects, Springer International Publishing
9. Abidin NZ (2010) Investigating the awareness and application of sustainable construction concept by Malaysian developers. Habitat Int 34(4):421–426
10. Akbiyikli R, Dikmen SU, Eaton D (2009) Sustainability and the Turkish construction cluster: a general overview. In: Proceedings of the construction and building research conference of the royal institution of chartered surveyors, University of Cape Town, 10–11 September
11. Aghimien DO, Awodele OA (2016) Sustainability of compressed stabilized interlocking earth block (CSIEB) for building construction in Nigeria. In: Ebohon OJ, Ayeni DA, Egbu CO, Omole FK (eds) Proceedings of the joint international conference (JIC) on 21st century human habitat: issues, sustainability and development, 21–24 March 2016, Akure, Nigeria, pp 783–792
12. Alabi AA (2012) Comparative study of environmental sustainability in building construction in Nigeria and Malaysia. J Emerg Trends Econ Manage Sci 3(6):951–961
13. Aje IO (2015) Effective financing options for sustainable construction in a developing economy. In: A paper delivered at a 2-Day national seminar on sustainable construction in a developing economy: factors and prospects organized by the Nigerian institute of quantity surveyors, ondo state chapter held at theodore idibiye francis auditorium, Federal University of Technology, Akure Ondo State from Thursday 8th–Friday 9th October
14. Davies OOA, Davies IO (2017) Barriers to implementation of sustainable construction techniques. MAYFEB J Environ Sci 2:1–9
15. Baron N, Donath D (2016) Learning from Ethiopia—a discussion on sustainable building. In: Proceedings of SBE16 hamburg international conference on sustainable built environment strategies—stakeholders—success factors, held from 7th to 11th march in Hamburg, Germany
16. Djokoto SD, Dadzie J, Ohemeng-Ababio E (2014) Barriers to sustainable construction in the Ghanaian construction industry: consultants' perspectives. J Sustain Develop 7(1):134–143
17. James P, Matipa WM (2004) Sustainable construction in a developing country: an assessment of how the professional's practice impacts the environment. In: Khosrowshahi F (ed), 20th Annual ARCOM conference, Heriot Watt University, 1–3 September

18. Aghimien DO, Aigbavboa CO (2019) Impediments of sustainable construction in the south african construction industry. In: environmental design and management international conference (EDMIC 2019), Obafemi Awolowo University, Ile-Ife, Nigeria, 20th–22nd May, pp 256–264
19. Saad MM (2016) Impediments to implementation of green buildings in South Africa. In: 9th CIDB postgraduate conference, Cape Town, 2–4 February
20. Dalibi SG, Fend JC, Shuangqin L, Sadiq A, Bello A, Danja II (2017) Hindrances to green building developments in nigeria's built environment: the project professionals' perspectives. In: IOP Conference series: earth and environmental science, 2017 international conference on environmental and energy engineering (IC3E 2017), vol 63. pp 1–8
21. Aigbavboa C, Ohiomah I, Zwane T (2017) Sustainable construction practices: a lazy view of construction professionals in the South Africa construction industry. In: The 8th International conference on applied energy procedia, vol 105. pp 3003–3010
22. Lowe DJ, Zhou L (2003) Economic factors of sustainable construction. In: RICS COBRA foundation construction and building research conference, University of Wolverhampton, The RICS Foundation, London, 1–2 September
23. Kibert CJ (2013) Sustainable construction–green building design and delivery. John Wiley , NJ
24. Ametepey O, Aigbavboa C, Ansah K (2015) Barriers to successful implementation of sustainable construction in the Ghanaian construction industry. In: 6th International conference on applied human factors and ergonomics and the affiliated conferences, procedia manufacturing, vol 3(1). pp 1682–1689
25. Häkkinen T, Belloni K (2011) Barriers and drivers for sustainable building. Build Res Inf 39(3):239–255
26. Shi Q, Zuo J, Huang R, Huang J, Pullen S (2013) Identifying the critical factors for green construction—an empirical study in China. Habitat Int 40(1):1–8
27. Powmya A, Abidin ZN (2014) The challenges of green construction in Oman. Int J Sustain Constr Eng Technol 5(1):33–41
28. Oke AE, Aghimien DO, Aigbavboa CO, Musenga C (2019) Drivers of sustainable construction practices in the zambian construction industry. In: 10th International conference on applied energy (ICAE2018), 22–25 August 2018, Hong Kong, China. Energy Procedia, vol 158. pp 3246–3252
29. Williams K, Dair CA (2007) framework for assessing the sustainability of brownfield developments. J Environ Planning Manage 50(1):23–34
30. Opoku A, Ahmed V (2015) Leadership and sustainability in the built environment. Taylor and Francis, London
31. Aghimien DO, Adegbemo TF, Aghimien IE, Awodele AO (2018) Challenges of sustainable construction: a study of educational buildings in Nigeria. Int J Built Environ Sustain 5(1):33–46
32. Al-Sanad S (2015) Awareness drivers actions and barriers of sustainable construction in Kuwait. Int Conf Sustain Design Eng Constr Procedia Eng 118(1):969–983
33. Pitt M, Tucker M, Riley M, Longden J (2009) Towards sustainable construction: promotion and best practice. Constr Innov 9(2):201–224
34. Mousa AA (2015) Business approach for transformation to sustainable construction: an implementation on a developing country. Resour Conserv Recycl 101:9–19
35. Al-Yami AM, Price ADF (2006) A framework for implementing sustainable construction in building briefing project. In: Boyd D (ed) Proceedings 22nd annual ARCOM conference, association of researchers in construction management, Birmingham, 4–6 September
36. Miranda L, Marulanda L (2001) Sustainable construction in developing countries a Peruvian perspective. In: Agenda 21 for sustainable construction in developing countries, Latin America Position Paper, (2001)
37. Isa M, Rahman MMGMA, Sipan I, Hwa TK (2013) Factors affecting green office building investment in Malaysia. Proc-Soc Behav Sci 105:138–148
38. Hatamleh MT, Hiyassat M, Sweis GJ, Sweis RJ (2018) Factors affecting the accuracy of cost estimate: case of Jordan. Eng Constr Architectural Manage 25(1):113–131

39. Kissi E, Abdulai-Sadick M, Agyemang DY (2018) Drivers militating against the pricing of sustainable construction materials: the Ghanaian quantity surveyors' perspective. Case Stud Constr Mater 8:507–551
40. Osaily NZ (2010) The key barriers to implementing sustainable construction in West Bank—Palestine, Robert Kennedy College/Zurich University of Wales/UK
41. Abidin ZN, Khalfan M, Kashyap M (2003) Moving towards more sustainable construction. In: Proceedings of the construction and building research conference of the royal institution of chartered surveyors. School of Engineering and the Built Environment University of Wolverhampton, 1st to 2nd September
42. Abisuga AO, Oyekanmi OO (2014) Organizational factors affecting the usage of sustainable building materials in the Nigerian construction industry. J Emerg Trends Econ Manage Sci 5(2):113–211
43. Ayarkwa J, Acheampong A, Wiafe F, Boateng BE (2017) Factors affecting the implementation of sustainable construction in ghana: the architect's perspective. In: ICIDA 2017—6th international conference on infrastructure development in Africa—12–14 April, Knust, Kumasi, Ghana, pp 377–386
44. Parkin S (2000) Context and drivers for operationalizing sustainable development. In: Proceedings of ICE, vol 138. pp 9–15
45. Azhar N, Farooqui RU, Ahmed SM (2008) Cost overrun factors in construction industry of Pakistan. In: First international conference on construction in developing countries (ICCIDC–I), advancing and integrating construction education, research and practice, 4–5 August
46. Ogunkah IC, Yang J (2013) Analysis of factors affecting the selection of low-cost green building materials in housing construction. Int J Sci 2(9):41–75
47. Babalola IH, Oluwatuyi OE, Akinloye L, Aiyewalehinmi E (2015) Factors influencing the performance of construction projects in Akure, Nigeria. Int J Civil Eng Constr Estate Manage 3(4):57–67
48. Ackroyd S, Hughes JA (1981) Data Collection in Context. Longman, London
49. Tan WCK (2011) Practical research methods. Pearson Custom, Singapore
50. Heckathorn DD (2011) Comments: snowballing versus respondent-driven sampling. Sociol Methodol 41(1):355–366
51. Atkinson R, Flint J (2001) Accessing hidden and hard-to-reach populations: snowball research strategies. Social Research Update, 33, University of Surrey, Guildford, pp 1–4
52. Rahman MM (2014) Barriers of implementing modern methods of construction. J Manage Eng 30(1):69–77
53. Chan APC, Darko A, Ameyaw EE (2017) Strategies for promoting green building technologies adoption in the construction industry—an international study. Sustainability 9(6):1–18
54. Hedaya AMA, Saad SMA (2017) Causes and effects of cost overrun on construction project in Bahrain: Part 2(PLS-SEM Path Modelling). Modern Appl Sci 11(7):28–37
55. Chin WW (1998) Issues and opinion on structural equation modeling. MIS Quarterly 22(1):7–16
56. Hwang H, Malhotra NK, Kim Y, Tomiuk MA, Hong S (2010) A comparative study on parameter recovery of three approaches to structural equation modeling. J Mark Res 47:699–712
57. Wong KK (2013) Partial least squares structural equation modeling (PLS-SEM) techniques using SmartPLS. Marketing Bulletin, vol 24. Technical Note 1. pp 1–32
58. Hair JF, Ringle CM, Sarstedt M (2011) PLS-SEM: indeed, a silver bullet. J Mark Theory Practice 19(2):139–151
59. Marcoulides GA, Saunders C (2006) Editor's Comments—PLS: a silver bullet? MIS Quarterly 30(2):3–9
60. Adeleke AQ, Bahaudin AY, Kamaruddeen AM (2015) A partial least square structural equation modeling (PLS SEM) preliminary analysis on organizational internal and external factors influencing effective construction risk management among nigerian construction industries. Rev Téc Ing Univ Zulia 38(3):143–155
61. Bamgbade JA, Kamaruddeen AM, Nawi MNM (2015) Factors influencing sustainable construction among construction firms in Malaysia: a preliminary study using PLS-SEM. Rev Téc Ing Univ Zulia 38(3):132–142

62. Maria S, Darma DC, Amalia S, Hakim YP, Pusriadi T (2019) Readiness to face industry 4.0. Int J Sci Technol Res 8(9):2363–2368
63. Wu Z, Jiang W, Cai Y, Wang H, Li S (2019) What hinders the development of green building? an investigation of China. Int J Environ Res Public Health 16:2–18
64. Shanmugapriya S, Subramanian K (2015) Partial least squares structural equation modelling to investigate the factors influencing productivity in indian construction projects. Aust J Basic Appl Sci 9(23):446–455
65. Memon AH, Rahman IA (2014) SEM-PLS analysis of inhibiting factors of cost performance for large construction projects in malaysia: perspective of clients and consultants. Sci World J 1–9
66. Durdyev S, Ismail S, Ihtiyar A, Abu Bakar NFS, Darko A (2018) A partial least square structural equation modeling (PLS-SEM) of barriers to sustainable construction in Malaysia. J Cleaner Prod 204:564–572
67. Li RY, Tang B, Chau KW (2019) Sustainable construction safety knowledge sharing: a partial least square-structural equation modeling and a feedforward neural network approach. Sustainability 11:1–18
68. Pallant J (2005) In: SPSS survival manual: a step by step guide to data analysis using SPSS for windows (Version 12). 2nd edn. Allen and Unwin, Crows Nest NSW 2065 Australia
69. Preacher KJ, MacCallum RC (2002) Exploratory factor analysis in behaviour genetics research: factor recovery with small sample sizes. Behav Genet 32:153–161
70. Tabachnick BG, Fidell LS (2007) Using multivariate statistics, 5th edn. Allyn and Bacon, Boston
71. Field A (2009) In: Discovering statistics using SPSS for windows. London—Thousand Oaks—New Delhi, Sage publications
72. Bagozzi RP, Yi Y (1988) On the evaluation of structural equation models. J Acad Mark Sci 16(1):74–94
73. Hair JF, Anderson RE, Tathan RL, Black WC (1998) Multivariate data analysis. Prentice Hall, Upper Saddle River, NJ
74. Hulland J (1999) Use of partial least squares (PLS) in strategic management research: a review of four recent studies. Strateg Manage J 20(2):195–204
75. Hair JF, Risher JJ, Sarstedt M, Ringle CM (2019) When to use and how to report the results of PLS-SEM. Eur Bus Rev 31(1):2–24
76. Henseler J, Hubona GS, Ray PA (2016) Using PLS path modeling in new technology research: updated guidelines. Indus Manage Data Syst 116(1):1–19
77. Franke GR, Sarstedt M (2019) Heuristics versus statistics in discriminant validity testing: a comparison of four procedures. Internet Res Forthcoming
78. Hair JF, Hult GTM, Ringle C, Sarstedt M (2016) A primer on partial least squares structural equation modeling (PLS-SEM). SAGE Publications, Incorporated
79. Henseler J, Dijkstra TK, Sarstedt M, Ringle CM, Diamantopoulos A, Straub DW, Ketchen DJ Jr, Hair JF, Hult GTM, Calantone RJ (2014) Common beliefs and reality about PLS: comments on Rönkkö & Evermann 2013. Organ Res Methods 17(2):182–209
80. Bagozzi RP, Yi Y (2012) Specification evaluation, and interpretation of structural equation models. J Acad Mark Sci 40:8–34
81. Singh R (2009) Does my structural model represent the real phenomenon?: a review of the appropriate use of structural equation modelling (SEM) model fit indices. Market Rev 9(3):199–212
82. Akter S, D'Ambra J, Ray P (2011) An evaluation of PLS based complex models: the roles of power analysis, predictive relevance and GoF index. In: Proceedings of the seventeenth americas conference on information systems, Detroit, Michigan August 4th–7th
83. Perera S, Karunasena G, Selvadurai K (2003) Application of value management in construction. Built-Environ Sri Lanka 4(1):3–12
84. Nguyen H, Skitmore M, Gray M, Zhang M, Olanipekun AO (2017) Will green building development take off? An exploratory study of barriers to green building in Vietnam. Resour Conserv Recycl 127:8–20

85. Hwang BG, Tan JS (2012) Green building management: obstacles and solutions for sustainable development. Sustain Develop 20(5):335–349
86. United Nations (2015) Sustainable development goals. Available on https://www.un.org/sustainabledevelopment/sustainable-development-goals/

Chapter 9
Barriers to the Adoption of Zero-Carbon Emissions in Buildings: The South African Narrative

Matthew Ikuabe, Douglas Aghimien, Clinton Aigbavboa, Ayodeji Oke, and Yambenu Ngaj

Abstract The earth is constantly faced by issues affecting the living conditions of its inhabitants and one of such challenges is the exacerbated effect of global warming which is largely attributed to the emissions of carbon (a major contributor to greenhouse gas) in the ecosystem. This has led to the call for a drastic abatement of operations that aid in the discharge of carbon emissions considering the harmful effect it has on the environment. Occupants of buildings engage in operations and activities that bring about the discharge of carbon, thus serving as an agent or contributor to the facing crisis of global warming. This study sets out to evaluate the barriers to the adoption of zero-carbon emissions in occupied buildings with a view to proffering ways to mitigate such practices. A comprehensive review of relevant literature was done which aided the identification of the barriers. Data for the study was elicited through a questionnaire survey from the built environment professionals. Methods of data analysis used were Percentage, Mean Item Score and Principal Component Analysis while Cronbach alpha was used in testing the reliability of the questionnaire. Findings from the study revealed that the hindering factors to the adoption of zero-carbon emissions in buildings and recommendations were made to help foster the adoption of zero-carbon emission processes in building operations and activities by its occupants.

1 Introduction

The twenty-first century is significantly and negatively impacted by climate change, which currently represents a global issue that needs to be urgently addressed. Tiwari [1] asserted that the level of emission of greenhouse gases is a huge influence on global warming and developing countries are the worst hit by the effect of these

M. Ikuabe (✉) · D. Aghimien · C. Aigbavboa · A. Oke · Y. Ngaj
Department of Construction Management and Quantity Surveying, University of Johannesburg, Johannesburg, South Africa
e-mail: ikuabematthew@gmail.com

© The Author(s), under exclusive license to Springer Nature Singapore Pte Ltd. 2021   135
R. J. Howlett et al. (eds.), *Emerging Research in Sustainable Energy and Buildings for a Low-Carbon Future*, Advances in Sustainability Science and Technology,
https://doi.org/10.1007/978-981-15-8775-7_9

gases. The activities and processes engaged by humans over the years have immensely contributed to the generation of these greenhouse gases. Carbon dioxide (CO_2) which is usually discharged as a result of the combustion of fossil fuels, wastes, wood and carbon is one of the major greenhouse gases [2]. Unfortunately, both developing and developed countries are contributors to the emission of carbon through the engagements of humans. Activities such as industrial discharge, combustion from vehicles, bush burning and domestic household processes are mostly the major culprit of the discharge of carbon to the ecosystem. There has been an increase in the global emission of CO_2 by almost 50% since 1990 [3]. Hence, Goal No 13 of the Sustainable Development Goals (SDGs) set by the United Nations aims at taking a holistic action against global warming.

Salagnac [4] stated that the building sector is one of the primary participants to greenhouse gas emission having unsustainable developments that have a negative impact on the social, environmental and economic life. Construction processes are usually very complex which involves the use of mechanical operations aided by the deployment of combustible fossil fuels. Equally, a huge chunk of wastes are always generated from construction processes, Son et al. [5] noted that 45–65% of waste deposited in landfills is generated by the construction industry. Furthermore, the activities of the occupants of completed buildings through their engagements also contribute to the emission of carbon to the immediate environment. Gill et al. [6] highlighted that cities are where the majority of the world's population resides, and it has been shown through statistics that an increase of population is expected to be 60% by 2030 and 70% by 2050, and regrettably, 80% of global greenhouse gas emissions are derived from cities. Consequently, the exponential increase of city dwellers over the years would have a huge implication on the discharge of carbon emission as a result of their activities. In South Africa, efforts have been made to push for the actualization of zero-carbon buildings. In 2018, the Mayors of Durban, Johannesburg, Tshwane and Cape Town formulated the requirements to ensure that new buildings are energy efficient including the reduction of electricity tariff and emission of greenhouse gas [7]. However, the attainment of these lofty policies and ideas seems to be posed by grave challenges.

In the face of the growing challenges facing the inhabitants of the earth, some of the issues that are clearly caused by the activities and engagements (carbon emission is part of it) should be given utmost consideration. The ways of mitigating or finding alternatives should be highly propagated. With the damaging effect caused in the ecosystem with regards to climate change and health challenges posed to humans, there is an urgent need to abate the discharge of carbon gases into the environment through the activities of the occupants of buildings. It is based on this premise this study is geared towards evaluating the barriers towards the adoption of zero-carbon emissions in buildings.

2 Review of Literature

2.1 Sustainability

Sustainability can be defined as the implementation of sustainable development methods and practices [8]. Also, Gladwin et al. [9] stated that sustainability is the process that provides a vision of community that respects the appropriate utilization of natural resources to make sure that the current generations attain a high level of economic security, achieve democracy and contribute in the control of their communities while preserving the integrity of the ecological systems and of life defined. Furthermore, Shurrab et al. [10] suggested that sustainability or the triple-bottom-line refers to the incorporation of environmental, economic, and social concepts. The terms "green construction" and "sustainable construction" are frequently used interchangeably [10]. In addition to other definitions, "Caring for the Earth" has defined sustainable construction as "development which improves the quality of human life while living within the carrying capacity of support ecosystems" [11].

According to [12], it is the responsibility of the building industry to take active measures towards sustainability. Equally, Kibert [8] argued that sustainable construction ought to be considered to be a subdivision of sustainable development. In the same light, CIB [13] asserted that sustainable development could be attained only if sustainable construction methods are deployed. In the housing, environmental, economics and spatial planning government policies, which are concerned about the improvement of the quality of the environment, building capacities, and sustainability, are considered to be the factors that affect sustainable construction and development in most cases. This is deemed very important considering the fact that it can directly improve the correlated developmental and construction industry problems. However, it is still debatable that the adoption of these policies could enhance the objective of sustainable construction [11]. Furthermore, it is suggested that in the drive for achieving sustainability in the African construction industry close attention should be given to the relationship between construction and development. It is much needed that the building sector brings more improved and specific solutions to alleviate the natural and environmental quality issues, spreading the word about development. This is not just saddled on the use of sustainable construction materials but also appropriate technologies that are more aware of the importance of energy-saving and its cost-effectiveness through the adoption of low carbon houses and environmentally friendly waste management methods. As stated by Al-Sanad [14], sustainability in construction aims to protect the environmental, social and economic well-being of the habitants.

2.2   Carbon Emissions

The most significant greenhouse gas in the atmosphere is carbon dioxide (CO_2), a trace gas. The concentration of CO_2 is measured in components per million, related to other gases. Carbon dioxide concentration has risen since the mid-nineteenth century and reached 410 ppm in June 2018. The annual rate of increase in CO_2 concentration was approximately 0.73 ppm per year in the late 1950s and around 2.11 ppm per year from 2005–2014. Its concentration increases because the rate of CO_2 emissions in the atmosphere is higher than the rate of its absorption, creating a carbon cycle imbalance [15]. In the current century, climate change is now recognized to be part of the environmental biodegradation, forming a major issue and raising a global concern. Trees and oceans absorb nearly the same quantity of complete CO_2 emissions and the remainder left in the atmosphere. The emission of CO_2 to the environment leads to an increase in atmospheric CO_2 concentration and also an increase in ocean and plant CO_2 absorption. The increased atmospheric CO_2 concentration results in a rise in temperature. Increasing ocean CO_2 absorption triggers ocean acidification and increased plant CO_2 absorption leads to carbon fertilization [15].

2.2.1   Effects of Carbon Emissions

The health and well-being of building occupants are the rudiments of productivity and are vigorous aspects of people's centric building design. Sick Building Syndrome (SBS) is a collection of factors that in numerous ways can have a harmful impact on physical health. In addition to physical health, psychological well-being is also associated because the human body is an interactive biological system. Meanwhile, the ideas of smart and viable structures have received significant attention in the latest decades [16]. Evidence of the hypothesis that building features and resulting indoor environmental quality continues to affect health outcomes has been proven and quite evident. CO_2-related SBS symptoms included headache, exhaustion, optical symptoms, nasal symptoms, allergy, asthma and diseases of the respiratory system. Additionally, Indoor air quality also seems to affect job efficiency and healthcare. CO_2 concentrations typically range from 350 to 2.500 ppm in office buildings. At levels in most indoor settings, CO_2 emissions can be regarded as a surrogate for other pollutants generated by the occupant, especially bio effluents. An assessment of the 94–96 BASE dataset discovered statistically significant dose–response relationships between CO_2 and the following symptoms: sore throat, irritated nose/sinus, a combination of mucous membrane diseases, narrow neck and wheeze; adjusted odds ratios for these symptoms ranged from 1.2 to 1.5 per 100 ppm rise in CO_2 [17].

One of the approaches in reducing carbon emission through buildings is through the implementation of zero-carbon buildings, which has been an attraction for policies in many countries [18]. Renukappa et al. [19] suggested that reducing carbon emission has the potential to reveal opportunities that would optimize the reduction of the rate at which materials are utilized, lower energy cost and lower cost of transportation. However, with the implementation of climate change policy such as the Kyoto Protocol, there is a wave of hesitation on the decision-makers' stance regarding the benefits that carry the reduction of carbon emissions due to the uncertain elements affecting the impact of carbon emissions reduction [19].

2.2.2   Barriers to the Attainment of Zero-Carbon Emission

According to [20], it is imperative to establish the barriers and challenges faced by developers and clients in the adoption of green buildings so as to enhance the discussion of its prominence and subsequent adoption. There are a number of potential obstacles to the attainment of Zero-Carbon Homes (ZCB) amongst which initial costs are perceived to be the key barrier, as many green buildings (GBs) are perceived to be more expensive than ordinary buildings [10]. Castillo and Chung [21]stated that ecological buildings escalate initial costs by an average of 2–7% over ordinary building costs. Furthermore, one of the challenges facing the implementation of sustainable developments is found in the negative perceptions regarding initial costs.

The design and construction of green buildings have been hindered by a wide range of factors which include cultural, financial, technical, design-related issues and legal [22]. Furthermore, Williams and Adair [23] argues that integrating renewable technologies in buildings is one of the major impediments in the adoption of zero-carbon homes, as it is seen that such technologies are currently untrustworthy by many. The technical and design barriers are quite similar to the cultural barriers standing against the achievement of zero-carbon homes construction. In addition, there are unwillingness from developers to adopt new techniques, methods and usage of products that are more sustainable stated [23]. This is a resultant of traditional attitudes from the building industry which has an unyielding stance towards innovations [22].

Amongst other barriers, there is the lack of adequate researches providing structured and complete work to serve as a bridge between the adoption of Zero-Carbon Homes in the construction industry to an actual adoption of the organizational business level[10]; Government incentives, builders unwillingness to adopt change [14]. There is a poor level of implementation of carbon reduction emissions initiatives in the building industry due to the absence of cognizance of its benefits. The reduction index statistics results show that there has been a reduction of 71% from the energy and utilities sector, 69% from the transport sector and only 30% from the construction industry [19].

CIB [13] reported that the most critical obstacle to low carbon houses in the building sector's absence of the ability to effectively enforce zero-carbon practices. This is further affirmed by [24] that ignorance or absence of common knowledge of sustainability can hinder the achievement of zero-carbon in buildings. While developers show trust in their capacity to access and use understanding in particular, this trust falls when problems of sustainable construction are discussed [25]. This presumes that in order to enforce its practice, experts in the construction sector need to be entirely accustomed to the principles of low carbon housing. Williams and Adair

[23] acknowledged that this was not the case, as professionals in construction need the best accessible data on goods and instruments to accomplish viable development. Due to a lack of data, proof of hindrance in their studies was a prevalent experience for most stakeholder organizations; Stakeholders have confessed in several instances that they are not conscious of sustainable policies or options that fall within their expertise [26]. Similarly, the installation of sustainable technologies and equipment needs fresh types of skills and knowledge [23]. Each industry's workforce is the backbone of the need to involve experts who are not only knowledgeable but who can encourage sustainable building as a team. If unattended, this obstacle will show a significant gap in expertise and abilities in the building industry [26].

3 Methodology

The study attempts to evaluate the barriers to the adoption of zero-carbon emissions in buildings with the use of a quantitative survey. The survey was carried out in Gauteng Province in South Africa. Responses were elicited from built environment professionals (Architects, Quantity Surveyors, Builders and Engineers) with the aid of a questionnaire, adopting convenience sampling while the study area is Gauteng Province, South Africa. The use of questionnaires as the instrument for collection of data is a result of its ease of usage and its propensity for wide coverage [27]. The number of questionnaire distributed totaled sixty-three (63) while fifty-nine (59) were retrieved and all deemed good for analysis. The period for data collection covered a period of one month. A portion of the questionnaires was self-administered while the remaining was carried out with help of field agents. The questionnaire was made up of two sections; the first section pertained to the demographic information of the respondents while the other section elicited responses from the respondents on the barriers to the adoption of zero-carbon emission in buildings. A provision of the list of barriers sourced from relevant literature was made and respondents were asked to rate their level of agreement. A Likert scale of 1–5 was adopted using 1as strongly disagree, 2 agree, 3 neutral, 4 agree and 5 as strongly agree. Cronbach alpha was used in ascertaining the reliability of the questions in the second section which gave a value of 0.899, thus indicating high reliability of the questions [28]. Mean Item Score was used in ranking the barriers based on their significance; and principal component analysis was used in converting similar related variables with linear correlated features into sets of components that retain the variation exhibited in the original variable [29].

4 Results

4.1 Barriers to the Adoption of Zero-Carbon in Buildings

Following the review of relevant literature and the subsequent identification of the barriers to the adoption of zero-carbon emission in buildings, thirty-six (36) barriers were presented to the respondents to rate them based on their level of agreement. Mean Item Score was deployed in ranking the barriers based on their level of significance.

In establishing the variables with similar underlying features, factor analysis (FA) was deployed in indicating constructs of subscales made up of correlated variables. Table 1 shows the communalities of the barriers and they all have a value above 0.6, thus suitable for the purpose.

In trying to establish the suitability of the data for factor analysis, Kaiser–Meyer–Olkin (KMO) measure of sampling adequacy and Bartlett's test of sphericity was used. Result from Table 2 shows a KMO value of 0.775 and Bartlett's significant value of 0.000, thus indicating a good value for factor analysis [30].

With the requirements for FA being met, principal component analysis with varimax rotation was deployed. Having set a cut off at 0.50, four components were extracted having eigenvalues greater than 1. Table 3 shows the extracted components with their factor loadings (Table 4).

5 Discussion of Findings

5.1 Lack of Knowledge and Expertise

The first principal component has twelve (12) factor loadings. This component has the highest number of factor loadings among the four principal components generated from the FA. This translates to the fact that this component serves as the major hindrance to the adoption of zero-carbon emission in buildings. The barriers in this component are lack of skills and knowledge, unfamiliarity of principles of low carbon housing, lack of customer prior knowledge, lack of willingness, ignorance or absence of common knowledge of sustainability, lack of demand, lack of experience and expertise, resistance to change and lack of manufacturer and vendor support. This component is labeled 'Lack of Knowledge and Expertise' as a result of the correlated features of the variables.

As a result of lot of people not being aware of the numerous benefits of adopting zero-carbon emissions in buildings, this has impeded the implementation of such practice in housing schemes. Likewise, not much emphasis is placed on the devastating effect carbon emission has on the ecosystem. This corroborated by Rydin et al. [25], observing that ignorance or absence of common knowledge of sustainability can

Table 1 Ranking of barriers to the adoption of zero-carbon emission

Barriers	MIS	STD	Rank
Lack of customer prior knowledge	4.38	0.607	1
Absence of data on the performance	4.37	0.655	2
Ignorance or absence of common knowledge of sustainability	4.32	0.668	3
Overestimation of the cost of capital	4.3	0.586	4
Lack of experience and expertise	4.29	0.728	5
Lack of demand	4.29	0.869	5
Underestimation of the potential cost savings	4.24	0.665	7
Lack of government initiatives	4.24	0.777	7
Fear of unforeseen expenses	4.22	0.75	9
Lack of skills and knowledge	4.21	0.986	10
Lack of tax incentives	4.19	0.47	11
Lack of zero-carbon methods adoption	4.19	0.47	11
Lack of willingness	4.14	0.8	13
Absence of techniques	4.13	0.729	14
Fear of greater zero-carbon building investment expenses	4.13	0.813	14
Resistance to change	4.1	0.756	16
Absence of ability to effectively enforce zero-carbon practices	4.08	0.655	17
Unconsciousness of zero-carbon emission policies	4.02	0.833	18
Lack of market-based instruments	3.98	0.635	19
Unfamiliarity principles of low carbon housing	3.98	0.833	19
Lack of zero-carbon regulation	3.95	0.718	21
Lack of zero-carbon technologies	3.95	1.038	21
Absence of measuring techniques	3.92	0.768	23
Absence of steering	3.92	0.867	23
Lack of zero-carbon equipment	3.9	0.928	25
Extra expenses for building testing and inspection	3.89	0.825	26
Greater consultant charges	3.86	0.82	27
Absence of building codes	3.78	0.812	28
Lack of manufacturer and vendor support	3.73	0.865	29

MIS mean item score; *STD* standard deviation

hinder the achievement of zero-carbon in buildings. Likewise, a major contributor to this is the lack of expertise. This view is supported by Djokoto et al. [26] stating that in order to enforce the practice, experts in the construction setting need to be entirely accustomed with the principles of low carbon housing. Similarly, Williams and Adair [23] noted that there is a significant gap in expertise in the building industry to propel the adoption of sustainable practices in buildings.

Table 2 Communalities of the barriers to the adoption of zero-carbon emission in buildings

	Initial	Extraction
Absence of ability to effectively enforce zero-carbon practices	1.000	0.626
Ignorance or absence of common knowledge of sustainability	1.000	0.725
Unfamiliarity principles of low carbon housing	1.000	0.709
Unconsciousness of zero-carbon emission policies	1.000	0.716
Lack of experience and expertise	1.000	0.757
Lack of zero-carbon technologies	1.000	0.798
Lack of zero-carbon equipment	1.000	0.840
Lack of skills and knowledge	1.000	0.718
Absence of data on performance	1.000	0.621
Lack of tax incentives	1.000	0.638
Lack of zero-carbon regulation	1.000	0.729
Lack of market-based instruments	1.000	0.814
Absence of building codes	1.000	0.782
Absence of measuring instruments	1.000	0.797
Absence of techniques	1.000	0.737
Absence of steering	1.000	0.820
Lack of manufacturer and vendor support	1.000	0.766
Lack of zero-carbon methods adoption	1.000	0.736
Fear of greater zero-carbon building investment expenses	1.000	0.732
Fear of unforeseen expenses	1.000	0.797
Extra expenses for building testing and inspection	1.000	0.748
Overestimation of the cost of capital	1.000	0.717
Underestimation of the potential cost savings	1.000	0.773
Greater consultant charges	1.000	0.743
Resistance to change	1.000	0.566
Lack of government initiatives	1.000	0.834
Lack of customer prior knowledge	1.000	0.671
Lack of willingness	1.000	0.788
Lack of demand	1.000	0.689

Extraction method: Principal Component Analysis

Table 3 KMO and Bartlett's test

Kaiser–Meyer–Olkin measure of sampling adequacy		0.775
Bartlett's Test of Sphericity	Approx. Chi-Square	1546.850
	Df	630
	Sig.	0.000

Table 4 Rotated component matrix[a]

	Component			
	1	2	3	4
Lack of skills and knowledge	0.812			
Unfamiliarity of principles of low carbon housing	0.806			
Lack of customer prior knowledge	0.746			
Lack of willingness	0.738			
Ignorance or absence of common knowledge of sustainability	0.732			
Lack of demand	0.622			
Lack of experience and expertise	0.617			
Resistance to change	0.611			
Lack of manufacturer and vendor support	0.605			
Lack of zero-carbon technologies		0.766		
Lack of zero-carbon equipment		0.757		
Absence of measuring technique		0.651		
Lack of market-based instrument		0.647		
Absence of data on performance		0.599		
Lack of zero-carbon methods adoption		0.570		
Absence of technique		0.558		
Absence of building codes			0.712	
Absence of steering			0.652	
Absence of ability to effectively enforce zero-carbon practices			0.632	
Lack of zero-carbon regulation			0.555	
Lack of tax incentives			0.536	
Lack of government initiatives			0.530	
Unconsciousness of zero-carbon emission policies			0.527	
Overestimation of the cost of capital				0.772
Fear of greater zero-carbon building investment expenses				0.771
Fear of unforeseen expenses				0.672
Extra expenses for building testing and inspection				0.662
Underestimation of the potential cost of savings				0.596
Greater consultant charges				0.564

Extraction method: Principal Component Analysis
[a]4 components extracted

5.2 Technological Hindrance

The second principal component has nine (9) factor loadings. The variables making up this component are lack of zero-carbon technologies, lack of zero-carbon equipment, lack of market-based instrument, absence of measuring technique, absence of data on performance, lack of zero-carbon methods adoption and absence of technique. This component is labeled 'Technological hindrance'. This finding is supported by Samari [31], stating that integrating renewable technologies in buildings is one of the major issues in the feasibility of zero-carbon homes. To further buttress this view, Djokoto et al. [26] stated that there is a gulf between technological needs and the adoption of zero-carbon homes. Thus, suggesting that a lot needs to be done to shore up the gap in technological applications in achieving zero-carbon emission in buildings.

5.3 Inadequacy of Government Policies and Initiatives

The third component is comprised of eight (8) factor loadings. This includes the absence of building codes, absence of steering, absence of the ability to effectively enforce zero-carbon practices, lack of zero-carbon regulation, lack of tax incentives, lack of government initiatives and unconsciousness of zero-carbon emission policies. The component is subsequently labeled 'Inadequacy of government policies and initiatives. The government's inability to steer the adoption of buildings with zero-carbon emission through the implementation of policies and legislation is very evident. Williams and Adair [23] noted that the absence of building codes and public policies has been a major impediment to sustainable buildings. Efforts should be geared by government in stimulating actions that encourage practices such as zero-carbon emission. Similarly, Castillo and Chung [21] made case for governments' participation in activities that propels sustainable practices. To this end, the government should ensure that policies that are in tandem with the goals of the UN's SDGs should be implemented. This is in line with the actions taken by the Mayors of Durban, Johannesburg, Tshwane and Cape Town by formulating the requirements to ensure that new buildings are energy efficient including the reduction of electricity tariff and emission of greenhouse gas [7].

5.4 Financial Impediments

The last component has seven (7) factor loadings which are overestimation of the cost of capital, fear of greater zero-carbon building investment expenses, fear of unforeseen expenses, extra expenses for building testing and inspection, underestimation of the potential cost of savings and greater consultant charges. The component is labeled

'Financial Impediments'. Castillo and Chung [21]stated that ecological buildings escalate initial costs by an average of 2–7% over ordinary building costs. With that in mind, there is setback established considering the comparative rise in the cost of putting up such buildings. Furthermore, Williams and Adair [23] noted that the high expense of putting up zero-carbon buildings is a major barrier.

6 Conclusion and Recommendations

The study focused on the barriers to the adoption of zero-carbon emission in buildings with professionals in the built environment as the respondents to questionnaire survey used for the study. Revealed from the study are the major barriers namely lack of knowledge and expertise, technological hindrance, inadequacy of government policies and initiatives and financial impediments. As a result, the study concludes that if visible improvement is to be made on the adoption of zero-carbon emission in buildings, a lot has to be done on enlightening the general populace of the dangers of the discharge of carbon in the ecosystem. Similarly, efforts need to be put into sensitizing the people of the benefits of such adoption. Equally, the provision of the requisite technologies for zero-carbon emission buildings should be readily available to foster the embracement of such practices. Government should as a matter of importance introduce policies and initiatives that would drive the acceptance of sustainable practices considering the glaring benefits that come with such. One notable aspect to be equally considered is the financial implication that comes along with the implementation of practices such as zero-carbon emission in buildings. However, any attempt to steer the general populace in imbibing such practices must be accompanied with assurances of regulated and flexible cost implications. As such, the government could introduce subsidy regimes for the procurement and installation of components that make up zero-carbon emission buildings.

References

1. Tiwari A (2011) Comparative performance of renewable and nonrenewable energy source on economic growth and CO_2 emissions of Europe and Eurasian countries: a PVAR approach. Econ Bull 31:2356–2372
2. Pervez M, Henebry G (2015) Spatial and seasonal responses of precipitation in the Ganges and Brahmaputra river basins to ENSO and Indian Ocean dipole modes: implications for flooding and drought. Nat Hazards Earth Syst Sci 15:147–162
3. United Nations (2015) Sustainable development goals. https://www.un.org/sustainabledevelopment/sustainable-development-goals/. Assessed 8 Feb 2020
4. Salagnac J (2012) Global political initiatives and overtones. In: Booth C, Hammond F, Lamond J, Proverbs D (eds) Solutions to climate change challenges in the built environment. Blackwell, London, pp 45–55

5. Son H, Kim C, Chong W, Chou J (2011) Implementing sustainable developmentin the construction industry: constructors'perspectives in the US and Korea. Sustain Develop 19(5):337–347
6. Gill E, Handley F, Ennos R, Pauleit S (2007) Adapting cities for climate change: the role of the green infrastructure. Built Environ. 3(1):115–133
7. C40 Energy (2018). Four South African cities strive to make all new buildings zero carbon. https://www.c40.org/press_releases/south-african-cities-make-all-new-buildings-zero-carbon. Assessed 8 Feb 2020
8. Kibert CJ (1994) Establishing principles and a model for sustainable construction. Paper presented at the proceedings of first international conference of CIB TG 16 on sustainable construction, Tampa, Florida
9. Gladwin T, Kennelly J, Krause T (1995) Shifting paradigms for sustainable development: Implications for management theory and research. Acad Manag Rev 20(4):874–907
10. Shurrab J, Hussain M, Khan M (2019) (2019) Green and sustainable practices in the construction industry: a confirmatory factor analysis approach. Eng Construct Architect Manage 26(6):1063–1086
11. Adebayo AA (2000) Sustainable construction in Africa. CIB agenda 21 for sustainable construction in developing countries (2000)
12. Bourdean L (1999) Sustainable development and the future of construction: a comparison of visions from various countries. Build Res Inform 27(6):345–366
13. CIB (1999) Agenda 21 on sustainable construction CIB report publication. Department for Communities and Local Government Zero carbon homes. https://assets.publishing.service.gov.uk/government/uploads/system/uploads/attachment_data/ile/6288/1905485.pdf. Accessed 8 May 2019
14. Al-Sanad S (2015) Awareness, drivers, actions, and barriers of sustainable construction in Kuwait. In: International conference on sustainable design, engineering and construction. Procedia Eng 118:969–983
15. Jain M (2018) Impact of increasing CO_2 emissions on environment. Science ABC. Available at https://www.scienceabc.com/social-science/greenhouse-gas-co2-emmission-effect-environment.html. Accessed on 21 July 2019
16. Ghaffarianhoseini A, Alwaer H, Omrany H, Alalouch C, Croome D, Tookey J (2018) Sick building syndrome: are we doing enough? Architect Sci Rev 61(3):99–121
17. Robertson DS (2006) Health effects of increase in concentration of carbon dioxide in the atmosphere. Curr Sci 90(12):1607–1609
18. Zhao X, Burnett JW, Lacombe DJ (2015) Province-level convergence of China's carbon dioxide emissions. Appl Energy 150:286–295
19. Renukappa S, Akintoye A, Egbu C, Goulding J (2013) Carbon emission reduction strategies in the UK industrial sectors: an empirical study. Int J Climate Change Strategies Manage 5(3):304–323
20. Mesthrige W, Kwong H (2018) Criteria and barriers for the application of green building features in Hong Kong. Smart Sustain Built Environ 7(3/4):251–276
21. Castillo R, Chung N (2005) The value of sustainability, Stanford, California, USA. Center for Integrated Facility Engineering (CIFE), Department of Energy, Washington DC
22. Osmani M, O'reilly A (2009) Feasibility of zero carbon homes in England by 2016: a house builder's perspective. Build Environ 44(9):1917–1924
23. Williams K, Adair C (2007) What is stopping sustainable building in England? Barriers experienced by stakeholders in delivering sustainable developments. Sustain Develop 15:135–147
24. Häkkinen T, Belloni K (2011) Barriers and drivers for sustainable building. Build Res Inform 39(3):239–255
25. Rydin Y, Amjad U, Moore S, Nye M, Withaker M (2006) Sustainable construction and planning. The academic report. Centre for Environmental Policy and Governance, The LSE SusCon Project, CEPG, London School of Economics, London
26. Djokoto S, Dadzie J, Ohemeng-Ababio E (2014) Barriers to sustainable construction in the ghanaian construction industry: consultants perspectives. J Sustain Develop 7(1):134–143

27. Tan P (2011) Towards a culturally sensitive and deeper understanding of "rote learning" and memorization of adult learners. J Stud Int Educ 15(2):124–145
28. Tavakol M, Dennick R (2011) Making sense of Cronbach's Alpha. Int J Med Educ 2:53–55
29. Jollife IT (2002) Principal component analysis, 2nd ed. Springer, New York
30. Tabachnick B, Fidell L (2007) Using multivariate statistics, 5th ed. Allyn and Bacon, New York
31. Samari M, Godrati N, Esmaeilifar R, Olfat P, Shafiei M (2013) The investigation of the barriers in developing green building in Malaysia. Mod Appl Sci 7(2). https://doi.org/10.5539/mas.v7n2p1

Part IV

Chapter 10
System Dynamics Analysis of Energy Policies on the building's Performance

Ibrahim Motawa and Michael Oladokun

Abstract Various policies have been formulated by governments to reduce energy consumption in buildings. Evaluating the effectiveness of a proposed policy requires consideration of complex interrelationships that exist among many variables. Therefore, this research aims to develop a dynamic model to analyse the impact of energy policies on buildings performance. The principle of socio-technical systems as an approach to model this complexity has been advocated in this research. A System Dynamics model has been developed to simulate the intrinsic interrelationship between the dwellings, occupants and environment systems. This chapter will analyse the impact of various policy scenarios on energy consumption in building towards achieving the UK national targets; namely: improvements in the uptake of dwelling insulation measures, occupants' behavioural changes, and policy change on energy prices. An integrated scenario has been also assumed to combine the effect of the first three ones. The main findings indicate that it is unlikely for anyone scenario alone to meet the required binding reductions unless an integrated solution is adopted. The developed model considers various qualitative conditions that are not usually simulated using the traditional regression-based forecasting of energy use in buildings. The developed model can be used to test various policies other than the UK context considering various data sets of the model variables.

Keywords Socio-technical systems · System dynamics · Buildings performance

I. Motawa (✉)
Belfast School of Architecture and the Built Environment, Ulster University, Londonderry BT37 0QB, UK
e-mail: i.motawa@ulster.ac.uk

M. Oladokun
Heriot Watt University, Edinburgh, UK

1 Introduction

There is number of policies/strategies like fabric insulation improvement, energy tariffs, alternative energy sources, micro-generation, energy subsidy for the uptake of technology like micro-generation, initiatives on fuel poverty, etc. which are initiated to reduce energy consumption in buildings. These various policies require a more comprehensive approach to evaluate their effectiveness and consider the complex interrelationships among many embedded variables. Such an approach should take into account the various qualitative conditions which are not usually considered when using the traditional regression-based forecasting of energy consumption in buildings. Therefore, this research aims to develop a dynamic model to analyse the impact of energy policies on buildings performance; advocating the principle of socio-technical systems as an approach to model this complexity.

The study focuses on domestic buildings in the UK and on analysing the impact of various policy scenarios on energy consumption in building towards achieving the UK national targets. However, the methodology can be applied in other countries considering relevant data sets are used. This study will contribute to the body of knowledge by modelling the interrelationships among the variables involved. Further, the study aims to provide a reliable tool that can serve as learning laboratory for policy/decision-makers to test different policies and run "what if" scenarios on energy consumption.

This chapter will discuss the relevant literature review, the methodology adopted to conduct this study, general overview of the concept of Socio-Technical Systems (STS) and its relevance to this research, the System Dynamics (SD) model developed to analyse the impact of energy policies on buildings performance, the model running and results, the study limitations and future work, and the study conclusions. First the relevant literature is discussed in the following section.

2 Literature Review

The effects of climate change due to carbon dioxide and other greenhouse gas emissions could see the global temperatures rise by up to 6 °C [42] thereby causing extremes in weather systems. Reduction of the energy consumption patterns in dwellings is, therefore, seen as one of the breakthroughs to curtail this threat. According to the Office of National Statistics [27], CO_2 emissions attributed to domestic buildings alone is around 26% of the total UK's carbon emissions. For this reason, the UK domestic sector is chosen as the centre of focus for both mitigation and adaptation agendas with a view to meeting the CO2 emissions reductions target of 80% by 2050 based on 1990 levels as laid down by the Climate Change Act of 2008. To meet the target of CO_2 emissions reduction, the UK's government has initiated a number of policies/strategies like fabric insulation improvement, energy tariffs, alternative energy sources, micro-generation, energy subsidy for the uptake of technology like micro-generation, initiatives on fuel poverty, etc. to serve as measures against this menace. In general, the influences of policy interventions on energy consumption/energy savings have been studied in different ways, such as the studies relevant to the environmental psychological theory which attempt to analyse the relation between policy instruments and energy consumption behaviour by households, such as the work of Abrahamse et al. [1]. Despite the various policies initiated to reduce energy consumption in buildings, a system-based approach to analyse the effectiveness of these policies is still needed.

Based on the outcomes of the research in the area of Post-Occupancy Evaluation (POE) of buildings which indicate that actual building performance is never the same as predictions [3, 39], buildings are becoming more complex and the technology, occupant behaviour and the environmental regulations increase the pressure for more accurate predictability of this end product, Way and Bordass [44]. Therefore, the relationship between buildings, occupants' behaviour, and the environment in and around buildings should be a key component for any policy analysis. Prior studies in the area of POE have concentrated more on evaluating occupants' satisfaction rather than modelling occupants' behaviour towards the use of the systems put in place in buildings. Stevenson and Rijal [39] argued that one area of uncertainty researchers are still struggling with is in finding means to establish a concrete methodology that links the technical assessment of building with that of human behaviour to capture the influence of occupants' behaviour on buildings performance. It is being argued that the way occupants behave towards buildings (thereby making buildings unsustainable) should not be used as an excuse for under-performance of buildings. This must, however, be understood and influenced appropriately [37, 38]. Therefore, the need to explore and model energy consumption in buildings should consider both the technology and the social aspects of buildings when analysing the effectiveness of energy policy.

There have been a number of studies on modelling approaches/techniques to capture domestic energy consumption, especially at the national level. Johnston [18] and Kavgic et al. [19] argue that these approaches/techniques vary tremendously in terms of requirements, assumptions made, and the predictive abilities of the models. According to the International Energy Agency [16], these approaches are top-down approaches and bottom-up approach. However, Böhringer and Rutherford [2], Kavgic et al. [19] and Kelly [20] acknowledge another modelling approach paradigm derived from both top-down and bottom-up approaches together. The top-down techniques rely on the kind of interaction subsisting between the energy sector and the economy in general at an aggregated level in order to predict the behaviour of energy consumption and carbon emissions at the household level. On the other hand, bottom-up techniques mainly focus on the energy sector utilising a disaggregated approach that contains a high level of details at the household level to model energy consumption and carbon emissions.

Among the examples of using the top-down modelling approach, several household energy consumptions and carbon emissions models have been developed such as the work of Hirst et al. [14], Haas and Schipper [13], FitzGerald [8], and Summerfield

[40]. On the other hand, quite a number of researchers (e.g. [18–21, 23, 41]) have adopted the bottom-up approach who basically used two major epistemic methods, statistics and building physics methods. These models are found to vary considerably based on the levels of disaggregation, complexity, resolution of output, output aggregation levels, scenario analysis performed, model validation, and their availability for the purpose of scrutiny. In general, there are a number of limitations in the existing modelling techniques: (1) lack of transparency in the model algorithms, (2) inability to account for the complex, interdependencies, and dynamic nature of the issue of energy consumption and carbon emissions, (3) limited evidence to show for the occupants-dwelling interactions, and (4) lack of enough capacity to accommodate qualitative data input. As such, there is a need to scout for more robust and accurate modelling approaches that take into consideration the kind of complexity involved due to high interdependencies, chaotic, non-linearity, and qualitative nature of some of the variables involved. Therefore, this research will adopt a modelling approach which will address the main research question of this study, "how could a system-based approach be designed and used to analyse the effect of an energy policy on building performance while taking into account both technical and social variables that are relevant to households' energy performance?".

The consideration of a system-based approach to evaluate how best an energy policy can be to reduce energy consumption in dwellings becomes more realistic and comprehensive over the traditional statistical or technological system approaches on their own. It is against this backdrop that this chapter intends to demonstrate a methodology that is capable of modelling the complex system of different policies regarding household energy consumption. The study focuses on domestic buildings in the UK. However, the methodology can be applied in other countries considering relevant data sets are used. This study will contribute to the body of knowledge by modelling the interrelationships among the variables involved. Further, the study provides a reliable tool that can serve as learning laboratory for policy/decision-makers to test different policies and run "what if" scenarios on energy consumption. The next section first introduces the adopted research methodology then carries out a general overview of the concept of Socio-Technical Systems (STS) and its relevance to this research.

3   Research Method

This research uses a mixed-method approach in order to achieve its objectives. The reasons for adopting this approach include the nature of the research problem, the data and the methods of collecting this data, and the purpose of the research. The research problem involves answering questions relating to '*what*' and '*how*', which means a single approach cannot be used to answer those questions. It is also evident that the nature of the research entails capturing both the qualitative and quantitative data, which by implication means the triangulation of data collection methods. The adopted STS as the modelling approach is a pluralistic approach that considers both qualitative and quantitative modelling. This research has been conducted over a three-year period, completed (as it stands) in 2016, and used the available UK data sets for domestic buildings.

The research method followed in this study is based on the concept of STS as a systems-based approach to scientific inquiry. STS emerged from the studies undertaken by the Tavistock Institute (London) during the post-war reconstruction of the industry. Cartelli [4] reported that the emergence of this concept is highly necessary for pursuit of a fit between the workforce and machine during the introduction of technological systems for work automation especially if those workers show resistance to technological innovation. Considering the different STS theories as described in the literature (such as Multilevel Perspective (MLP), Innovation System, Strategic Niche Management, Transition Management, etc. as noted by Shove [36] and Geels [10], this research focuses on modelling the impact of policy instruments (such as: incentives) on energy consumption by households considering the technological and social behaviour in conjunction with these policies. STS as a concept is founded on two main principles. The first one is the interaction between two sub-systems (Social and Technical) that set the conditions for successful (or unsuccessful) system performance. Walker et al. [43] argued that the interactions are comprised partly of linear "cause and effect" relationships which can normally be "designed", and partly from non-linear, complex, even unpredictable relationships which may be often unexpected. Soft, which is a socio, does not necessarily behave like the hard, which is technical [43]. That is, people are not machines and their behaviour grows in much more complexity and interdependency. Therefore, STS as a methodology of systems-based approach of scientific inquiry is used to handle this kind of complexity. The second of the two main principles is that "optimisation" of the two sub-systems must be sought. That is, the need for 'joint optimisation' of the two sub-systems.

Based on these principles, STS has been implemented in many fields such as organisational development [6], innovation development and diffusion (Rohracher [34] and Geels and Kemp [11]), energy supply and demand (Shipworth [35], Motawa and Banfill [25]), communication and telecommunication engineering [31], water management [17] and agriculture and food [22].

For the study on energy consumption, the STS is comprised of an interplay between the building technology systems, the external environment, and the occupants. In a previous study Oladokun [29] the modelling techniques of STSs have been reviewed, which included: Actor-Network Theory, Agent-Based Modelling technique, Bayesian Belief Network, Configuration modelling, Fuzzy Logic, Morphological Analysis, Social Network Analysis, and System Dynamics (SD). A number of criteria were set upon to select which technique is more suitable to model the problem of energy consumption in buildings. The available modelling techniques are compared to one another based on if the technique considers: (1) transparency, (2) multiple interdependencies, (3) dynamic behaviours, (4) feedback processes, (5) non-linear relationships, (6) modelling *hard* and *soft* data, (7) uncertainties of the variables involved, (8) chaotic assumptions, and (9) the use of the model as learning laboratory. Table 1 summarises the conducted comparison for the available STS modelling techniques.

Table 1 Comparative analysis of STS modelling techniques

Criteria	ANT	ABM	BBN	CM	FL	MA	SNA	SD
Transparency	✓	✓	✓				✓	✓
Multiple interdependencies	✓	✓	✓				✓	✓
Dynamic behaviours		✓						✓
Feedback processes								✓
Non-linear relationships	✓	✓	✓	✓	✓	✓	✓	✓
Modelling "hard" and "soft" data	✓	✓	✓	✓	✓	✓	✓	✓
Chaotic assumptions	✓	✓	✓	✓	✓	✓		✓
Uncertainties		✓	✓	✓	✓	✓		✓
Learning laboratory tool			✓					✓

After a careful appraisal of all the techniques, the study recommended System Dynamics (SD) as a suitable technique to simulate the effect of energy policies on household energy consumption. SD was specifically introduced by Forrester [9] in order to handle complex problems that have multiple interdependencies and are dynamic in nature with many feedback structures. The tools for this technique have built-in functions to capture the non-linear relationships among variables with the capability of processing both qualitative and quantitative data. The transparency aspect of it has been greatly enhanced which means that all the model variables including the algorithms can be assessed and scrutinised by third parties. The following sections will illustrate the structure of the proposed SD tool.

ANT: Actor-Network Theory, ABM: Agent-Based Model, BBN: Bayesian Belief Network, CM: Configuration Modelling, FL: Fuzzy Logic, MA: Morphological Analysis, SNA: Social Network Analysis, SD: System Dynamics.

The system dynamics methodology involves interrelated and linked steps towards the 'understanding of the system'. This reveals that at any point in time in each stage of system dynamics, a better insight into the problem is gotten, which eventually leads to a better understanding of the system under study. The methodology indicates that the first stage is to identify the problem in question and properly define it through the situation analysis. This involves identifying the variables in the problem and relates them to one another in order to find out the causal relationships and feedbacks in the system. Based on Ranganath and Rodrigues [32], the most important aspect of problem identification is to identify the time-based policy parameters, which influence the dynamics of the system under study.

The second stage is the system conceptualisation. This involves representing the 'cause and effect' relationship between the variables in the system pictorially and this is called Causal Loop Diagram (CLD). Stage three involves formulating the model by representing the model using the Stock and Flow Diagram (SFD). SFD diagrams are the pictorial representation of the behaviour of the system in the form of accumulation (stock) and flow (rate). This automatically leads to stage four where the SFD is turned into a simulation model. It must be emphasised that mere CLD or SFD do not constitute the system dynamics. It is when the variables in the model are related together in terms of equations before it can be said that it forms the simulation model. Once the model is validated accordingly, the simulation is then run and the output of the simulation is presented in the form of graphs. These graphs reveal the pattern exhibit by the variables under study over a period of time. Based on these outputs, policy analysis and improvement is carried out by the decision-makers and this is stage five of the methodology. Implementation of the policy improvement (stage six) then concludes the system dynamics methodology. This method demonstrates system dynamics as a powerful analysis tool for use by decision-makers. It is worth mentioning that system dynamics has the ability to be used as a learning laboratory tool [33] in conjunction with other traditional decision making techniques.

The application of system dynamics methodology to model the impact of different UK policies on household energy consumption relies first on the identification of the relevant variables and the development of the relationships among them. For the problem under investigation in this chapter, the proposed model is comprised of an interplay between the dwelling system (dwellings' physical parameters and dwellings' dynamic variables) with the occupants system (biophysical variables, behavioural variables and household characteristics) and the environment (climatic variables, economic variables and policy/regulations variables). The physical parameters of the dwelling system include the dwelling size, materials, heating system and stock of appliances; and the dynamic variables include dwelling internal temperature, ventilation rates, amount of hot water and appliances use. The biophysical variables of the occupants system consist of variables such as occupants' thermal comfort in the form of metabolic rate, respiration and clothing. The behavioural variables include household income, socio status and number of occupants; whereas household characteristics relate to the individual beliefs, attitudes, knowledge and personalities. The climatic variables of the environment system include external temperatures, insulation and wind levels. The economic variables involve variables such as energy prices and energy tax; whereas the policy variables embrace the governmental and society influences. All these variables seamlessly work together to influence household energy consumption. The full list of the considered variables for the proposed model and the methodology adopted to identify them can be found elsewhere [24].

The second most important aspect for the application of system dynamics methodology to model the impact of different UK policies on household energy consumption relates to the validation of the identified CLDs and SFDs to ensure that the model produces realistic and accurate results. This research firm up a rigorous validation process for the outputs of the model. Two different approaches were used: model structure validation and model behaviour validation. In the case of the first approach, a number of tests (as indicated by Groesser and Schwaninger [12]) were used to test the model's equations technically. The tests are mainly for parameter verification, structure verification, parameter adequacy, dimensional analysis, integration error and sensitivity analysis tests. All these tests were conducted and approved the model validity,details on these tests can be found elsewhere, Motawa and Oladokun [24]. The model behaviour validation has been conducted by both unstructured and semi-structured interviews with industry practitioners. The selection of the industry practitioners has utilised the database of the Scottish Statistic Register for professionals in energy and environment to contact with energy experts in buildings. The most qualified ten experts were selected to partake in the study. The unstructured interview method was used to elicit interviewees' mental knowledge of the model structure and variables. This is basically to ascertain the correctness of the initial causal loop diagrams (CLDs) drawn based on the authors' knowledge of the system under study as elicited through the review of literature and government documents. Also, at this stage respondents have helped in formulating some of the relationships between certain variables in the model that are with a lack of empirical data and/or evidence of existing relationships among them. This method is in line with the approach of Coyle [5] regarding establishing the causal relationships among the variables in SD models.

Whereas the semi-structured interview approach was used for validating the model results and application based on the scoring method. For these interviews, 15 respondents took part in the exercise which included three experts with SD background. At this stage, experts and industry practitioners have tested the behaviour of the model outputs to ensure that it reflects their expectations or otherwise and to advance reasons for any plausible behaviour noticed.

The developed model has been coded using Vensim SD environment and includes over 90 variables with over 400 equations. Due to space limitation, this chapter explains a few of the key components of the model. The authors welcome any communication with the interested readers for further details on any part of the model. The following sections present the main model components.

4   The Main Components of the Proposed SD Model of Household Energy Consumption

The developed model consists of six modules: population/household, dwelling internal heat, occupants' thermal comfort, climatic-economic-energy efficiency interaction, household energy consumption, and household CO_2 emissions. The following sections will illustrate the key components of these modules in the form of the developed "Stock and Flow Diagrams" (SFDs). SFDs simulate the model variables and the relationships among them into controlling equations.

4.1   SFD for the Population/Household Module

The research utilised the profile of population growth to estimate the number of households in the UK he key input to estimate the amount of energy consumption. Population is being influenced by a number of variables which include: birth and death rates, reproductive time, population equilibrium time, average total fertility rate, mortality and average life expectancy, as shown in Fig. 1.

Population is modelled as 'stock'. This is being controlled by an inflow (births) and an outflow (deaths). 'Births' is influenced by 'reproductive time', 'population equilibrium time', 'average total fertility rate', and 'deaths'. On the other hand, 'deaths' are determined primarily by 'mortality' rate. 'Mortality' is generated based on 'average life expectancy' and 'mortality lookup' profile. 'Mortality lookup' is one of the model variables which is qualitative and based on experts' judgement as well as information from the literature. The relationships among these variables are developed by regression analysis and certain SD functions. For example, Eq. 1 uses the SD software 'Vensim' to interpret 'Population' as the stock with 'births' as inflow (rate) and 'deaths' as outflow (rate). However, the equation for 'births' (Eq. 2) is a modified formula to the equation developed by Forrester [9] for his World Dynamics model.

'Mortality' in Eq. 4 is determined based on the profile of mortality rate that is qualitatively captured as 'mortality lookup' according to available data from the Office of National Statistics [27] as shown in Fig. 2. The regression analysis relationship between population and household is developed (as shown in Eq. 5) based on the available data from ONS [27].

$$\text{Population}(t) = \text{INTEGRAL}\left[\text{births} - \text{deaths, population } (t_0)\right] \qquad (1)$$

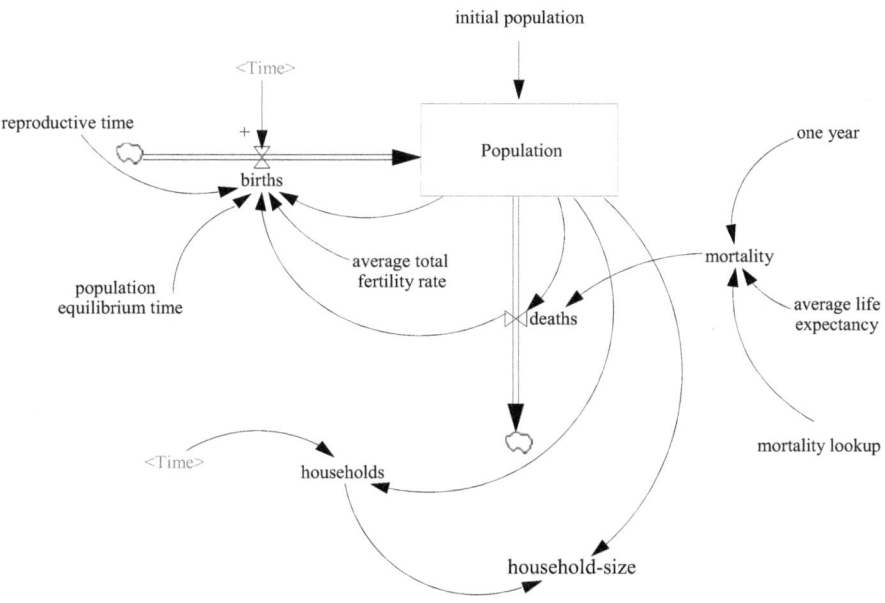

Fig. 1 SFD for population/household module

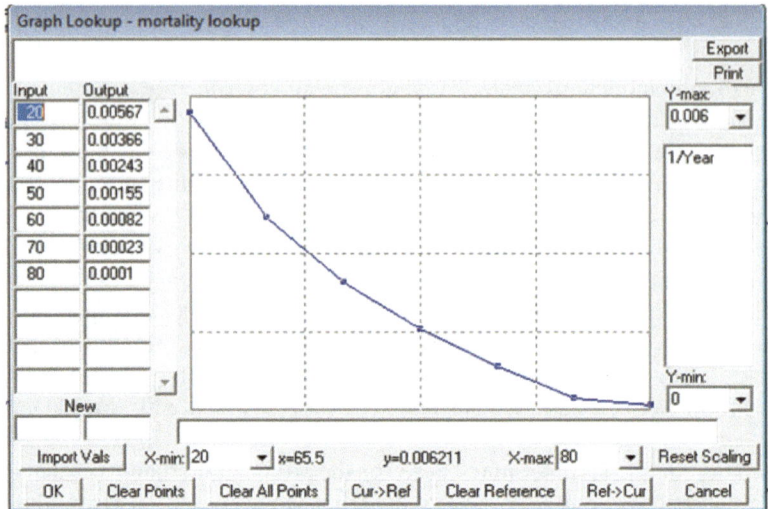

Fig. 2 Mortality lookup

$$
\text{births} = \begin{cases} \text{IF THEN ELSE (Time} = \text{population equilibrium time, deaths,} \\ \text{IF THEN ELSE (Time} \leq 2011, \text{ average total fertility rate*Population} \\ *0.08/\text{reproductive time,} \\ \text{FORECAST (average total fertility rate} * \text{Population} * 0.08/\text{reproductive time, 39, 100)))} \end{cases} \quad (2)
$$

$$
\text{deaths } = \text{Population} * \text{mortality} \quad (3)
$$

$$
\text{mortality} = \text{mortality lookup(average life expectancy/one year)} \quad (4)
$$

$$
\text{households } = -3.436e008 + 182058 * \text{Time } + 0.067 * \text{Population} \quad (5)
$$

4.2 SFD for the Climatic—Economic—Energy Efficiency Interaction Module

This module simulates the interactions of the energy efficiency, economic, and climatic variables that are included in the model. As shown in Fig. 3, the effect of the household energy efficiency measures on household energy bills is highlighted. Also, the effects of the unfavourable climatic conditions on international energy prices and consequently on household energy bills are elaborated. All these work together seamlessly as a system to give the effect of energy bills on energy consumption, which ultimately have effects on carbon emissions. An example of sample data driving this

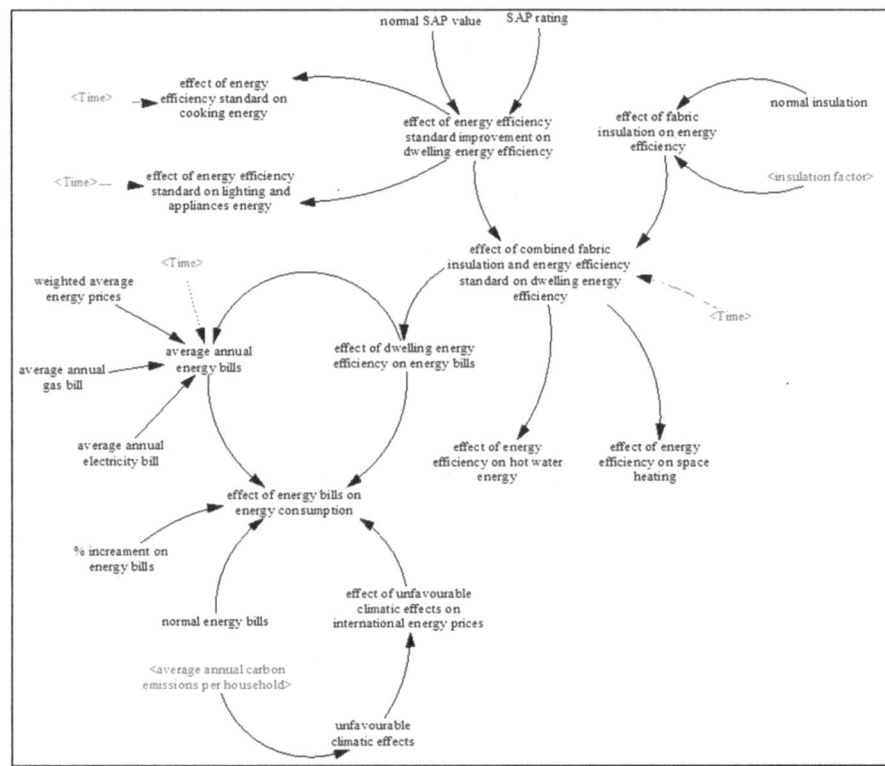

Fig. 3   SFD for economic—climatic—energy efficiency interaction module

module is given in Table 2. Examples of the main equations of this module are given in Eqs. 6–10.

Table 2   Sample data under climatic-economic-energy efficiency interaction module

Variable	Unit of measurement	Minimum	Maximum	Mean	Standard error	Standard deviation
Average annual gas bill	£	372	659	542.07	12.44	79.66
Average annual electricity bill	£	378	578	490.43	8.43	53.96
Weighted average energy prices	–	3.45	6.01	4.74	0.10	1.61
SAP rating	–	17.60	55.00	37.86	1.61	10.31

Source [30]

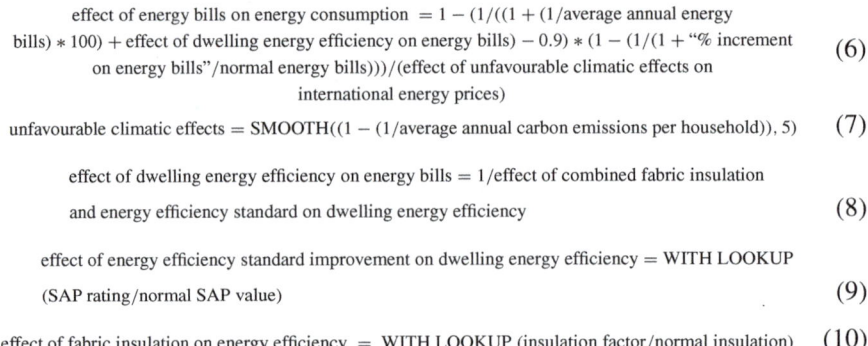

$$\text{effect of energy bills on energy consumption} = 1 - (1/((1 + (1/\text{average annual energy}$$
$$\text{bills}) * 100) + \text{effect of dwelling energy efficiency on energy bills}) - 0.9) * (1 - (1/(1 + \text{``\% increment}$$
$$\text{on energy bills''}/\text{normal energy bills})))/(\text{effect of unfavourable climatic effects on}$$
$$\text{international energy prices)} \tag{6}$$

$$\text{unfavourable climatic effects} = \text{SMOOTH}((1 - (1/\text{average annual carbon emissions per household})), 5) \tag{7}$$

$$\text{effect of dwelling energy efficiency on energy bills} = 1/\text{effect of combined fabric insulation}$$
$$\text{and energy efficiency standard on dwelling energy efficiency} \tag{8}$$

$$\text{effect of energy efficiency standard improvement on dwelling energy efficiency} = \text{WITH LOOKUP}$$
$$\text{(SAP rating/normal SAP value)} \tag{9}$$

$$\text{effect of fabric insulation on energy efficiency} = \text{WITH LOOKUP (insulation factor/normal insulation)} \tag{10}$$

Figure 3 shows the relationship between the two variables 'The effect of dwelling energy efficiency on energy bills' and 'average annual energy bills'. 'The effect of dwelling energy efficiency on energy bills' on energy bills was qualitatively modelled to show how the model will respond to changes in energy efficiency improvements. As shown in Fig. 4, the model behaviour suggests that there will be an incline in the bills' values over time due to these improvements. The model uses the available data about the 'average annual energy bills' to develop a future profile as shown in Fig. 5. It is clear that 'The effect of dwelling energy efficiency on energy bills' is balancing 'The effect of dwelling energy efficiency on energy bills' as shown in Fig. 5 which shows a gradual increase.

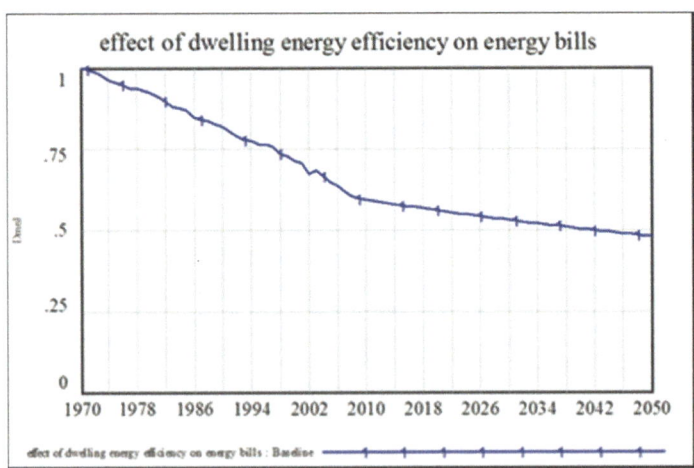

Fig. 4 The effect of dwelling energy efficiency on energy bills

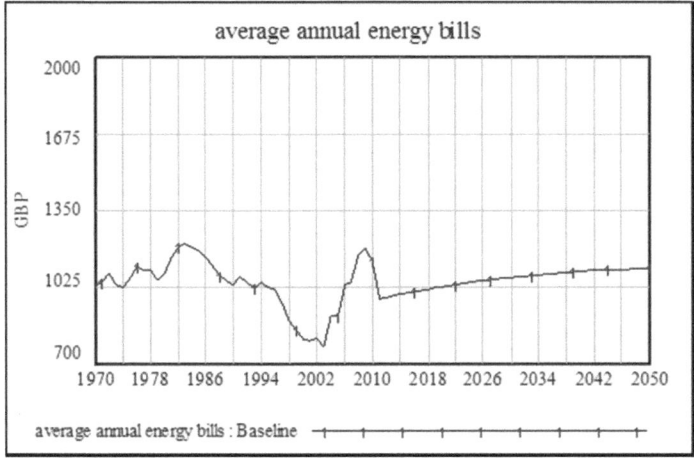

Fig. 5 Average annual energy bills

4.3   SFD for Household Energy Consumption Module

In this module, household energy consumption is modelled based on the five different end uses of energy (space heating energy consumption, hot water energy consumption, cooking energy consumption, lighting energy consumption, and appliances energy consumption). Carbon emissions are then modelled by converting the total energy consumption to carbon emissions through the use of a conversion factor termed 'energy to carbon conversion'. An example of the developed SFDs for 'lighting energy consumption' is shown in Fig. 6. The 'Average annual household energy consumption' (shown in Fig. 7) is therefore determined by adding all the household energy consumption stocks of the five end uses of energy. An example of data driving the module is shown in Table 3.

The 'energy to carbon conversion factor' used in this model is assumed to be the conversion factor of energy to carbon conversion factor of energy from electricity source. This is done for simplicity sake. Ideally, energy conversion factor of different fuels (i.e. gas, oil, electricity, etc.) to meet household energy consumption by end uses needs to be determined and applied appropriately. The average annual carbon emissions per household is determined by the same approach as described under household energy consumption.

The basic conditions affecting the identified variables have been simulated under 'Baseline scenario'. The scenario assumes the current energy efficiency effects and behaviour will continue at the same trends. The scenario assumes the dwelling internal temperature of householders having a set-point of 19 °C. Further, the scenario assumes that any change to energy bills will not significantly affect the energy consumption behaviour of the householders as the 'standard' consumption behaviour

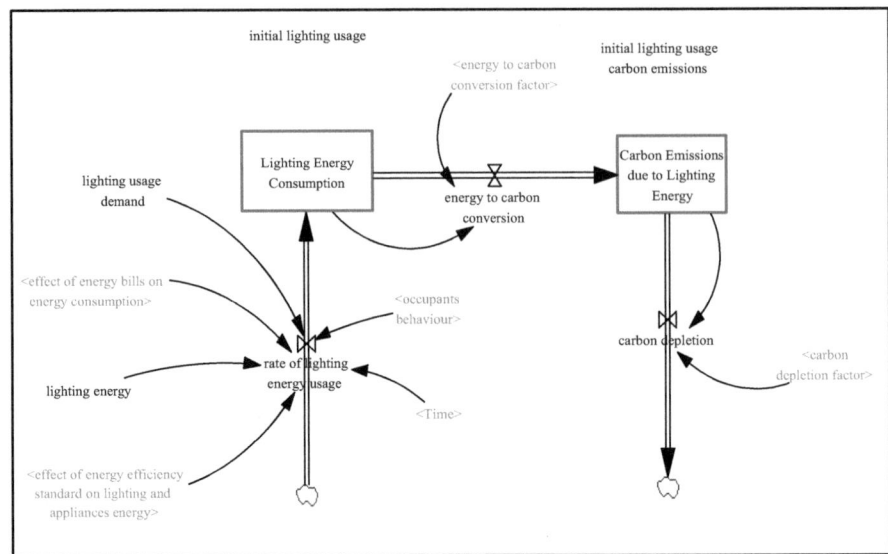

Fig. 6 SFD for lighting energy consumption and carbon emissions

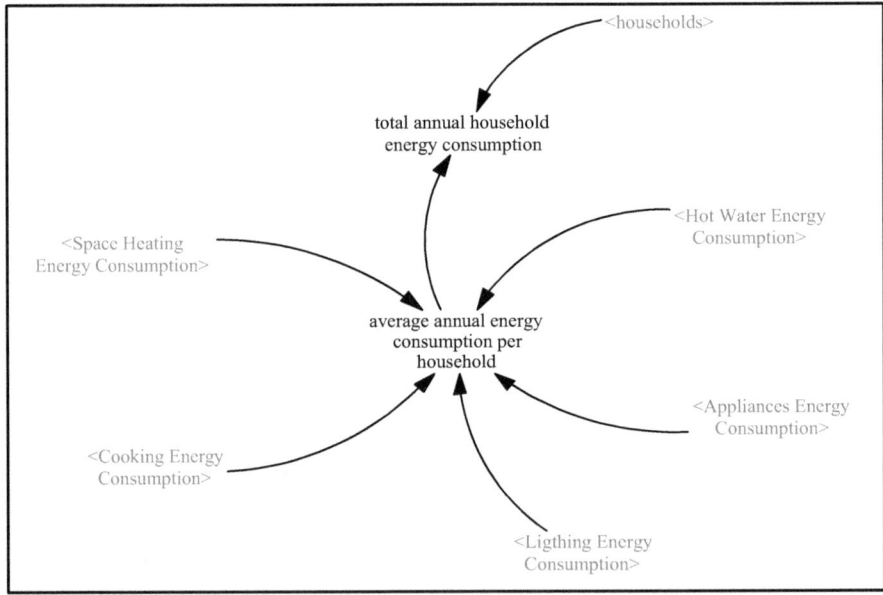

Fig. 7 SFD for household energy consumption

Table 3 Sample data for household energy by end uses (Lighting data)

Variable	Unit of measurement	Minimum	Maximum	Mean	Standard error	Standard deviation
Lighting	MWh	0.55	0.69	0.65	0.01	0.04

Source [30]

Table 4 Sample data to initiate the Baseline Scenario for the year 2005

"Average annual energy consumption per household" (MWh)	21.91	SAP Rating	48.1
Households	25,375,000	Average annual gas bill	448
Population	60,236,000	Average annual electricity bill	444
"Space heating energy consumption" (MWh)	14.38	Weighted average energy prices	4.19
"Hot water energy consumption" (MWh)	3.62	average annual carbon emissions per household	6.1
"Cooking energy consumption" (MWh)	0.61	total annual household carbon emissions	154.59
"Lighting energy consumption" (MWh)	0.66	External air temp	10.5
"Appliances energy consumption" (MWh)	2.33	Relative humidity	89

will be maintained. A sample data used to initiate the baseline scenario is shown in Table 4, for the year 2005. The results were then compared with the available historical data in order to predict the consumption within the targeted time horizon. The following sections discuss the model results.

5   Results of the Model Application

The key results of the proposed model for the population/Household module and the Climatic—Economic—Energy Efficiency Interaction module will be presented in this section.

5.1   Behaviour Analysis of the Population/Household Module

The number of households plays a major role in accurately estimating the amount of energy consumption in the entire UK's housing stock. Household energy consumption is strongly influenced by the population, the number of households and the average household size. For example, the amount of household energy consumption attributable to hot water consumption and usage of some appliances is greatly influenced by the household size. However, there is a minimum level of household energy consumption applicable to each household as the operation of some energy-consuming appliances like fridge, freezer, etc. don't depend on the household size.

An output of this module is presented in Fig. 8 with a comparison against the reference model. The model shows that the total UK population is on an upward trend till 2050. The model indicates that the UK population of 55.63 million in 1970 will grow to 69.78 million by the year 2050. This figure shows a yearly increase of 0.31% on the average. The Office of National Statistics [28] suggests that the UK population receive an annual growth of 0.28% averagely between the year 1970 and 2010, while the model output for the same period shows an average of yearly growth of 0.26%. The slight difference in the two estimates can be attributed to the consideration of the mortality factor in the proposed model which includes some subjective data.

Furthermore, in order to gain more insight into the behaviour of the number of households as produced by the model, further analysis was conducted at an interval of ten years from 1970. The results of this study are compared with that of Johnston [18] and Palmer and Cooper [30] as shown in Table 5. The behaviour of this model indicates that an increase in the number of households on a yearly basis with a decline in the level of this growth until 2050. The result shows almost the same downward trend as the output of Johnston [18] model. The major difference in the two models lies in the period of 2040–2050, where Johnston [18] specifically fixed the trend of this period (and not as the result of analysis) based on the assumption that the number of households will not change in those years. This assumption may explain the difference between the two models. Correspondingly, when the output of this model is compared with that of Palmer and Cooper [30], which is based on the ONS available data till 2010, the results follow a 'lumpy' trend with a combination of peaks and troughs, but overall the number of households grow at 0.99% yearly, whereas the results of this model show a growth of 1.02% yearly.

5.2 Behaviour Analysis of the Climatic—Economic—Energy Efficiency Interaction Module

Space heating energy has been moving in an upward direction until the year 2004 when it begins to fall apart due to bad weather conditions in 2010, as shown in Fig. 9. The behavioural attitude of occupants in seeking more thermal comfort at home and home extension might explain this trend. The model estimates the space heating energy to follow a downward trend from the year 2004 until 2050 due to the improvements in energy efficiency (SAP rating) as a result of stringent building regulations.

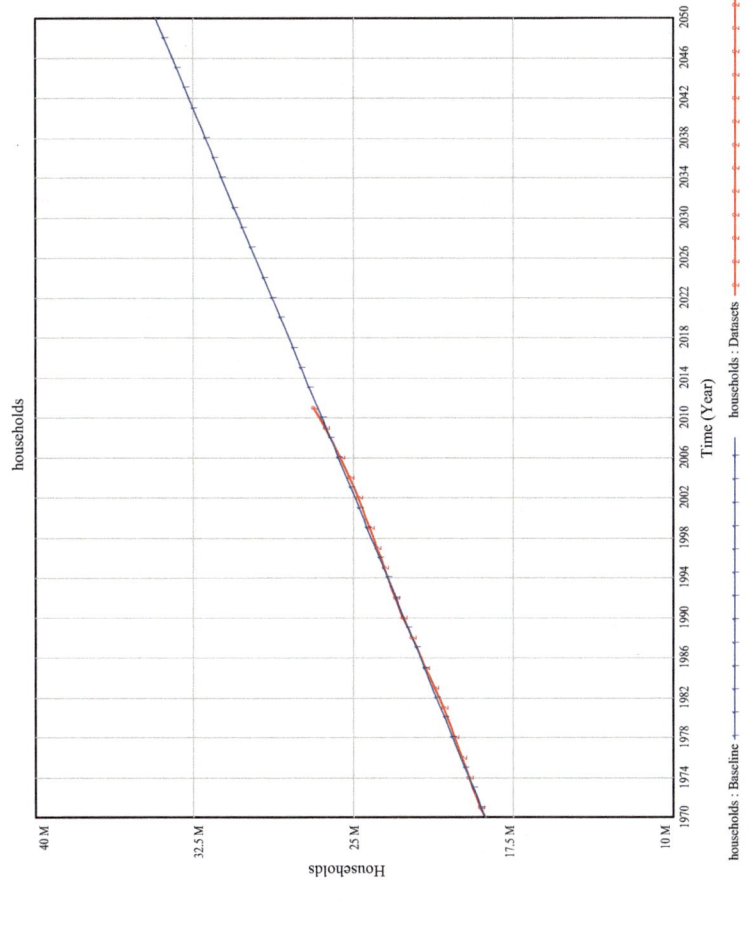

Fig. 8 Projected total UK population under the 'baseline' scenario

Year	Johnston [18] (%)	Palmer and Cooper [30] (%)	This study (%)
1970–1980	–	0.82	0.93
1980–1990	–	0.88	0.84
1990–2000	0.76*	0.72	0.77
2000–2010	0.70	0.81	0.71
2010–2020	0.68	-	0.67
2020–2030	0.53	-	0.63
2030–2040	0.37	-	0.59
2040–2050	-0.09	-	0.55
Overall yearly average	0.56*[†]	0.99[†]	1.02[†]

Table 5 Average yearly percentage increase/decrease in the number of households

* The year starts from 1996; [†] was computed based on [(final year value – base year value)/number of years*100%]

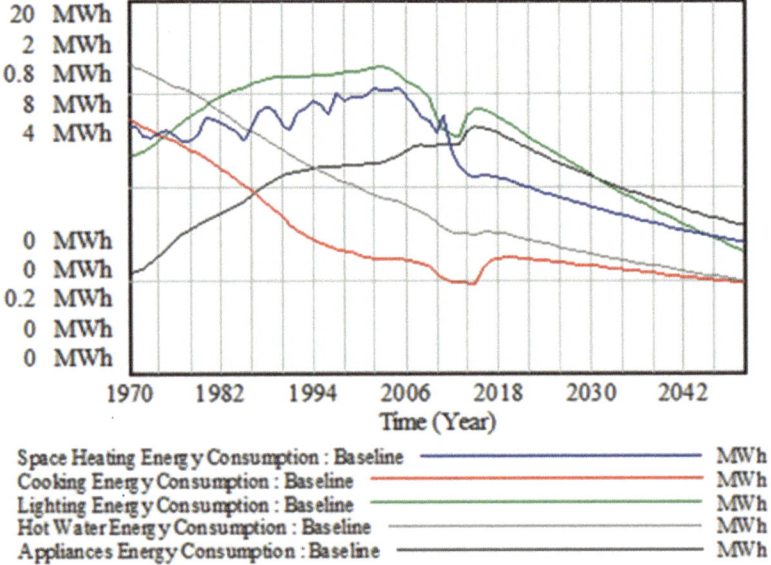

Fig. 9 Average Energy Consumption for the five end uses per household

The behaviour pattern for hot water usage and cooking per household shows a downward trend since 1970, as shown in Fig. 9. The reason for the trend of hot water usage may be connected to reduction in heat loss (in terms of improvement in energy efficiency 'SAP rating') due to improved lagging of hot water pipes and tanks coupled with improvements in household heating systems that is being witnessed due to changes to building regulations. A further probe into the behaviour of the model indicates that the slope of the trend slightly changed around the year 2014 and follows this new trend until the year 2050. Should the trend follow the slope of the graph since 1970 until 2014, it may mean that by 2040, the average energy consumption for hot water would have net-zero, which is practically impossible. It needs to note that irrespective of the demand for cut in household energy consumption, it will not translate to mean that no hot water would be required at homes in years to come as there will be a minimum amount of hot water energy required for each household.

Generally, the energy consumption for cooking has a steep slope till the 1990s and the downward trend seems levelling for a short period since the year 2000 apart from a short period between 2008 and 2016. The general downward trend may be due to changes in lifestyle through saving in household cooking energy as most families eat in eateries, which consequently reduces the rate of cooking at home. However, the trend levelling up is more pronounced around the year 2016 until 2050. This saw the slope of the trend of average household cooking energy to be gentler compared to the preceding years. The reason for this trend could be explained as a result of a decline in the size of households. This is due to the fact that cooking energy per head is claimed to be higher in single-person households [7].

Household lighting energy remains a small fraction of the total household energy. The behaviour exhibited by the output of the model Fig. 9 shows that the average lighting energy consumption per household remarkably follows an upward trend since 1970 until 2004 when begins to gradually come down. This decline may be as a result of the Government's policy of the Carbon Emissions Reduction Target (CERT), which ensures that energy-consuming incandescent bulbs are replaced in homes with energy-efficient ones. However, the simulation result suggests that the rate of decline of household lighting energy consumption would decrease as of 2016 as against the trend witnessed between 2004 and 2016. This may be a result of the likely increase in the lighting points in homes especially in kitchens and bathrooms with higher specifications. This may therefore likely reduce the rate of decline by offsetting the savings that would have recorded should the trend of decline between 2004 and 2016 maintained.

The model suggests that household appliances energy use has been on the increase since 1970. This result is consistent with historical data [30]. The reasons for this trend are explained based on three factors. Firstly, the trend may be due to the fact that many homes now acquire electric gadgets more than before, which continue to grow, based on changes in occupants' lifestyle and their access to more disposable income. Secondly, owing to these gadgets alone may not result in a surge in household appliances energy if they are not put into use. So, the rate at which these gadgets are being put into use has been on increase. This may probably due to changes in lifestyle as previously argued. Additionally, changing to the use of energy-consuming appliances for some tasks that were previously or traditionally completed manually as well as the increasing trend of working from home may be responsible for this surge. Thirdly, the results of the study conducted by EST et al. [7] indicate that the use of cold appliances like freezer and large fridges has been on the increase and they constitute about 50% of the household appliances energy use. Further, there has been growth in the use of microwaves to thaw out frozen food. Combining all these together have seen household appliances energy on the increase. However, there is an event overturn in and around 2016 as dictated by the result of the simulation that household appliances energy will follow a gentle decline till 2050. This output may explain the optimistic view regarding different on-going research efforts directed at improving the energy efficiency of cold appliances. This hopefully would see the deployment of even more energy-efficient cold appliances in the coming years as they have a share in the household appliances energy consumption.

6 Discussion on the Model Results and Use

The household energy consumption results have been compared with the historical data [30]. The statistical 'goodness of fit' test was conducted for the main four output variables of the model to compare the results with the historical data as shown in Table 6. The R-square values for all variables are acceptable at the 99% confidence level.

Further analysis of the results has been conducted to illustrate the expected reductions in household energy consumption for the years 2020 and 2050 relative to the year 1990 based on the 'baseline' scenario, as shown in Table 6. To this end, the model results suggest that the energy consumption ascribable to the UK households are expected to decline by about 9.35% by 2020 and about 26.60% by the year 2050. The implication of this result is that under the 'baseline' scenario, it is unlikely to meet the target reductions of 34 and 80% relative to 1990 by the years 2020 and 2050 respectively as enshrined in the Climate Change Act of 2008. The proposed model has been run to evaluate new proposed scenarios that may be introduced to help achieve the targeted energy consumption, which will be discussed next.

The discussion below will consider how the results are sensitive to new energy policies initiated to reduce carbon emissions. This will also enable testing these policies before the actual implementation.

Table 6 Model validation based on statistical significance

Variable	R	R-square	P-value
Average annual energy consumption per household	0.967	0.935	0.000
Total annual household energy consumption	0.953	0.908	0.000
Average annual carbon emissions per household	0.940	0.884	0.000
Total annual household carbon emissions	0.895	0.801	0.001

6.1 Policy Formulation and Analysis

The 'baseline' scenario presented communicates the most probable way in which the household energy consumption of the UK housing stock will evolve over the years starting from 1970 until 2050. This is based on the assumption that the trends depicted by historical data will continue in that way. Evidently, if there is any policy change in the future that is clearly different from the current ones, the profile of energy consumption could be altered. According to Murphy et al. [26], there could be many policies that affect household energy consumption such as subsidies, tax incentives, awareness-raising campaigns, energy labels, smart meters instalment, energy audits, etc. More recently, there have been also few community-led initiatives to improve energy efficiency in household, examples of the lessons learnt can be found elsewhere, Hoppe et al. [15]. Based on the approach adopted for the development of the proposed model of this research, these policies can be classified as technology, economics, and occupant related. The developed model can accommodate these three categories based on the main variables that may be affected by a certain policy. For this paper, four hypothetical scenarios are assumed that include 'efficiency', 'behavioural change', 'economic', and 'integrated' scenarios. In all of these scenarios, the future of energy consumption for the entire UK housing stock is explored. The following section explains the underlay assumptions for each of the scenarios and discusses the results emanating from them.

6.2 'Efficiency' Scenario

This scenario assumes a situation whereby more stringent energy efficiency measures is put in place to deeply cut carbon emissions. Therefore, the uptake of improved dwelling insulation measures will increase to improve the energy efficiency rating of dwellings which will lead to more airtightness of dwellings. Two cases were made and the model was run to explore the results of each one. The first case (Efficiency 1) assumes that Fabric insulation depicted as 'insulation factor' in the model is assumed to increase by 25% beyond the levels set under the 'baseline' scenario and all other related energy consumption activities are assumed as the standard maintained by the householders for the baseline scenario. The second efficiency case (Efficiency 2) assumes that the installation of the improvement measures may result also in rebound effects (as a form of occupants' behaviour change) which makes occupants seek more thermal comfort. Thereby the dwellings internal temperature is assumed to change the set-point from 19 to 21 °C. All other variables are set at the same level as the 'baseline' scenario. Table 7 shows the results for both cases.

For the first case, the behaviour of efficiency scenario based on end uses shows a downward trend for energy consumption for all the end uses. However, for appliances, the expectation is mixed as the results show that no reductions in energy consumption below the 1990 levels are anticipated. Although there are technological improvements

Table 7 Percentage (%) reductions in household energy consumption for all scenarios for the year 2020 and 2050 relative to 1990 base as enshrined in Climate Change Act of 2008

	Household energy consumption (TWh) (1990)		Baseline scenario	Efficiency scenario		Behaviour change scenario	Economic scenario		Integrated scenario
				(1)	(2)		(1)	(2)	
Space heating	300.92	For year 2020	−7.99	−16.12	−10.66	−23.99	−13.01	−12.45	−26.36
Hot Water	108.20		−26.61	−43.99	−29.08	−42.38	−23.19	−22.19	−44.71
Cooking	18.88		−44.28	−71.38	−47.19	−37.08	−45.34	−43.38	−38.24
Lighting	15.29		−10.92	−23.05	−15.24	−25.18	−10.04	−9.61	−26.88
Appliances	47.93		+35.34	+44.85	+29.65	13.85	38.72	37.05	11.60
Total	491.22		−9.35	−18.65	−12.33	−24.89	−11.36	−10.87	−27.17
Space heating	300.92	For year 2050	−25.38	−36.13	−31.53	−38.88	−30.91	−20.66	−43.72
Hot Water	108.20		−41.10	−53.30	−46.51	−51.70	−55.88	−37.35	−55.90
Cooking	18.88		−44.92	−61.79	−53.92	−36.12	−65.77	−43.96	−42.11
Lighting	15.29		−25.18	−44.08	−38.46	−37.67	−35.52	−23.74	−46.50
Appliances	47.93		+5.21	−14.04	−12.25	−12.23	10.61	7.09	−23.81
Total	491.22		−26.60	−39.00	−34.03	−38.96	−33.84	−22.62	−44.49

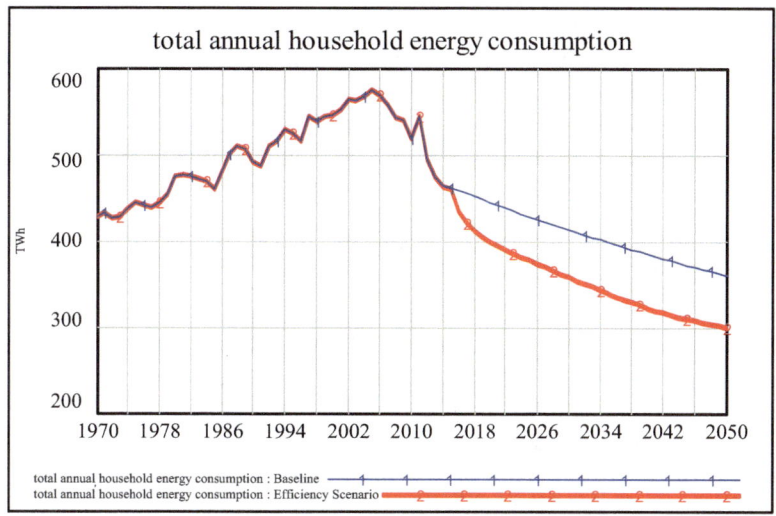

Fig. 10   Total annual Household Energy Consumption for Efficiency verses Baseline scenario

in home appliances in terms of energy efficiency, but this advancement could not be immediately translated into much savings. However, by 2050, some savings are expected. The results of this case are shown in Fig. 10 in comparison to the baseline scenario which shows a total reduction in energy consumption by 18.65 and 39% for the year 2020 and 2050 respectively as against minimum targets of 34 and 80% respectively. The implication of this is that laying much of the emphasis on energy efficiency improvements alone without corresponding efforts on other aspects of policy target is unlikely to yield the required level of savings. Table 7 also shows the results of efficiency case 2 where the reduction in energy consumption has been affected by the rebound effect to be 12.33% and 34.03% for the year 2020 and 2050 respectively. This indicates how the behavioural change may affect the total energy consumption which should be considered when analysing any future policies.

6.3   'Behavioural Change' Scenario

The effect of occupants' behavioural change on energy consumption is tested by this scenario. Frugal consumption behaviour is assumed, that is the daily habitual behaviours of occupants tend towards energy saving in their homes which will be reflected in a set-point of dwelling internal temperature to be 18.5 °C instead of 19 °C for the baseline scenario. Other indicators associated with the frugal behaviour have been made to reflect the difference in energy requirements for this behavioural category. As shown in Table 7, a significant reduction is observed when the 'Behavioural Change' scenario is implemented. Like 'efficiency' scenario, the appliances energy consumption is expected to show no reductions in energy consumption below the 1990 levels, however with much less consumption. The total results indicate it is unlikely to meet the reduction targets of 80% by 2050. As suggested under the 'efficiency' scenario, the Government policy should target other policy areas in addition to the 'behavioural change' in order to meet the required level of reductions.

6.4 'Economic' Scenario

The 'Economic scenario' has been also considered in two cases. The first case (Economic 1) considers the effects of energy bills on energy consumption. The scenario assumes that the energy prices will be frozen as a result of governmental initiative (instead of the assumed annual increase) which will provide more disposable income for householders in an attempt to lower the number of those in fuel poverty. The second case (Economic 2) considers the rebound effects that may also result in the householders seeking more thermal comfort by increasing the dwelling internal temperature set-point from 19 to 20 °C. All other variables are kept as they were for the 'baseline' scenario. It is evident from the results shown in Table 7 that by adopting this scenario alone it is unlikely to meet the reduction targets as enshrined in the Climate Change Act of 2008. The rebound effects have also shown some reductions in the total savings of consumption.

6.5 'Integrated' Scenario

The 'integrated' scenario combines the assumptions made under the 'efficiency', 'behavioural change', and 'economic' scenarios. The scenario assumes that the energy efficiency improvements as described and emphasised under the 'efficiency' scenario will be maintained. Further, the scenario assumes that householders will display frugal energy consumption behaviour. That is, they exercise some behavioural habits aiming at saving energy consumption at home like turning down heating in vacant rooms, washing at lower temperatures. With all these, they are however assumed to maintain a dwelling internal temperature set-point of 20 °C. Additionally, energy prices are expected to be frozen as explained under the 'economic' scenario.

   Generally, it is apparent from the results of the 'Integrated' scenario as shown in Table 7 that the efficiency of a vigorous behavioural change in addition to the improvements in energy efficiency measures through stringent building regulations and other UK Government's policy frameworks display the capability of achieving the legally binding reduction targets better. However, it is unlikely to meet the reduction targets of 80% by the middle of this century, only about 44.49% reductions are likely to be achieved.

7 Conclusions and Policy Implications

The previous empirical studies relating to the socio-technical systems of household energy consumption and carbon emissions have established the theoretical underpinning of frameworks used in analysing the impact of energy policies on carbon emissions. Among the frameworks available, an integrated and holistic approach is required to combine a number of disciplines together and provide a framework capable of shaping the impact of energy policies on energy consumption and carbon emissions. The interacting systems affecting this framework comprise of the dwellings, occupants, and external environment systems. The variables identified within the dwellings system are related to dwellings' physical characteristics. The variables related to occupants system are in terms of household characteristics, occupants' thermal comfort, and occupants' behaviour. The variables related to external environment system are in terms of climatic, economic, and cultural influences.

In order to simulate the impact of energy policies on these variables, there are quite a number of energy and carbon emissions models that have evolved over the years with the capability of estimating energy consumption and carbon emissions, especially in the housing sector of the economy. The available models vary considerably based on the levels of disaggregation, complexity, resolution of output, output aggregation levels, scenario analysis performed, model validation, and their availability to the members of public for scrutiny.

A careful appraisal of the existing modelling approaches suggests that there are a number of limitations in the existing modelling techniques, which are (1) lack of transparency in the model algorithms, (2) inability to account for the complex, interdependencies, and dynamic nature of the issue of energy consumption and carbon emissions, (3) limited evidence to show for the occupants-dwelling interactions, and (4) lack of enough capacity to accommodate qualitative data input. And as such, there is the need to scout for more robust modelling approaches that take into consideration the kind of complexity involved and bedevilling the issue of household energy consumption and carbon emissions due to high interdependencies, chaotic, non-linearity, and qualitative nature of some of the variables involved.

After considering various techniques, the system dynamics methodology was considered the most suitable technique in conceptualising the problem under investigation. This research aimed to develop a dynamic model to analyse the impact of energy policies on buildings performance. Therefore, this chapter introduced a dynamic model to evaluate the effectiveness of proposed energy policies while considering the complexity of household energy consumption. The developed System Dynamics (SD) model simulates the interrelationship between the dwellings, occupants and environment systems. The developed model has the advantage of simulating the various qualitative conditions incorporated with energy consumption in addition to the traditional technological conditions that are lonely considered by regression-based forecasting.

The developed SD model includes these modules: population/household, dwelling internal heat, occupants thermal comfort, climatic-economic-energy efficiency inter-action, household energy consumption, and household CO_2 emissions. Some scenarios were formulated in order to illustrate the use of the model to simulate the impact of certain policies on carbon emissions reduction in the housing sector. The developed scenarios demonstrate the way by which household energy consumption and carbon emissions attributable to the UK housing stock would evolve over the years under different assumptions. Four scenarios were illustrated to include 'efficiency', 'behavioural change', 'economic', and 'integrated' scenarios.

The 'efficiency' scenario considers the effects of improvements in energy efficiency measures on household energy consumption. The 'behavioural change' scenario simulates the effects of occupants' behaviour change on household energy consumption and carbon emissions profile. Furthermore, the 'economic' scenario assumes a case of policy change by Government favouring energy prices reduction, thereby reducing the energy bills payable by the householders and its consequences on household energy consumption and carbon emissions. Lastly, an 'integrated' scenario combines the assumptions in the first three scenarios were considered.

As the main conclusion from the scenarios tested above is that more combined efforts should be made in terms of technology installation, behaviour change, and policy incentives in order to have a significant impact on the household energy consumption. It is unlikely for any of the scenarios to individually meet the required legally binding reductions of 80% cut in energy consumption.

For all the scenarios, the insights from the model show that the greatest reductions in both household energy consumption and carbon emissions are expected from both the space and water heating. It is also concluded that the behavioural change represented by the daily habitual behaviours of occupants towards energy saving has a significant effect on the amount of reduction. On the other hand, the rebound effect has a considerable negative effect on reducing the total energy consumption that should be always considered when analysing any future policies.

The developed model builds on the existing modelling efforts, which are traditionally restricted to building physics and regression-based forecasting, in order to generate new insights into the future using a non-deterministic systems approach. The developed model is highly transparent as all the variables and algorithms developed can be scrutinised, unlike many of the existing models.

The developed model has the capability of producing a clear understanding of household energy consumption and carbon emissions associated with it. This can serve as a decision making policy tool with the capability to direct policy decisions by testing the effect different policy scenarios such as energy efficiency improvements and behavioural change likely to have on household energy consumption and carbon emissions. The insights generated will allow policy makers to make informed decisions regarding any future policy formulations concerning energy and carbon emissions within the UK housing sector.

8   Model Limitations and Further Research Work

While the model is specifically developed for the UK housing sector, the model structure can be used for other countries considering other necessary data sets to run the model algorithms.

The current version of the developed model is an aggregated model of the entire UK households. To provide more accurate results for each type of dwellings, another line of research can be followed by disaggregating the model structure based on different dwelling types such as detached, semi-detached dwellings (with varying in size, age, thermal insulation standards) and different occupancy patterns (e.g. pensioners, singles, family with/without children).

A research on the energy supply side of the housing sector based on the system dynamics approach is encouraged. This is considered important because of the need to highlight the effect of clean energy supply, due to technological advancement, on the household energy consumption and carbon emissions. The output of such a research can be linked to the developed model in this research to form an improved socio-technical model of energy and carbon emissions of the UK housing sector.

This research can be replicated with an expanded model boundary to accommodate other variables that are excluded in this current version. This will include incorporation of the carbon emission factors of different fuels used for different end uses of household energy consumption, dwelling exposure, dwelling orientation, occupants' social class, and the likes. As indicated in literature for this kind of complex approach in research, STS models are a broad unit of analysis and have a difficulty in drawing precise boundaries. However, it is also recommended that the impact of major global variables such as oil crises, geopolitical events, and disruptive events can be considered for further model development. Inclusion of such variables would improve the accuracy of the model results and allow further options of different policies to be explored.

References

1. Abrahamse W, Steg L, Vlek C, Rothengatter T (2005) A review of intervention studies aimed at household energy conservation. J Environ Psychol 25(3):273–291
2. Böhringer C, Rutherford TF (2009) Integrated assessment of energy policies: decomposing top-down and bottom-up. J Econ Dyn Control 33(9):1648–1661
3. Bordass W, Cohen R, Field J (2004) Energy performance in non-domestic buildings: closing the credibility gap.Paper presented at the Building Performance Congress, Frankfurt, Germany
4. Cartelli A (2007) Socio-technical theory and knowledge construction: towards new pedagogical paradigms. Issues Inform Sci Inform Technol 4:1–14
5. Coyle RG (1997) System dynamics modelling: a practical approach. Chapman and Hall, London
6. De-Greene KB (1988) Long wave cycles of sociotechnical change and innovation: a macropsychological perspective. J Occup Psychol 61(1):7–23
7. EST, DECC, and Department of Environment, Food and Rural Affairs (DEFRA) (2012) Powering the nation: household electricity-using habits revealed. EST/DECC/DEFRA, London

8. FitzGerald J (2002) A model for forecasting energy demand and greenhouse gas emissions in Ireland. The Economic and Social Research Institute, Dublin. Working Paper No 146
9. Forrester JW (1971) World dynamics. Wright-Allen Press, Cambridge
10. Geels FW (2002) Technological transitions as evolutionary reconfiguration processes: a multi-level perspective and a case-study. Energy Policy 31:1257–1274
11. Geels FW, Kemp R (2007) Dynamics in socio-technical systems: typology of change processes and contrasting case studies. Technol Soc 29(4):441–455
12. Groesser SN, Schwaninger M (2012) Contribution to model validation: hierarchy, process, and cessation. Syst Dynamics Rev 28(2):157–181
13. Haas R, Schipper L (1998) Residential energy demand in OECD-countries and the role of irreversible efficiency improvements. Energy Econ 20(4):421–442
14. Hirst E, Lin W, Cope J (1977) A residential energy use model sensitive to demographic, economic, and technological factors. Quart Rev Econ Finance 17(2):7–22
15. Hoppe T, Graf A, Warbroek B, Lammers I, Lepping I (2015) Local governments supporting local energy initiatives: lessons from the best practices of Saerbeck (Germany) and Lochem (The Netherlands). Sustainability. 7(2):1900–1931
16. International Energy Agency (1998) Mapping the energy future: energy modelling and climate change policy, Paris
17. Jayanesa HAH, Selker JS (2004) Thousand years of hydraulic civilization: some sociotechnical aspects of water management. World Water Council, pp 225–262
18. Johnston D (2003) A physically based energy and carbon dioxide emission model of the UK housing stock. Ph.D. thesis, Leeds Metropolitan University, UK
19. Kavgic M, Mavrogianni A, Mumovic D, Summerfield A, Stevanovic Z, Djurovic-Petrovic M (2010) A review of bottom-up building stock models for energy consumption in the residential sector. Build Environ 45:1683–1697
20. Kelly S (2011) Do homes that are more energy efficient consume less energy? A structural equation model for England's residential sector. Electricity Policy Research Group, EPRG Working Paper, University of Cambridge
21. Lee T, Yao R (2013) Incorporating technology buying behaviour into UK-based long term domestic stock energy models to provide improved policy analysis. Energy Policy 52:363–372
22. Marques FC, Dal Soglio FK, Ploeg JD (2010) Constructing sociotechnical transitions towards sustainable agriculture: lessons from ecological production of medicinal plants in Southern Brazil. In: Proceeding of innovation and sustainable development in agriculture and food, Montpellier, 28–30 June 2010.
23. Mhlas A, Kassem M, Crosbie T, Dawood N (2013) A visual energy performance assessment and decision support tool for dwellings. Visual Eng 1:7
24. Motawa I, Oladokun M (2015) A model for the complexity of household energy consumption. J Energy Build 87:313–323
25. Motawa IA, Banfill PF. Energy-efficient practices in housing—a system dynamics approach. In: Proceedings of 18th CIB World Building Congress: TG62—built environment complexity, Salford, UK, pp 44–56
26. Murphy L, Meijer F, Visscher H (2012) A qualitative evaluation of policy instruments used to improve energy performance of existing private dwellings in the Netherlands. Energy Policy 45:459–468
27. Office for National Statistics (2009) Digest of UK energy statistics. TSO, London
28. Office for National Statistics (2013) Households. Available at https://www.ons.gov.uk/ons/tax onomy/search/index.html?nscl=Households&nscl-orig=Households&content-type=Dataset& content-type=Reference+table&sortDirection=DESCENDING&sortBy=pubdate. Assessed on 09 Dec 2013
29. Oladokun MG (2014) Dynamic modelling of the socio-technical systems of household energy consumption and carbon emissions. Ph.D. thesis, Heriot-Watt University, UK
30. Palmer J, Cooper I (2012) United Kingdom housing energy fact file. London, DECC
31. Patnayakuni R, Ruppel C (2010) A socio-technical approach to improving the systems development process. Inform Syst Front 12(2):219–234

32. Ranganath BJ, Rodrigues LLR (2008) System dynamics: theory and case studies. International Publishing House Pvt. Limited, New Delhi
33. Rodrigues A, Bowers J (1996) The role of system dynamics in project management. Int J Project Manage 14(4):213–220
34. Rohracher H (2003) The role of users in the social shaping of environmental technologies. Eur J Social Sci 16(2):177–192
35. Shipworth D (2006) Qualitative modelling of sustainable energy scenarios: an extension of the Bon qualitative input–output model. Construct Manage Econ 24(7):695–703
36. Shove E (1998) Gaps, barriers and conceptual chasms: theories of technology transfer and energy in buildings. Energy Policy. 26(15):1105–1112
37. Soldaat K (2006) Interaction between occupants and sustainable building techniques. In: ENHR conference: housing in an expanding Europe: theory, policy, participation and implementation, Ljubljana, Slovenia, 2–5 July 2006, pp 1–15
38. Stevenson F, Leaman A (2010) Evaluating housing performance in relation to human behaviour: new challenges. Build Res Inform 38(5):437–441
39. Stevenson F, Rijal HB (2010) Developing occupancy feedback from a prototype to improve housing production. Build Res Inform. 38(5):549–563
40. Summerfield A (2010) Two models for benchmarking UK domestic delivered energy. Build Res Inform 38(1):12–24
41. Swan LG, Ugursal VI (2009) Modelling of end-use energy consumption in the residential sector: a review of modelling techniques. Renew Sustain Energy Rev 13:1819–1835
42. United Nations—Department of Economic and Social Affairs (2010) Buildings and construction as tools for promoting more sustainable patterns of consumption and production. Sustain Develop Innov Briefs 9. https://www.un.org/esa/sustdev/publications/innovationbriefs/index.htm. Viewed: 23 May 2011
43. Walker GH, Stanton NA, Salmon PM, Jenkins DP (2008) A review of sociotechnical systems theory: a classic concept for new command and control paradigms. Theor Issues Ergon Sci 9(6):479–499
44. Way M, Bordass B (2005) Making feedback and post-occupancy evaluation routine 2: Soft landings—involving design and building teams in improving performance. Build Res Inform 33(4):353–360

Chapter 11
Investigating the Application of LEED and BREEAM Certification Schemes for Buildings in Kazakhstan

Serik Tokbolat and Farnush Nazipov

Abstract Kazakhstan is one of the leading economies of the Central Asian region. With the ambitious goal of being among the fifty most developed countries in the world, Kazakhstan is rigorously attempting to boost its infrastructure development. Among others, significant resources are invested in the construction of both residential and non-residential buildings. As a proponent of sustainable development, the government of Kazakhstan is seeking ways of reducing both energy use and greenhouse gas emissions. As a part of this agenda, Kazakhstan hosted EXPO2017 «Future Energy» exhibition which triggered the integration of sustainability principles into all areas of the economy including the construction sector. Among others, the construction industry started adopting widely recognized environmental assessment certification schemes such as the Leadership in Energy & Environmental Design (LEED) and the Building Research Establishment Environmental Assessment Method (BREEAM). Up to this day, more than 50 buildings, especially from rapidly expanding cities such as Nur-Sultan and Almaty, have obtained LEED and BREEAM certificates and were recognized as green buildings. This study investigates the nature of adopting these methods in the context of Kazakhstan with the aim of understanding the driving factors of such application, characteristics of the certified buildings, and the potentials of promoting the certification schemes at a wider scale.

Keywords LEED · BREEAM · Green building

1 Introduction

Buildings account for significant energy consumption globally and cause considerable impact on the planet's ecosystem by emitting greenhouse gases, consuming

S. Tokbolat (✉) · F. Nazipov
School of Engineering and Digital Sciences, Nazarbayev University, Nur-Sultan 010000, Kazakhstan
e-mail: stokbolat@gmail.com

© The Author(s), under exclusive license to Springer Nature Singapore Pte Ltd. 2021   181
R. J. Howlett et al. (eds.), *Emerging Research in Sustainable Energy and Buildings for a Low-Carbon Future*, Advances in Sustainability Science and Technology,
https://doi.org/10.1007/978-981-15-8775-7_11

tremendous amounts of resources and producing wastes sent to landfills [1]. The status quo in this regard is not acceptable and requires dramatic paradigm change towards sustainability in all sectors including construction industry. The philosophy and practice of sustainable construction are not as advanced in developing countries such as Kazakhstan as in developed ones like Scandinavian states, Canada, South Korea [2]. In Kazakhstan, challenges associated with energy deficit and carbon emissions could be resolved, among others, by modernization of existing infrastructure, designing and constructing low energy buildings, integrating renewable energy-related technologies, strengthening existing and developing new building codes and standards [3]. In addition, a strong impetus towards construction sustainability, in general, and green buildings, in particular, could be created by introducing rating and certification schemes to assist guide, demonstrate, and document efforts to provide environmentally and socio-economically high-performance buildings. The aim of this study is to look into such rating and certification schemes and to investigate some of the first "pioneering" buildings that undergone certification process in Kazakhstan.

2 Background

At a global scale, certification efforts were significantly advanced in the 1990–2000s with the creation of a number of voluntary rating programs that have been recognized by governments, construction industries and professionals around the world [4]. One of them is the first and most popular green building rating systems in the UK - Building Research Establishment's Environmental Assessment Method (BREEAM). BREEAM is a system of evaluating, rating and certifying a range of sustainability parameters of buildings, including their energy performance. In BREEAM, the project performance is assessed against best practices in the following categories: (1) management; (2) health and well-being; (3) energy; (4) transport; (5) water; (6) materials; (7) waste; (8) land use and ecology; (9) pollution [5].

Another widely recognized rating system, Leadership in Energy and Environmental Design (LEED), was developed by the US Green Building Council (USGBC) to set criteria for assessment of environmental performance of new and existing buildings as well as entire neighborhoods. LEED-certified buildings are considered as the ones that generate less emission compared to standards buildings, reduce energy consumptions, promote water efficiency, etc. In LEED, Performance and parameters of projects are compared against best practices in 9 main areas: (1) location and transport; (2) sustainable sites; (3) water efficiency; (4) energy and atmosphere; (5) materials and resources; (6) indoor environmental quality; (7) innovation; (8) regional priority; (9) integrative process [6].

In the last decade, many countries started to develop so-called "domestic building environmental assessment methodologies" in correlation to their climatic conditions and culture [7]. However, the global trend goes to standardization of assessment methods and tends to establish some global ranking systems such as LEED and

BREEAM. Both LEED and BREEAM certification takes into account many parameters when assessing the energy performance of the building, hence globally recognized as a leading and widely imported rating system [8]. According to Kawazu if both systems are compared by six common categories: 'Indoor Environment' category, 'Quality of Service' category, 'Energy' category, 'Resource and Materials' category, and 'Off-site Environment' category, output results show identical trend and same qualitative, comprehensive rating [9].

First building attempted to get LEED certification in Kazakhstan was KBTU institute mixed-use complex in January 2012. Until 2016 only a few buildings were attempting to get LEED certification and none of them was able to score more than 40 in order to get at least registered certification. However, after 2016 many buildings got silver and gold certification. Moreover, this trend keeps ascending.

First introduction of BREEAM certification was in 2015 and for 2 years until 2017 more than 55 buildings and complexes received Pass, Good, Very good, and Excellent rating. However, after 2017 no buildings were applied for BREEAM certification.

It can be seen, that Kazakhstan's certification trend mostly inclines towards getting LEED certification as opposed to BREEAM certification. Possible reasons for this could be the brand name since LEED is slightly more recognized in the world. Another reason is the motivation of other owners to make their certification globally accepted where LEED is more established in comparison to BREEAM. Since Kazakhstan is only at the beginning of the wide implementation of green building certification and this trend has started to expand relatively recently, LEED certification seems to be a better start.

3   Results and Discussion

This paper provides an overview of the first implementation of LEED and BREEAM certification schemes in Kazakhstan. Most of the buildings are located in two major cities of the country: Nur-Sultan, capital city, and Almaty, the largest city in Kazakhstan. The types of certified buildings can be divided into two categories: (1) commercial buildings (office buildings/business centers) and (2) residential buildings. Most of the certified buildings were built prior to or after EXPO2017 «Future Energy» exhibition. Among reasons for certification could be the desire to attract foreign companies which tend to rent "green" offices/buildings, incentives from the government to create a "green" image of the country in the lead up to the EXPO exhibition, and, finally, willingness of construction companies to increase market competitiveness. It is not clear whether the ultimate goal was to build sustainable buildings in the first place. Nevertheless, it is certain that having several dozens of certified buildings will create new trends and practices in the construction sector. The quality and performance of certified buildings might change the construction standards and norms as other companies will have examples to follow. Most of the buildings covered in this review were LEED certified. Data collection was mainly done online [10].

Fig. 1 Talan Towers

3.1 Talan Towers (Gold)

Talan Tower is a luxury structure which consists of two towers (Fig. 1). The first tower is 26-storey Ritz-Carlton hotel and the second one is 30-storey high-class office complex. Towers are bridged by a three-level podium building. Totally, building occupies 1.05 million square feet. Construction duration is 5 years. Considering the harsh climate conditions of Nur-Sultan city ranging from −35 to +40 °C, it is notable that the building uses 20% less energy compared to similar buildings in the city. LEED certification for this structure was issued in 2018. The building scored the highest score in Kazakhstan (70 out of 110) and received LEED Core & Shell 2009 Gold Rating Level. In Energy and Atmosphere category, it scored 23 out of 37. In Indoor Environmental Quality category, it scored 9 out of 12. In Sustainable Sites category it scored 23 out of 28. In Water Efficiency category, it scored 12 out of 10. In Innovation Design category, it scored 2 out of 6.

3.2 Esentai Tower, Almaty (Silver)

Esentai Tower is one of the first high-rise buildings in Almaty which is 162 m high (Fig. 2). Construction of tower was started in 2006 and successfully finished in 2008. Totally, building occupies 667,368 ft^2. LEED certification for this structure was issued in March 9, 2016. Building scored 53 out of 110 possible points and took LEED for Existing Buildings 2009 Silver Rating Level. In Energy and Atmosphere category, it scored 16 out of 35. In Indoor Environmental Quality category, it scored

Fig. 2 Esentai Tower

4 out of 15. In Sustainable Sites category, it scored 17 out of 26. In Materials and Resources category, it scored 6 out of 10. In Water Efficiency category, it scored 6 out of 14. In Innovation Design category, it scored 4 out of 6. In addition to this, in terms of Energy and Atmosphere structure has 81 Energy Star Performance Rating; in terms of Materials and Resources the structure has 40% sustainable purchasing of electric equipment, 50% reuse, recycle or compost of ongoing consumables, 75% reuse or recycle of durable goods, 90% sustainable purchasing of reduced mercury lamps, 70% diversion of waste from facility alteration and additions; in terms of Sustainable Sites, the structure has 75% reduction in conventional commuting trips and in terms of Water Efficiency, the structure has 20% reduction in indoor potable water use.

3.3　Wilo Kazakhstan, Almaty (Gold)

The Wilo Campus is a Swedish HVAC manufacturer which is the first "green" factory in Kazakhstan (Fig. 3). The Wilo campus was officially opened and started functioning in 2017. Totally, building occupies 26,177 ft^2. LEED certification for this building was issued in 2018. Building scored 63 out of 110 possible points and gained LEED Core & Shell 2009 Gold Rating Level. In the Energy and Atmosphere category it scored 19 out of 37; Indoor Environmental Quality—8 out of 12; Sustainable Sites—18 out of 28; Water—10 out of 10; Innovation Design—5 out of 6. In addition to this, in terms of Energy and Atmosphere, the building has 22% improvement. In terms of Indoor Environmental Quality, the structure has 90% of occupied space has high-quality views. In terms of Water Efficiency, the structure has 40% reduction in baseline indoor water use and 100% reduction in potable landscape water use.

Fig. 3 Wilo Kazakhstan

3.4 Commercial Building T5, Expo Village (Gold)

The main concept of EXPO2017 exhibition was "future green energy" in the context of sustainable development. The Expo City has a dozen buildings that earned LEED certification of different ranking levels (Fig. 4). The case study office building in Expo City occupies 342,052 ft^2. LEED certification for this building was issued in 2019. Building scored 61 out of 110 possible points and obtained LEED for New Construction 2009 Gold Rating Level. In the Energy and Atmosphere category it scored 17 out of 35; Indoor Environmental Quality—12 out of 15; Sustainable Sites—15 out of 26; Water Efficiency—10 out of 10; Innovation Design—5 out of 6. In addition to this, in Energy and Atmosphere category, the structure has 26% improvement; Indoor Environmental Quality—90% of occupied space has quality views; Water Efficiency—100% reduction in potable landscape water use and 40% reduction in baseline indoor water use.

Fig. 4 Expo Village (T5)

3.5 *Commercial Building T4, Expo Village (Gold)*

It is office building which totally occupies 384,559 ft^2. LEED certification for this structure was issued on January 29, 2019. Building scored 62 out of 110 possible points and took LEED for New Construction 2009 Gold Rating Level. In Energy and Atmosphere category, it scored 18 out of 35. In Materials and Resources category, it scored 2 out of 14. In Indoor Environmental Quality category, it scored 8 out of 12. In Sustainable Sites category, it scored 15 out of 26. In Water Efficiency category, it scored 10 out of 10. In Innovation Design category, it scored 5 out of 6. In addition to this, in terms of Energy and Atmosphere structure has a 28% improvement on baseline building performance rating; in terms of Indoor Environmental Quality, the structure has 90% of occupied space has quality views; in terms of Water Efficiency, the structure has 100% reduction in potable landscape water use and 40% reduction in baseline indoor water use; in terms of Materials and Resources, the structure has 75% diversion of construction and demolition debris.

3.6 *Commercial Building B11, Expo Village (Gold)*

It is office building which totally occupies 120,977 ft^2. LEED certification for this structure was issued on January 4, 2019. Building scored 60 out of 110 possible points and took LEED for New Construction 2009 Gold Rating Level. In Energy and Atmosphere category, it scored 19 out of 35. In Materials and Resources category, it scored 2 out of 14. In Indoor Environmental Quality category it scored 10 out of 15. In Sustainable Sites category it scored 15 out of 26. In Water Efficiency category it scored 9 out of 10. In Innovation Design category it scored 5 out of 6. In addition to this, in terms of Energy and Atmosphere structure has 30% improvement on baseline building performance rating; in terms of Indoor Environmental Quality, the structure has 90% of occupied space has quality views; in terms of Water Efficiency, the structure has 100% reduction in potable landscape water use and 40% reduction in baseline indoor water use; in terms of Materials and Resources, the structure has 75% diversion of construction and demolition debris.

3.7 *Residential B4-B7, Expo Village (Certified)*

It is a Multi-family residential building that totally occupies 373,194 ft^2. LEED certification for this structure was issued in 2019. Building scored 45 out of 110 possible points and obtained LEED Core & Shell 2009 Certification (Fig. 5). In the Energy and Atmosphere category it scored 12 out of 37. In Indoor Environmental Quality category it scored 4 out of 12. In Sustainable Sites category, it scored 16 out of 28. In Water Efficiency category, it scored 10 out of 10. In Innovation Design

Fig. 5 Expo Village
(B4-B7)

category it scored 3 out of 6. In addition to this, in terms of Energy and Atmosphere, the structure has 16% improvement; in terms of Indoor Environmental Quality, the structure has 90% of occupied space has quality views; in terms of Water Efficiency, the structure has 100% reduction in potable landscape water use and 40% reduction in baseline indoor water use.

3.8   Residential B1, Expo Village (Certified)

It is a Multi-family residential building that totally occupies 177,904 ft². LEED certification for this structure was issued in November 8, 2018. Building scored 47 out of 110 possible points and took LEED Core & Shell 2009 Certification. In Energy and Atmosphere category it scored 14 out of 37. In Indoor Environmental Quality category, it scored 4 out of 12. In Sustainable Sites category, it scored 16 out of 28. In Water Efficiency category, it scored 10 out of 10. In Innovation Design category, it scored 3 out of 6. In addition to this, in terms of Energy and Atmosphere structure has 20% improvement on baseline building performance rating; in terms of Indoor Environmental Quality structure has 90% of occupied space has quality views; in terms of Water Efficiency structure has 100% reduction in potable landscape water use and 40% reduction in baseline indoor water use.

3.9   Residential B12, B13, Expo Village (Certified)

It is Multi-family residential building that totally occupies 141,421 ft². LEED certification for this structure was issued in October 18, 2018. Building scored 45 out of 110 possible points and took LEED Core & Shell 2009 Certification. In Energy and Atmosphere category it scored 12 out of 37. In Indoor Environmental Quality category it scored 4 out of 12. In Sustainable Sites category, it scored 16 out of 28. In Water Efficiency category it scored 10 out of 10. In Innovation Design category it scored 3 out of 6. In addition to this, in terms of Energy and Atmosphere structure has 16% improvement on baseline building performance rating; in terms of Indoor Environmental Quality structure has 90% of occupied space has quality views; in terms of Water Efficiency structure has 100% reduction in potable landscape water use and 40% reduction in baseline indoor water use.

Except for construction projects which earned a particular LEED Rating Level there are 7 buildings that are registered. Additionally, there are 50 buildings in Kazakhstan which passed and earned BREEAM certification.

3.10   MEGA Silk Way Shopping and Entertainment Center (In-Use)

MEGA Silk Way shopping mall is a newly built BREEAM certified center of shopping and entertainment (Fig. 6). It was designed by English architects from Chapman Taylor Company. The total floor area of the building is 140,000 m^2, with a length of 500 m. and width of 160 m. The sizes and the space are quite remarkable. It has the height of 8 floors building, 9 entrances and parking for 2100 cars outdoors and 400 indoors. The building obtained BREEAM In-Use International certification. There was no available data in terms of energy, water and resources efficiency. However, the fact that the building has a "green" certification can be felt considering various parameters.

Fig. 6 MEGA Silk Way

It should be mentioned that these buildings have various features that make them more sustainable than other typical buildings of the same class. For example, public transportation has good accessibility. The certified residential buildings have very safe and pedestrian-oriented yards. Cars are allowed to park only in underground parking areas. Parks, shops and amenities are in close proximity. Various sustainable measures can be observed in and around the buildings such as, for example, solar panels, greenery, passive design elements, etc. Certified business centers and office buildings have various sustainability elements incorporated into architectural elements as well as the way indoor environments are designed. Considering the provided limited information regarding the energy and water use as well as the points the certified buildings obtained can be a strong indication that these buildings are successful in meeting sustainability requirements. In the country with construction norms and practices based on outdated Soviet-era construction standards, having LEED and BREEAM certified buildings is a strong signal that the policies, standards and norms as well as construction practices should be upgraded and made "green". The construction industry, associated governmental agencies, academics and all other related stakeholders could benefit from having certified buildings in the country to use this fact as a driver for making construction industry more sustainable. The limitation of this study can be linked to the fact that the economic aspect of obtaining certification is not considered. All these buildings are new and assumed to have considerable resources invested both from the government and affiliated organizations. Future studies should take this aspect into consideration.

4 Conclusion

This paper has found that Kazakhstan, in total, has about 50 LEED and BREEAM certified buildings. Most of them are located in Nur-Sultan and Almaty. Majority of the certified buildings are commercial or residential. The case study buildings performed very well and obtained certifications with some of them being of the highest level. The construction industry and government should use the certified buildings to upgrade the existing standards and norms to the level of "green" buildings. The financial aspect should be considered at a deeper level to understand possible barriers to certifying more buildings in the future.

References

1. Portnov BA, Trop T, Svechkina A, Ofek S, Akron S, Ghermandi A (2018) Factors affecting homebuyers' willingness to pay green building price premium: Evidence from a nationwide survey in Israel. Build Environ 137:280–291
2. Tokbolat S, Karaca F, Durdyev S, Nazipov N, Aidyngaliyev A (2018) Assessment of green practices in residential buildings: a survey-based empirical study of residents in Kazakhstan. Sustainability 10(12):4383

3. International Environmental Agency (IEA) (2016) Energy statistics of OECD countries
4. Vierra S (2014) Green building standards and certification systems. National Institute of Building Sciences
5. BRE Global Ltd. (2014) BREEM UK New Construction. Non-domestic buildings. Technical Manual. Draft 0.1
6. U.S. Green Building Council (USGBC) (2016) Profile
7. Roderick Y, McEwan D, Wheatley C, Alonso C (2009) Comparison of energy performance assessment between LEED, BREEAM and Green Star. In: Eleventh international IBPSA conference, pp 27–30
8. Cole RJ, Valdebenito MJ (2013) The importation of building environmental certification systems: international usages of BREEAM and LEED. Build Res Inform 41(6):662–676
9. Kawazu Y, Shimada N, Yokoo N, Oka T (2005) Comparison of the assessment results of BREEAM, LEED, GBTool and CASBEE. In: Proceedings of international conference on the sustainable building, pp 1700–1705 (2005)
10. The Green Building Information Gateway. GBIG (2018)

Part V

Chapter 12
Examining Undergraduate Courses Relevant to the Built Environment in the 4IR Era: A Delphi Study Approach

John Aliu, Clinton Aigbavboa, and Ayodeji Oke

Abstract The fourth industrial revolution (4IR) marks a significant period in world history as this innovative era has the potential to change the way we think and execute ideas in every area of our lives. The emerging technologies coupled with its rapid impacts prompt a swift and urgent revamp of higher education curricula worldwide. This is to provide present-day built environment students with an opportunity to comprehend the intrinsic details and applications of this latest disruptor. The aim of this research is to determine undergraduate courses that will be relevant to the built environment in the nearest future. A qualitative Delphi approach was adopted to validate these courses as institutions of higher learning prepare students for this latest wave of innovation. Fourteen experts completed a two-stage iterative Delphi study process and reached consensus on all 29 technological areas identified. This study found that concepts such as data analytics, artificial intelligence, computer programming, computer coding and data mining should be integrated into the curricula of universities to ease the transition of students from the lecture-room to the world of work. It is recommended that universities across South Africa and beyond integrate innovation-driven courses to provide an understanding of the technologies accompanying the 4IR era to not only produce graduates who comprehend these applications but also to build on these concepts to address socio-economic problems plaguing the continent.

J. Aliu (✉) · C. Aigbavboa
Cidb Centre of Excellence, Faculty of Engineering and the Built Environment, University of Johannesburg, Johannesburg, South Africa
e-mail: Ajseries77@gmail.com

C. Aigbavboa
e-mail: caigbavboa@uj.ac.za

A. Oke
SARChl in Sustainable Construction Management and Leadership in the Built Environment, University of Johannesburg, Johannesburg, South Africa
e-mail: emayok@gmail.com

© The Author(s), under exclusive license to Springer Nature Singapore Pte Ltd. 2021   195
R. J. Howlett et al. (eds.), *Emerging Research in Sustainable Energy and Buildings for a Low-Carbon Future*, Advances in Sustainability Science and Technology, https://doi.org/10.1007/978-981-15-8775-7_12

Keywords Built environment · Digitalization · Employability · Fourth industrial revolution · Skills revolution

1 Introduction

The fourth industrial revolution (4IR) has gathered momentum across the globe as it marks a new chapter in human development. This era represents a significant transformation in the way we think, work, live, relate with others and execute ideas in every area of our lives. In recent times, the era has been celebrated as a revolutionary accomplishment and has been hailed as one that will introduce algorithmic precision and automation to the decisions we make from switching on the television to predicting future behaviours. In fact, a few years ago, storing documents in the cloud sounded like fiction, not anymore! Unlike the previous industrial revolutions, the 4IR era is renowned for its vibrant, unpredictable and extraordinary technologies that have the potential for both promise and peril [27]. This disruptive era merges the physical, digital and biological worlds to solve socio-economic problems in a way that has never been imagined. The accuracy, speed, scope and complexity of this revolution have prompted nations to rethink sustainability policies, prompting organisations to seek ways to harness converging technologies to address arising global issues.

With the adoption of robotic technologies coupled with the integration of Building Information Modelling (BIM) into construction activities, the industry is captivated with the adoption of 4IR elements. The efficiency and quality output derived by adopting 4IR elements into its processes has been identified as one of the best approaches to boost the sector, which currently trails other sectors in terms of productivity [31]. The emerging 4IR technologies in the construction industry places significant pressures on educational systems globally, to develop built environment students who are capable of handling industry positions in various capacities after graduation. This study focuses on South Africa and seeks to determine the undergraduate courses that will become relevant to the built environment in the nearest future due to the advent of the fourth industrial revolution.

2 Review of Literature

2.1 Digitalisation in developing countries

While the world basks in the euphoria of the 4IR, Africa is seemingly gathering momentum for this innovative age, with South Africa as one of its leading voices. Under the leadership of President Cyril Ramaphosa, the '4IR' theme has become the leading mantra of new policy initiatives designed to move the nation forward and out of the economic crisis. One such example is the inaugural Digital Economy Summit which was held at the Gallagher Convention Centre, Johannesburg on 5th July 2019, under the theme "Advancing the African Agenda on the Fourth Industrial Revolution through Digital Transformation." During the summit, President Ramaphosa highlighted seven major priorities to be achieved through the 4IR namely "enhancing economic transformation and job creation; improving education outcomes and skills revolution and ensuring healthy nation; consolidating the social wage through reliable and quality basic services and enhancing spatial integration; human settlements and local government; advancing social cohesion and safe communities; creating a capable, ethical and developmental state; and working for better Africa and world". In achieving these priorities, the President asserts that "one million young people will be trained in data science and related skills by 2030". In realising this long-term vision, President Ramaphosa tasked the educational systems (both primary schools and universities) to revisit and revamp their curricula in what he referred to as "skills revolution". In achieving this, the President further stated that "…we are introducing subjects such as coding and data science and analytics at the primary school level to prepare our young people for the jobs of the future"… [54]. Subsequently, the President established a 30-member "Presidential Commission on the Fourth Industrial Revolution" to advise the government on leveraging opportunities presented by 4IR. Professor Tshilidzi Marwala who is an internationally recognised scholar in the artificial intelligence aspect of 4IR and the Vice-Chancellor of the University of Johannesburg was appointed as the commission's deputy chair.

This Presidential mandate aligns with the "Building a Capable 4IR Army" capacity development programme which was announced by the nation's Communications Minister Stella Tembisa Ndabeni-Abrahams last year to digitally equip communities to foster economic competitiveness across the nation. The Minister states that is "our aim is to take a completely different direction and government's programmes will enable people to program the machines, including making them". These initiatives by the South African government highlights its readiness to become a major player in the 4IR era. However, in achieving this 2030 mandate as stated by President Ramaphosa, the role of education cannot be overstated as this was the second of the seven listed priorities to be addressed by the 4IR which is "improving education outcomes and skills revolution and ensuring healthy nation". While President Ramaphosa continues to advocate for South Africa's 4IR march, the previous President, Jacob Zuma was also at the forefront of this innovative discussion. In fact, between 2016 and 2019, the terms 'fourth industrial revolution' and '4IR' were frequently used in government speeches and initiatives. According to Mail & Guardian Data Desk [45], the top 12 words that were frequently used in the same sentence as 4IR are—education, people, economy, government, need, new, skills, digital, country, development, technology and world. This illustrates how South Africa continues to be one of the leading voices on the continent regarding 4IR discussions. This resonates with President Ramphosa's mandate of proposing "new subjects to be introduced at schools to prepare students for 4IR careers which includes data science and analytics, artificial intelligence (AI), blockchain, additive manufacturing, robotics and quantum computing". It is against this backdrop that this article highlights the 4IR courses that will become relevant to the built environment in the nearest future.

The objectives of this study align with South Africa's long-term National Development Plan (NDP) which seeks to develop the skills, knowledge and capabilities of its citizens through the provision of quality education across all levels by the year 2030. This mandate also resonates with the Sustainable Development Goals (SDGs) proposed by the United Nations which plans to ensure quality and inclusive education for all by 2030. In achieving the objectives, a qualitative Delphi approach was adopted to validate these courses as universities across South Africa and beyond prepare students for this latest wave of innovation.

2.2 *The Introduction of 4IR to Present-Day Curricula*

Globally, the quest to develop the future workforce with 4IR knowledge is well and truly underway as educational sectors seek ways to expand interdisciplinary knowledge among students. This can be achieved by introducing 4IR elements into existing learning content or as stand-alone modules.

The main aim of adopting and adapting 4IR components into the built environment space is to improve the existing curricula and pedagogy which will result in developing graduates who are competent to handle the drastic changes that accompany advance digitalisation[5, 6]. From the main features of 4IR which includes big data analysis, artificial intelligence, cloud computing, augmented reality, industrial internet amongst many others, several literatures have suggested possible educational courses that will become relevant to the built environment in present-day terms or in the nearest future. These can be found in Table 1. It is pertinent to note that 4IR technologies are numerous and exceeds those listed in Table 1. However, for the sake of this study, these 29 concepts were considered.

3   Research methodology

3.1 *The Delphi Process*

For this research study, a qualitative Delphi approach was conducted to validate the 4IR courses identified from literature. The Delphi technique gained popularity during the beginning of the Cold War in the 1950s. During the period of geopolitical tension between the Soviet Union and its satellite states, the Delphi technique was used as a forecasting tool to ascertain the impact of warfare on enemy territory [10]. Historically, the word "*Delphi*" is coined from the term "*Oracle of Delphi*" - an ancient Greek temple where the oracle was situated. The Greek mythology states that the "*Oracle of Delphi*" was frequently consulted to predict the outcomes of

Table 1 Review of undergraduate courses

Undergraduate courses/technologies	Literature sources
Principles of Robotics	Boden et al. [9]
Networking and its applications	Hermann et al. [27], Hernández-Muñoz et al. [28]
Data analytics	Wamba et al. [71]
Cybersecurity	Blazek et al. [8]
Artificial intelligence courses	Hermann et al. [27]
Computer programming	Chookaew et al. [13]
Computer coding	Nager and Atkinson [48]
Data mining courses	Qi et al. [55]
Computational knowledge	Beg et al. [7]
Concepts of Algorithms	Lateef and Hota [39]
Database analytics	Sellers-Blais [60]
Database systems design	Hermann et al. [27]
Cloud-based technology	Sharma et al. [61]
Cloud computing	Gbadegesin et al. [19]
Technology innovation processes	Audretsch et al. [4]
Information Technology modules	Hermann et al. [27], Hernández-Muñoz et al. [28]
Principles of Automation	Neelakandan et al. [49]
Emotional intelligence modules	Jameson et al. [36]
Critical thinking courses	Conrad and Newberry [14]
Multidisciplinary collaboration	Hesjedal et al. [29]
Multidisciplinary collaboration Management	Hesjedal et al. [29]
Blockchain concept and application modules	Hughes et al. [33]
Cryptocurrency understanding module	Hughes et al. [33]
Mechatronics engineering	Mutambara [47]
Photonics and Quantum Materials	Xu et al. [73]
Biotechnology courses	Moo-Young [46]
Energy Systems courses	Gazafroudi et al. [18]
Digital control systems	Hermann et al. [27]
3D printing and application	Hermann et al. [27]

Source Author's compilation

battles which helped troops to perfect their strategies before embarking on warfare. This technique allowed experts to offer their opinions on probabilities and the likelihoods of possible enemy surprise attacks while planning. The aim of this predictive process was to achieve a consensus of ideas after several rounds of expert deliberations (iterations). Based on the successful application of this format, the Delphi technique gained prominence and has since been introduced into academic research. In recent times, it has become a qualitative methodology that seeks to get the opinions of experts through multi-round surveys with the aim of reaching a consensus. According to [24], the Delphi technique is designed to get experts' opinions on matters that are still gaining recognition, such as the objectives of this study. In a typical Delphi approach, experts respond to questions and submit their responses to the researcher who collates and analyses the information to determine the central and extreme tendencies [21]. For this study, the experts were anonymous throughout the Delphi process and were allowed to resubmit their opinions as many times as possible until consensus was reached. Similar to how it was used during the Cold War, the Delphi process helps to make predictions and projections concerning a specific subject matter. In this study, due to the advent of the fourth industrial revolution, the pressures on universities to integrate technological-driven courses into their curricula have gained momentum, hence this research predicts the modules that will become relevant to the built environment in the nearest future. Over time, the Delphi method has become a systematic thinking tool and has gained recognition from experts across several disciplines for generating credible outcomes and provoke relevant discussions [44]. This research is expected to stimulate extensive studies on graduate employability (Employability 4.0) in South Africa and beyond, hence, the Delphi method is relevant to the realisation of the study's objectives.

One of the advantages of the Delphi technique is that it stems from the constructivist paradigm by cutting across both the quantitative and qualitative methods of data collection and analysis [20, 59]. This allows for the results and conclusions of the research to be generally represented to the wider population [64]. More so, the opinions of experts make the Delphi method a robust and credible one as professional opinions are delivered which ensures the reliability of the study [23]. The strength of the Delphi method also lies in the various rounds (iterations) which helps in initial feedback and dissemination of subsequent rounds for further review and opinions [65]. From the foregoing discussions, a Delphi technique was conducted to obtain the opinions of experts on the possible 4IR courses that will become relevant to the built environment in the nearest future. For this study, two rounds of the Delphi process were conducted before experts reached a consensus on the various courses. Generally, the strength of qualitative research such as the Delphi technique is based on the methodological rigour in conducting such a process [41]. For this study, a detailed procedure was adopted as shown in Fig. 1.

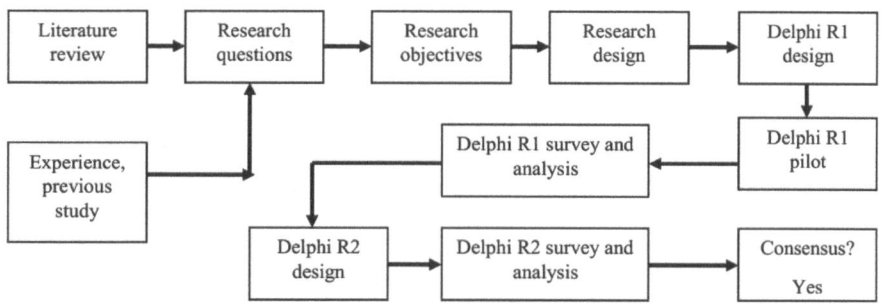

Fig. 1 Delphi design for this study (Author's compilation)

3.2 Conducting the Delphi study

Having established the roles of the Delphi process in obtaining the opinions of experts through multi-round surveys with the aim of reaching a consensus, it is pertinent to state its critical features which include the anonymity of experts; statistical responses obtainable and repetitive process with controlled responses [43]. The Delphi method is also an interactive research technique used to obtain experts' judgement on a specific subject matter [25]. Through the series of rounds, experts are allowed to modify their responses, maintain their opinions or agree with the rankings of the previous round. Hence, the aim of the Delphi process is to eliminate the variability of responses to reach a consensus on a particular issue or question [25, 32]. For this study, the aim was to attain consensus on the possible 4IR courses that will become relevant to the built environment in the nearest future to ensure the employability of built environment graduates. While achieving consensus is the main aim of every Delphi study, the choices of experts play a pivotal role in the success of the research [26]. It is also critical to note that during the Delphi process, more attention should be given to the group responses (convergence of ideas) rather than the individual response. To this end, several parameters (median and interquartile deviation (IQD)) were adopted to measure consistencies, central tendencies and hence, determine consensus.

3.3 Selection of Delphi Experts for the Study

According to Okoli and Pawlowski [50], Delphi experts refer to a panel of knowledgeable individuals who are selected for a specific study based on certain criteria (criterion sampling). Based on the fact that the quality of opinions depends on experts' knowledgeability, this research presents the demographic characteristics of the selected panel to portray their knowledge and expertise to justify their selection. Over time, several researchers have provided several criteria in deciding who should be called an expert. Most of the studies suggest that to be called an expert, an individual should possess some of the following including understanding the subject matter in question,author of publications relating to the subject in question; attendance of conference and workshop relating to the subject in question; willingness to participate in the study at that point in time; willingness to partake in several rounds when required; good communicators to avoid a slow process; employed with an accredited institution of higher education; satisfactory number of years of experience in the construction industry; member of an academic committee; possess an academic degree in a related discipline; registered with an accredited professional body and several others, [1, 25, 57, 69]. However, across various studies, it has been agreed that the researcher has the autonomy to decide who should be called an expert based on any of the above-listed criteria or as stated by the researcher as it suits the study. For this study, an expert was required to satisfy 3 or more of the following criteria: possess at least a Bachelor's degree in a built environment discipline,employed with either a construction firm or an accredited university in South Africa; possess at least 5 years working experience within a construction firm or an accredited university in South Africa; registered with an accredited professional body and an author of publications relating to the built environment as shown in Table 2. From Table 2, the 14 experts who completed the Delphi process satisfied 3 more of the stated criteria.

3.4 Determination of the Delphi Panel Size

Due to the fact that the Delphi approach is more qualitative than quantitative, fewer number of participants or experts are required for the process. Over the years, numerous researchers have recommended several sample sizes required for a Delphi process. Dalkey and Helmer [16] and Linstone and Turoff [42] both suggest 7 experts,Cavalli-Sforza and Ortolano [12] recommend 8–12 experts,Phillips [53] states between 7 and 12 experts while Andranovich [3] and Skulmoski et al. [63] propose 10–15 participants if the panel of experts are of a homogeneous sample and possess similar backgrounds. Due to time constraints and an unpredictable schedule of experts, a small sample (14 experts from both the construction industry and academics) was considered for this research.

The fourteen (14) experts agreed to participate in the Delphi study after they received a comprehensive description (requirements or instructions) of the process. Before the study began, the selected experts were requested to forward their curriculum vitae (CV) to ascertain if they meet the qualification threshold for this study. Thereafter, they received the first-round questionnaire survey which comprised both open-ended and closed questions with the option of rankings and stating their opinions. Both the first-round and second-round Delphi questionnaires can be found in Appendices 2 and 3. For this study, experts were required to possess 3 or more of the outlined eligibility criteria. As shown in Table 2, all 14 experts satisfied the criteria. 5 experts met all the parameters, 8 experts satisfied 4 of the criteria while 1 expert met 3 of the criteria.

Table 2 Assessment of Delphi expert qualifications

S/N	Eligibility criteria for Experts	E1	E2	E3	E4	E5	E6	E7	E8	E9	E10	E11	E12	E13	E14
1	Possess at least a Bachelor's degree	X	X	X	X	X	X	X	X	X	X	X	X	X	X
2	Currently employed with a tertiary institution or professional in the construction industry	X	X		X			X	X	X	X	X		X	X
3	At least 5 years of working experience with a tertiary institution or construction industry	X	X	X	X	X	X	X	X	X	X	X	X	X	X
4	Affiliated with professional bodies	X	X	X	X	X	X	X		X	X	X	X	X	
5	Author or co-author of a peer-reviewed publication	X	X	X	X		X	X	X		X		X		X
	Total	5	5	4	5	3	4	5	4	4	5	4	4	4	4

Source Author's compilation

3.5 Delphi Experts' Information

As stated earlier, one of the criteria to be an expert was to possess at least a Bachelor's degree in a discipline within the built environment. Following an analysis of the experts' CV, all 14 experts possessed at least a Bachelor's degree as shown in Table 3.

Based on their academic qualifications, 5 of the experts possessed a Doctorate; 7 of the experts had a Master's Degree while 2 of the experts possessed a Bachelor's Degree. This shows that around 86 percent of the experts possessed post-graduate degrees which highlight the credibility of this study. More so, these high academic qualifications of the experts boost the quality and reputability of this study's Delphi process. Secondly, the experts were required to possess a built environment background. As seen in Table 4, the 14 experts represent most of the various disciplines in the built environment.

Based on their academic qualifications, 2 of the experts possessed a background in Architecture; 1 expert was from Quantity Surveying; 2 of the experts where from Construction Project Management; while 9 of the experts were of Engineering background (Civil, Mechanical and Electrical). Further analysis of experts' CV showed that 8 of the experts were from higher institutions in South Africa while 6 experts were from the construction industry. Experts were also required to have at least 5 years of working experience with an academic institution or the construction industry. As seen from Table 5, the 14 experts all possessed a significant number of work experience in their spheres of influence.

Based on their years of experience 1 expert had 5 years of experience; 7 had 6–10 years of experience; 3 had 11–20 years of experience; 2 had 21–30 years of experience and 1 expert had above 31 years of working experience. Additionally, experts were required to be members of recognised and accredited professional

Table 3 Panel of experts' qualifications

Highest qualification	Number of experts
Doctor of Philosophy (Ph.D.)	5
Master's Degree (M.Sc. and M.Eng.)	7
Bachelor's Degree (B.eng.)	2
Total	14

Table 4 Panel of experts' field of specialisation

Field of specialisation	Number of experts
Architecture	2
Quantity surveying	1
Construction project management	2
Engineering (civil, mechanical, electrical)	9
Total	14

Table 5 Panel of experts' years of experience

Years of experience	Number of experts
5	1
6–10	7
11–20	3
21–30	2
Over 31 years	1
Total	14

bodies. Hence, 5 of the experts were registered with the Engineering Council of South Africa (ECSA), 2 were registered with the South African Council for the Project and Construction Management Professions (SACPCMP), 2 were registered with the South African Institution of Civil Engineering (SAICE) and 1 was registered with Project Management South Africa (PMSA). Finally, most of the authors were authors and co-authors of peer-reviewed publications within the built environment disciplines and have presented at academic conferences, webinars and seminars.

3.6 Determining consensus from the Delphi Study

As stated earlier, the main aim of the Delphi study is to attain consensus among the selected experts. Over time, many researchers have established several parameters for reaching consensus. According to Holey et al. [30], a consensus is achieved when there is a convergence (agreements) of opinions among experts. [56] suggest that consensus is achieved by recording the median and standard deviations (SD) values where a decrease in SD between rounds indicate higher levels of agreement among experts. In addition, [56] propose that for consensus to be reached, the inter-quartile deviation (IQD) should be less than or equal to 1, signifying that over 60% of the experts were largely positive or largely negative in their responses towards a specific issue. For this study, a consensus was achieved when the IQD = 0.00 or ≤1. The consensus scales adopted for this study is shown in Table 6.

Table 6 Consensus scales for this study

S/N	Consensus strength	Median	Mean	Interquartile deviation (IQD)
1	Strong	9–10	8–10	≤1 and ≥ 80% (8–10)
2	Good	7–8.99	6–7.99	≥1.1 ≤ 2 and ≥ 60% ≤ 79% (6–7.99)
3	Weak	≤6.99	≤5.99	≥2.1 ≤ 3 and ≤ 59% (5.99)

3.7   *Reliability and Validity of the Delphi Study*

The reliability and validity of the Delphi study deal with the rigorousness of the methodological process, expert panel determination, panel size determination and consensus determination [15, 74]. One of the ways to ensure the reliability of the Delphi study for this research was to ensure the clarity of the process to the 14 experts involved. In achieving this, detailed instructions were included in both rounds of the Delphi questionnaire to eliminate doubts and ambiguity (as shown in Appendix section). By selecting experts from a similar background (built environment), reliability was also ensured. To obtain a balance of opinions, experts selected for this study were obtained from both academics and the construction industry. The study also adhered strictly to the qualification criteria to further enhance the reliability of the Delphi study. The validity of this research was enhanced by ensuring the anonymity of experts from each other to eradicate the 'bandwagon' effect - tendency to make decisions and give opinions based on the choices of others [2]. The experts were provided with the option of willing participation to further enhance the internal validity of the Delphi study. This research also adopted multiple rounds which allowed experts to make modifications and suggestions where necessary and providing reasons for their additions, hence boosting the internal validity of the study.

4   Presentation of Findings

4.1   *Delphi Round 1*

The aim of this research is to determine the undergraduate courses that will become relevant to the built environment in the nearest future due to the advent of the fourth industrial revolution. Ratings were achieved using a 10-point Likert scale of 'no significance', 'low significance', 'medium significance', 'high significance' and 'very high significance'. While 'very high significance' had the highest weighting (9 and 10), 'no significance' was assigned the lowest weighting (1 and 2). Twenty-nine (29) undergraduate courses were identified from literature and government reports as shown in Table 1. More so, 10 of the courses - data analytics, critical thinking courses, technology innovation processes, concepts of algorithms, computational analysis, data mining courses, computer coding, computer programming, artificial intelligence (AI) courses and data analytics were highly selected by the experts based on the median score of 9.0 as shown in Fig. 2. From the 10 courses, only technology innovation processes and critical thinking courses achieved consensus based on IQD scores of 1.0 each. Likewise, principles of robotics, networking and its applications, cybersecurity studies, database analytics, database systems design, cloud-based technology, cloud computing, information technology modules, principles of automation, emotional intelligence modules, multidisciplinary collaboration modules, multidisciplinary collaboration management, blockchain concept

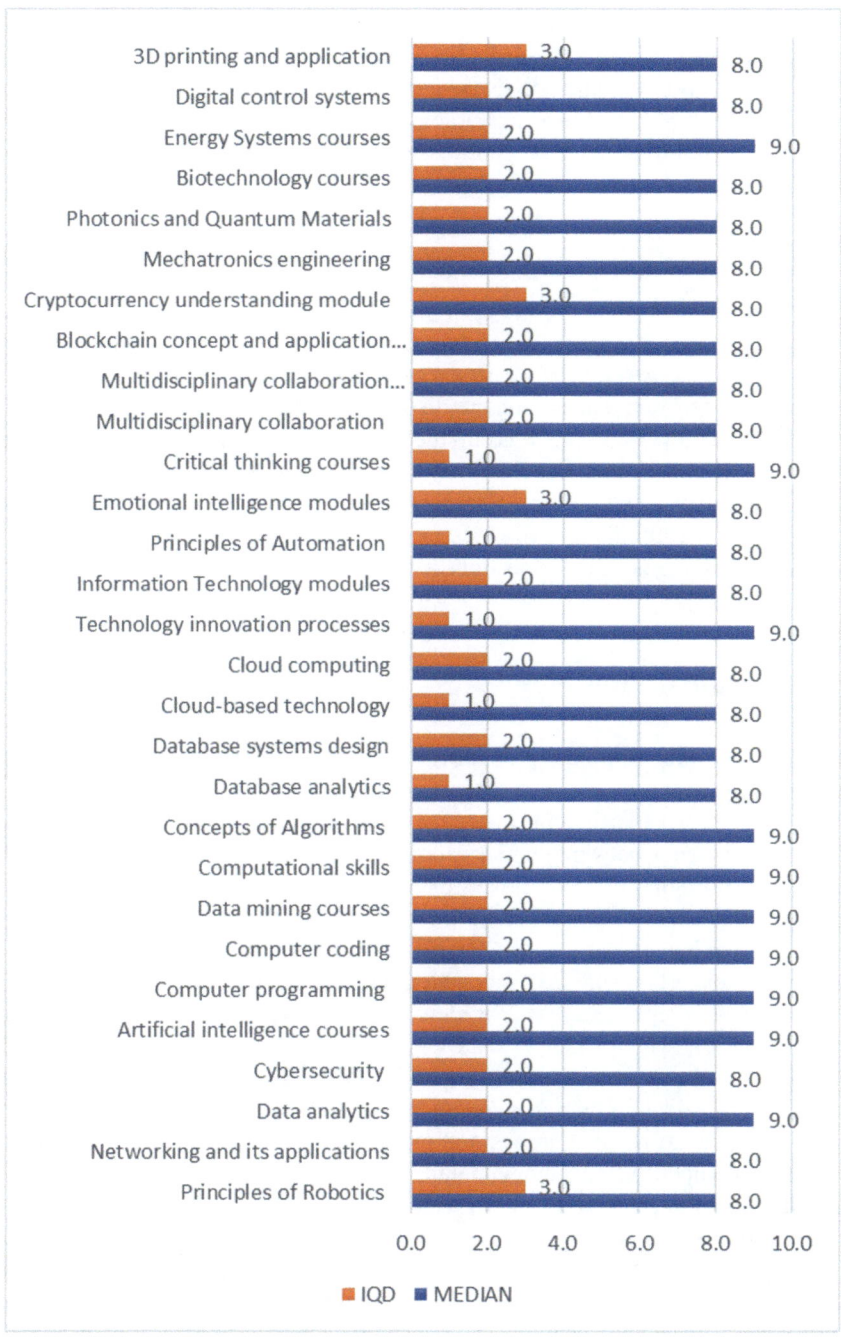

Fig. 2 List of undergraduate 4IR courses

and application modules, cryptocurrency understanding module, mechatronics engineering, photonics and quantum materials, biotechnology courses, digital control systems and 3D printing and application recorded median scores of 8.0 each. Only database analytics, cloud-based technology and principles of automation recorded an IQD score of 1.0 as the rest recorded IQD scores above 1.0. More so, no new courses were introduced by the experts.

4.2 Delphi Round 2

Shortly after the completion of the first round, the researcher analysed the data and sent out the second-round questionnaire. In this round, the experts were required to do one of the following—accept the group media value by returning the questionnaires back to the researcher without making any changes; maintain their original responses and choosing a new response (and providing explanations for making those changes). Table 7 presents a summary of the second-round responses with the median, mean, standard deviation and IQD scores duly recorded.

5 Discussion of Findings

After the successful completion of the second-round Delphi questionnaire, it was found that all 29 undergraduate courses achieved good consensus. Hence, there was no need for a third or fourth round process. Findings emanating from the Delphi study suggest that the courses which underwent both rounds resonate with what has been proposed by the South African government during recent 4IR-driven initiatives as well as several existing literature [17, 28, 38, 52, 70]. As observed from Table 6, all 29 courses had median scores between 8.0 and 9.0, indicating a very high significance rating by the experts. Data analytics which had a median value of 9.0 and a mean value of 8.36 was highlighted by President Ramaphosa during the inaugural Digital Economy Summit which was held at the Gallagher Convention Centre, Johannesburg on 5th July 2019. According to Wang et al. [72], data analytics refers to the science of analyzing raw data for human consumption. In the world today, organisations rely on data analytics to monitor trends and make better business decisions, providing them with a competitive advantage in an ever-changing world [71].

Artificial intelligence (AI) courses with a median value of 9.0 and a mean value of 8.14 were also ranked highly by experts and its importance in the curricula of educational institutions cannot be overstated. Like data analytics, the applications of AI across several fields and industries today are astonishing [22]. Known as the intelligence displayed by machines, AI is currently being employed globally to tackle economic and social challenges in fields such as agriculture, aviation, education, finance, healthcare, marketing, military, music, sensors, and transportation amongst others. It is against this backdrop, that the South African government is working round

Table 7 Future undergraduate courses

Future undergraduate courses	Median	Mean ($\bar{x}$)	SD (σX)	IQD
Principles of robotics	8.0	7.43	2.44	0.00
Networking and its applications	8.0	7.57	1.55	0.00
Data analytics	9.0	8.36	2.10	0.00
Cybersecurity	8.0	7.43	2.17	0.00
Artificial intelligence courses	9.0	8.14	2.63	0.00
Computer programming	9.0	8.43	1.95	0.00
Computer coding	9.0	8.29	2.27	0.00
Data mining courses	9.0	8.43	1.95	0.00
Computational knowledge	9.0	8.50	1.83	0.00
Concepts of algorithms	9.0	8.50	1.99	0.00
Database analytics	8.0	7.64	1.82	0.00
Database systems design	8.0	7.64	1.82	0.00
Cloud-based technology	8.0	7.57	2.06	0.00
Cloud computing	8.0	7.64	1.91	0.00
Technology innovation processes	9.0	8.50	1.29	0.00
Information technology modules	8.0	7.64	1.50	0.00
Principles of automation	8.0	7.57	1.91	0.00
Emotional intelligence modules	8.0	7.57	1.91	0.00
Critical thinking courses	9.0	8.57	1.34	0.00
Multidisciplinary collaboration	8.0	7.86	0.86	0.00
Multidisciplinary collaboration management	8.0	7.64	1.39	0.00
Blockchain concept and application modules	8.0	7.64	1.50	0.00
Cryptocurrency understanding module	8.0	7.21	2.26	0.00
Mechatronics engineering	8.0	7.36	2.13	0.00
Photonics and quantum materials	8.0	7.29	2.30	0.00
Biotechnology courses	8.0	8.07	1.07	0.00
Energy systems courses	9.0	8.86	0.66	0.00
Digital control systems	8.0	7.79	1.19	0.00
3D printing and application	8.0	7.79	1.25	0.00

the clock to ensure AI courses are introduced into the educational curricula to develop graduates who can utilise this knowledge for the greater good and benefit of the nation and beyond [54]. Computer programming with a median value of 9.0 and a mean value of 8.43 was also ranked significantly which resonates with the works of [13]. Through computer programming knowledge, students are offered the opportunity to learn database programs, software applications, image editing tools, word processors and several programming packages and languages to solve computer tasks. Like AI, computer programming is the heartbeat of several industries globally as they continue to develop products and services based on written computer programs. Consensus was also achieved in computer coding courses with a median value of 9.0 and a mean value of 8.29. This is another important 4IR course that provides knowledge of cryptography, data transmission, and a broader understanding of how computers generally operate. Like AI, coding is used almost every time because computer devices, cars and gadgets we use today require codes to function effectively [48]. Hence, its inclusion into the educational curricula is paramount.

Another course highly ranked is Data mining which has a median value of 9.0 and a mean value of 8.43. The use of data mining in modern businesses is one of the strong motivation for its inclusion into present-day educational curricula. Businesses of today use data mining to analyse large volume of data to obtain patterns of certain information that can improve productivity and efficiency, thus gaining competitive advantage. In recent times, several industries such as healthcare, engineering, banking and criminal investigation have adopted data mining to improve their service delivery [55]. For example, data mining has been used in the educational sector to predict future learning behaviour and performance outcomes of students [35, 62, 66]. It is based on those data mining applications that even the South African government has proposed its inclusion into the educational curricula. Computational knowledge was another course that gained high rankings with a median value of 9.0 and a mean value of 8.50. According to Thieret [67], computational knowledge provides information on data interpretation for the purpose of monitoring trends, testing relationships and hypotheses and developing models to solve complex problems in various fields [7]. Like computational knowledge, understanding the concepts of algorithm was another critical course ranked highly by the experts with a median value of 9.0 and a mean value of 8.50. In modern-day, understanding algorithm is simply the key to solving complex problems and processing data [39]. In recent times, young students around the continent have used algorithms for crazy inventions such as aeroplanes, generators and appliances that run on unorthodox fuels. Hence, the introduction of algorithm into the lecture room will stimulate problem-solving tendencies among students as Africa seeks solutions to its increasing problems [58].

Technology innovation processes with a median value of 9.0 and also a mean value of 8.50 was ranked highly significant by experts and understandably why. This area of 4IR knowledge deals with understanding how computer tools and devices interact during complex problem-solving [4]. Due to the importance of critical thinking skills in graduate employability and technologies that accompany the fourth industrial revolution, the inclusion of critical thinking courses into the educational curriculum is necessary. Experts for the study agreed with this assertion, hence, critical thinking courses were ranked with a median value of 9.0 and also a mean value of 8.57. These set of skills allows students to intellectually conceptualise, analyse, evaluate and solve problems by observing, reflecting and reasoning [51]. Through traditional teaching and academic exercises, educators play a key role in developing these set of skills among students. Finally, energy system courses were also highly ranked with a median value of 9.0 and a mean value of 8.86. Considering the way technology is changing the forms of energy production globally, it is necessary for students to understand the societal and political implications of these innovations [68]. With numerous challenges facing the energy sector on the continent, equipping young learners on energy systems is highly critical as African nations continue to seek renewable energies for a sustainable future.

Furthermore, there exists a strong consensus among experts on the remaining undergraduate courses with median values of 8. These include principles of robotics, networking and its applications, cybersecurity, database analytics, database systems design, cloud-based technology, cloud computing, information technology modules, principles of automation, emotional intelligence modules and multidisciplinary collaboration. Others include multidisciplinary collaboration management, blockchain concept and application modules, cryptocurrency understanding module, mechatronics engineering, photonics and quantum materials, biotechnology courses, digital control systems, 3D printing and application. These outlined 4IR courses were resonated in existing literature such as [28, 34, 40]. It is expected that these modules will develop young learners to not only understand the emerging 4IR technologies but also to apply them in addressing socio-economic problems facing the continent and beyond, especially as it relates to engineering, construction and architectural industry.

6 Conclusion

The fourth industrial revolution has gathered momentum across the globe and South Africa has continued to outline its vision to lead this march on the African continent. This innovative era represents a significant transformation as the interaction between humans and machines takes on a radical dimension. In recent times, the South African government has embarked on several initiatives across the nation, which highlights its readiness to become a major player in the 4IR era. One such initiative was the inaugural Digital Economy Summit which was held at the Gallagher Convention Centre, Johannesburg on 5th July 2019, where President Ramaphosa discussed the nation's commitment to skills and technology revolution that will boost the nation's human capital required in the 4IR era. During the summit, a live 3D hologram of the President was transmitted to the Rustenburg Civic Centre in the North West, showcasing some of the components of the 4IR elements. These emerging technologies have increased the pressure on educational institutions to introduce technology-driven courses to equip students with the required knowledge need to thrive with the advent of the latest disruptor. In achieving the objectives of this study, a qualitative Delphi approach was adopted to validate these courses as universities across South Africa and beyond take measures in preparing students for this latest wave of innovation. Built environment experts who satisfied the listed criteria for participation were drawn from both academics and industry practice. After two rounds of the Delphi process, a very high consensus was reached on all 29 4IR courses identified with an IQD value of 0.00. The study, therefore, achieved its stated objectives of validating these courses among experts after two Delphi iterations. One of the implications of this study is that the various 4IR initiatives proposed by the South African government signal its intention of embarking on digitalisation as a nation. This places significant pressures on the educational sector to revamp its curricula to accommodate several 4IR components as recommended by President Ramaphosa. To this end, the President inaugurated a 30-member committee that have been tasked with leveraging opportunities presented by the fourth industrial revolution to solve contemporary issues. AI scholar, Professor Tshilidzi Marwala, Vice-Chancellor of the University of Johannesburg was appointed as the deputy chair of the committee. Therefore, one of the recommendations for this study is that educational institutions should employ more innovative ways of teaching 4IR components. As earlier discussed, the 4IR era combines both physical and virtual reality, hence, students are to be immersed in virtual learning environments to fully grasp the concept of the innovativeness surrounding the fourth industrial revolution. While the nation is at the forefront of the 4IR discussion on the continent, it is necessary to note that this disruptive era is set to trigger drastic changes to the job market as AI and automation becomes the order of the day. According to recent statistics from McKinsey Global Institute, AI and automation are set to displace around one-fifth of the global workforce. Alarmingly, robots are projected to replace close to a billion workers by 2030 (McKinsey Global Institute, 2019)! Therefore, this age of digitalisation and automation calls for graduates who are digitally-savvy and multi-skilled to handle job positions in various sectors after graduation. Hence, follow-up research should be conducted to predict the various employability skills (Employability 4.0) that employers will seek in the nearest future due to the advent of the fourth industrial revolution.

References

1. Adler M, Ziglio E (1996) Gazing into the oracle: The Delphi method and its application to social policy and public health. Kingsley Publishers, London
2. Aigbavboa C (2013) An integrated beneficiary centred satisfaction model for publicly funded housing schemes in South Africa (Ph.D.). University of Johannesburg, Johannesburg
3. Andranovich G (1995) Developing community participation and consensus: The Delphi technique
4. Audretsch DB, Link AN, Scott JT (2019) Public/private technology partnerships: evaluating SBIR-supported research. In: The social value of new technology. Edward Elgar Publishing
5. Baena F, Guarin A, Mora J, Sauza J, Retat S (2017) Learning factory: the path to industry 4.0. Procedia Manuf 9:73–80
6. Baygin M, Yetis H, Karakose M, Akin E (2016) An effect analysis of industry 4.0 to higher education. In: 2016 15th international conference on information technology based higher education and training (ITHET). IEEE, pp 1–4
7. Beg M, Pepper RA, Fangohr H (2017) User interfaces for computational science: a domain specific language for OOMMF embedded in Python. AIP Adv 7(5):056025
8. Blazek P, Fujdiak R, Mlynek P, Misurec J (2019) Development of cyber-physical security testbed based on IEC 61850 architecture. Elektronika Ir Elektrotechnika 25(5):82–87
9. Boden M, Bryson J, Caldwell D, Dautenhahn K, Edwards L, Kember S, Newman P, Parry V, Pegman G, Rodden T, Sorrell T (2017) Principles of robotics: regulating robots in the real world. Connect Sci 29(2):124–129
10. Buckley C (1995) Delphi: a methodology for preferences more than predictions. Library Manage 16(7):16–19

11. Buhler D, Kuchlin W, Grubler G, Nusser G (2000) The virtual automation lab-web based teaching of automation engineering concepts. In: Proceedings of the seventh IEEE international conference and workshop on the engineering of computer-based systems (ECBS 2000), Edinburgh, UK, 3–7 April 2000, pp 156–164
12. Cavalli-Sforza V, Ortolano L (1984) Delphi forecasts of land use—transportation interactions. J Transport Eng 110(3):324–339
13. Chookaew S, Wanichsan D, Hwang GJ, Panjaburee P (2015) Effects of a personalised ubiquitous learning support system on university students' learning performance and attitudes in computer-programming courses. Int J Mob Learn Organisation 9(3):240–257
14. Conrad D, Newberry R (2012) Identification and instruction of important business communication skills for graduate business education. J Educ Bus 87(2):112–120
15. Creswell JW (2014) Research design. Qualitative, quantitative and mixed methods approaches, 4th edn. Sage Publications, Lincoln
16. Dalkey NC, Helmer O (1963) An experimental application of the Delphi method to the use of experts. J Inst Manage Sci 9:458–467
17. Eder A (2015) Acceptance of educational technologies in industrial-technical vocational training against the background of Industry 4.0. J Tech Educ JOTED 3(2):19–44
18. Gazafroudi AS, Soares J, Ghazvini MAF, Pinto T, Vale Z, Corchado JM (2019) Stochastic interval-based optimal offering model for residential energy management systems by household owners. Int J Electr Power Energy Syst 105:201–219
19. Gbadegesin A, Batoukov R, Reed DR, Microsoft Corp. (2013) Flexible scalable application authorization for cloud computing environments. U.S. Patent 8,418,222
20. Green RA (2014) The Delphi technique in educational research. SAGE Open 4(2):1–9. https://doi.org/10.1177/2158244014529773
21. Grisham T (2008) Cross cultural leadership. Doctor of Project Management, School of Property, Construction and Project Management, RMIT, Melbourne
22. Gupta R, Gupta R (2016) ABC of internet of things: advancements, benefits, challenges, enablers and facilities of IoT. In: 2016 symposium on colossal data analysis and networking (CDAN). IEEE, pp 1–5
23. Habibi A, Sarafrazi A, Izadyar S (2014) Delphi technique theoretical framework in qualitative research. Int J Eng Sci. 3(4):8–13
24. Häder M, Häder S (1995) Delphi and cognitive psychology: an approach to the theoretical foundation of the Delphi method. ZUMA News 19(37):8–34
25. Hallowell M, Gambatese J (2010) Qualitative research: application of the Delphi method to CEM research. J Construct Eng Manage 136(Special Issue: Research Methodologies in Construction Engineering and Management):99–107
26. Hasson F, Keeney S, McKenna H (2000) Research guidelines for the Delphi survey technique. J Adv Nurs 32(4):1008–1015
27. Hermann M, Pentek T, Otto B (2016) Design principles for Industrie 4.0 scenarios. In: 49th Hawaii international conference on system sciences (HICSS). IEEE, pp 3928–3937
28. Hernández-Muñoz GM, Habib-Mireles L, García-Castillo FA, Montemayor-Ibarra F (2019) Industry 4.0 and engineering education: an analysis of nine technological pillars inclusion in higher educational curriculum. In: Best practices in manufacturing processes. Springer, Cham, pp 525–543
29. Hesjedal E, Hetland H, Iversen AC (2015) Interprofessional collaboration: self-reported successful collaboration by teachers and social workers in multidisciplinary teams. Child Fam Soc Work 20(4):437–445
30. Holey EA, Feeley JL, Dixon J, Whittaker VJ (2007) An exploration of the use of simple statistics to measure consensus and stability in Delphi studies. BMC Med Res Methodol 7(52):1–10
31. Holt PEA, Kearney N (2015) Emerging technology in the construction industry: perceptions from construction industry professionals emerging technology in the construction industry: perceptions from. In: 122nd ASSEE annual conference & exposition
32. Hsu CC, Sandford BA (2007) The Delphi technique: making sense of consensus. Practical Assess Res Eval 12(10):1–8

33. Hughes L, Dwivedi YK, Misra SK, Rana NP, Raghavan V, Akella V (2019) Blockchain research, practice and policy: applications, benefits, limitations, emerging research themes and research agenda. Int J Inf Manage 49:114–129

34. Irfan MM, Rajamallaiah A, Ahmad SM (2018) Paradigm shift in the engineering curriculum: design thinking. J Eng Educ Transform

35. Jalota C, Agrawal R (2019) Analysis of educational data mining using classification. In: 2019 international conference on machine learning, big data, cloud and parallel computing (COMITCon). IEEE, pp 243–247

36. Jameson A, Carthy A, McGuinness C, McSweeney F (2016) Emotional intelligence and graduates–employers' perspectives. Procedia-Social Behav Sci 228:515–522

37. Kagermann H, Helbig J, Hellinger A, Wahlster W (2013) Recommendations for implementing the strategic initiative INDUSTRIE 4.0: securing the future of German manufacturing industry; final report of the Industrie 4.0 Working Group. Forschungsunion

38. Kalantari F, Mohd Tahir O, Mahmoudi Lahijani A, Kalantari S (2017) A review of vertical farming technology: a guide for implementation of building integrated agriculture in cities. In: Advanced engineering forum, vol 24. Trans Tech Publications, pp 76–91

39. Lateef O, Hota B (2019) Opportune acquisition LLC, 2019. Algorithm, data pipeline, and method to detect inaccuracies in comorbidity documentation. U.S. Patent 10,269,447

40. Lawler J, Molluzzo JC (2015) A proposed concentration curriculum design for big data analytics for information systems students. Inform Syst Educ Journal 13(1):45

41. Leung L (2015) Validity, reliability, and generalizability in qualitative research. J Family Med Prim Care 4(3):324–327. https://doi.org/10.4103/2249-4863.161306

42. Linstone A, Turoff M (1978) The Delphi method: techniques and applications. Addison-Wesley, Reading

43. Linstone, H.A. & Turoff, M. (2002). *The Delphi method: techniques and applications.* Available from: from www.is.njit.edu/pubs.php. (Accessed 12 December 2011).

44. Loo R (2002) The Delphi method: A powerful tool for strategic management, policing. Int J Police Strategies Manage 25(4):762–769

45. Mail & Guardian Data Desk (2019) What does "fourth industrial revolution' even mean? https://mg.co.za/article/2019-09-05-what-does-fourth-industrial-revolution-even-mean

46. Moo-Young M (2019) Comprehensive biotechnology. Elsevier, Amsterdam

47. Mutambara AG (2019) Decentralized estimation and control for multisensor systems. Routledge, London

48. Nager A, Atkinson RD (2016) The case for improving US computer science education. Available at SSRN 3066335

49. Neelakandan S, Tyagi A, Nagalkar D, Global eProcure (2019) Robotic process automation for supply chain management operations. U.S. Patent 10,324,457

50. Okoli C, Pawlowski S (2004) The Delphi method as a research tool: an example, design considerations and applications. Inform Manage 42:15–29

51. Paul R, Elder L (2019) A guide for educators to critical thinking competency standards: Standards, principles, performance indicators, and outcomes with a critical thinking master rubric. Rowman & Littlefield

52. Penprase BE (2018) The fourth industrial revolution and higher education. In: Higher education in the era of the fourth industrial revolution. Palgrave Macmillan, Singapore, pp 207–229

53. Phillips R (2000) New applications for the Delphi technique. Pfeiffer and Company, Annual San Diego

54. President Cyril Ramaphosa (2019) Digital economy summit, Gallagher Convention Centre, Johannesburg on 5th July 2019, Advancing the African Agenda on the Fourth Industrial Revolution through Digital Transformation. https://www.gov.za/speeches/president-cyril-ramaphosa-south-african-digital-economy-summit-5-jul-2019-0000

55. Qi C, Chen Q, Fourie A, Tang X, Zhang Q, Dong X, Feng Y (2019) Constitutive modelling of cemented paste backfill: a data-mining approach. Constr Build Mater 197:262–270

56. Rayens MK, Hahn EJ (2000) Building consensus using the policy Delphi method. Policy Politics Nursing Practice 1(2):308–315

57. Rogers MR, Lopez EC (2002) Identifying critical cross-cultural school psychology competencies. J Sch Psychol 40(2):115–141
58. Roteta E, Bastarrika A, Padilla M, Storm T, Chuvieco E (2019) Development of a Sentinel-2 burned area algorithm: generation of a small fire database for sub-Saharan Africa. Remote Sens Environ 222:1–17
59. Sarantakos S (2005) Social research, 3rd edn. Palgrave Macmillan, Basingstoke
60. Sellers-Blais DS, IfWizard Corp (2018) Automatically optimizing analytics database server. U.S. Patent 10,089,080
61. Sharma P, Swaminathan A, Brunner C, Chari MR, Qualcomm Inc. (2019) Sensor API framework for cloud based applications. U.S. Patent 10,248,184
62. Shruthi, P, Chaitra BP (2016) Student performance prediction in education sector using data mining
63. Skulmoski GJ, Hartman FT, Krahn J (2007) The Delphi method for graduate research. J Inform Technol Educ 6:1–21
64. Stewart J (2001) Is the Delphi technique a qualitative method? Med Educ 35(10):922
65. Stitt-Gohdes WL, Crews TB (2004) The Delphi technique: a research strategy for career and technical education. J Career Tech Educ 20(2):55–67
66. Sumitha R, Vinothkumar ES, Scholar P (2016) Prediction of students outcome using data mining techniques. Int J Sci Eng Appl Sci 2(6):132–139
67. Thieret TE, Xerox Corp (2015) System and method for applying computational knowledge to device data. U.S. Patent 8,965,949
68. Torrie RD, Stone C, Layzell DB (2016) Understanding energy systems change in Canada: 1. Decomposition of total energy intensity. Energy Econ 56:101–106
69. Veltri AT (1986) Expected use of management principles for safety function management. Ph.D. dissertation, West Virginia University, Morgantown
70. Wallner T, Wagner G, Costa YJ, Pell A, Lengauer E, Halmerbauer G, Seher F, Staberhofer F, Lienhardt CA (2016) Academic education 4.0. In: Proceedings of the international conference on education and new developments, Ljubljana, Slovenia, 12–14 June 2016, pp 155–159
71. Wamba SF, Gunasekaran A, Akter S, Ren SJF, Dubey R, Childe SJ (2017) Big data analytics and firm performance: effects of dynamic capabilities. J of Bus Res 70:356–365
72. Wang Y, Kung L, Byrd TA (2018) Big data analytics: Understanding its capabilities and potential benefits for healthcare organizations. Technol Forecast Soc Chang 126:3–13
73. Xu Y, Wang W, Ge Y, Guo H, Zhang X, Chen S, Deng Y, Lu Z, Zhang H (2017) Stabilization of black phosphorous quantum dots in PMMA nanofiber film and broadband nonlinear optics and ultrafast photonics application. Adv Func Mater 27(32):1702437
74. Yousuf MI (2007) Using experts' opinions through Delphi technique. Practical Assess Res Eval 12(4):1–8

Chapter 13
An Appraisal of the Level of Awareness and Adoption of Insurance Policies for Sustainable Construction

Ayodeji Oke, Oriabure Ijieh, and Olanrewaju Ogunniyi

Abstract Sustainable Construction projects are precarious by nature and a lot of accidents occur in the process and therefore there is a need for insurance policies. The purpose of this study is to appraise the level of awareness and adoption of insurance policies on the sustainability of construction projects in Nigeria. The objective of the study is to examine the various insurance policies available to sustainable Construction Projects. This study was achieved through a questionnaire survey method that was self-administered to Professional in the construction industry; insurer and Client Representatives, the data collected were analyzed using statistical software (SPSS 23.0) Statistical Package for Social Sciences which encompasses tool such as mean, frequency, percentile, standard deviation and ranking. The findings of the study indicate the types of insurance policies available to construction industry, concluded that Contractors' All Risks (CAR) insurance is an all-inclusive insurance cover used in construction contracts, most effective and adopted and it has been realized that the risk involved in engaging insurance policies are numerous. Thus, insurance policies have positive impacts highly beneficent to the performance of construction projects.

Keywords Accidents · Insurance policies · Risks · Construction project

1 Introduction

The construction industry is full of hazards, This is because from inception through the completion is made up of complex activities with various phases, for construction to take place various equipments has to be used in construction and in other to ensure uninterrupted activities on-site there has to be proper coordination [1]. Although the construction sector has been in the business of constructing houses since the cave era [8]. Hence there is a need to cover for unforeseen circumstances that may arise as a result of these activities. According to Emmett and Therese [5], the transfer of

A. Oke (✉) · O. Ijieh · O. Ogunniyi
The Federal University of Technology Akure, Akure, Nigeria
e-mail: emayok@gmail.com

© The Author(s), under exclusive license to Springer Nature Singapore Pte Ltd. 2021   217
R. J. Howlett et al. (eds.), *Emerging Research in Sustainable Energy and Buildings for a Low-Carbon Future*, Advances in Sustainability Science and Technology,
https://doi.org/10.1007/978-981-15-8775-7_13

uncertainty can be done through insurance, the main function is to a counterpart risk through transferring or shifting risk from one individual to a group of individuals.

However, the construction industry has an important part to play in the competitive delivery of goods and services, it is full of many known and unknown risk occurrence due to various source of uncertainty, which includes the performance of construction parties, resources availability, environmental conditions, the involvement of other parties, and contractual relations [6]. Liyadu [9] postulated that a contractor can prevent the risk of been financially crippled by substituting a small definite cost, i.e. the premium for the variability of construction. However Transferring risks in construction is an acceptable practice worldwide as the construction environment is full of many factors (activities) that generate risks and can't be controlled [13]. Thus insurance is a wise move for managing construction risk [15].

Although there are many studies on insurance bonds and policies in Nigeria, research of the level of awareness and adoption of these insurance policies is scarce, Therefore, the objectives of this paper are to assess the level of awareness of various insurable policies on sustainable construction projects and examine the level of adoption of these insurance policies available in the construction industry. This is with a view to identifying insurable risks and the types of insurance policies typically involved in the construction industry. This study will contribute to raising the awareness and the benefits of the insurable risks and policies which project participants are exposed to, in the contracts they enter into. It will also provide a tool for decision-making in contract formation especially in insurance policies available for the sustainability of construction projects.

2   Literature Review

2.1   Insurance Policies Available for Sustainable Construction.

In relation to construction operation, there are quite a number of insurance policies available to cover for losses, nevertheless, policy are usually taken in all construction project. Thus, any uncertainty that can be measured can be prevented. Special kinds of risk can give rise to claims known as 'perils'. However, an insurance policy will show clearly which claims are included in the insurance included or not. These types of insurance policy available for construction projects [4]. The following are various forms of insurance policies available for sustainable constructions; Contractor's All Risk Policy (CAR) Suwisai and Clinton [17], Rameezdeenet al. [16], Worker's Compensation policy [2, 11], Professional indemnity insurance policy Baartz and Longley [2], Looi [10], etc. According to Suwisai and Clinton [17], contractor's all-risk insurance policy is an all-inclusive insurance cover used in construction. The definition of all risk insurance in clause 22.2 of the JCT [7] defines the risks for which insurance is required. Payment of premium may be paid in quarter installment in a situation where the period of policy is more than 12 months, the first payment of installment is more by 5% while the final payment is paid 6 months before the policy expires. Also damages insured ranges from proper insurance is to protect employees/workers against injuries or occupational diseases received or incurred in the course of their work on a particular project. It is a mandatory type of insurance in many countries; the insurance covers medical costs and a portion of lost wages. It also covers the services needed to help an injured worker recover and return to work [11]. It is important to note that various countries and states around the globe have regulatory requirements for clients to affect employees compensation insurance for the advantage of their workers [2]. Professional indemnity insurance policy as opined by Baartz and Longley [2] as its name suggests, is a form of insurance that allows for an amount which the insured becomes legally liable to pay as a result of errors or omission while carrying out their duties. The cost expended while doing an investigation, defending or settling of uncertainties that have occurred is also in addition to policy limit. According to Looi [10] describe Professional insurance is a 'claim made' insurance, this means that the policy only responds to claims first made against the organisation during the policy period, irrespective of when the act of negligence actually occurred. Basic policies cover the structure and weather-proofing but this can be extended to include non-structural elements and mechanical and electrical services (such as heating, ventilating, air-conditioning, water systems, lifts, escalators, electrical distribution systems, building management systems, and so on), and some policies will provide cover for loss of rent, loss of profit or revenue, and the costs of working from alternative premises [10]. The policy, therefore, has, as most policies do, a number of exclusion and these usually include war, radiation, wear and tear, poor materials or poor workmanship, design error, erosion, force majeure, kidnapping, bombing, theft, earthquakes, flood. More so, as referred earlier it is expedient to manage and maintain the interface between the works and existing structures and the transition of a specific element under construction from the cover of contract work policy to cover under principal's industrial special risk, thereby ensuring there is a limited gap to cover, having regard to the fact that a principal will usually elect for its industrial special risk policy to commence only at the same time as commencement of operation which is commercially inclined [2]. Insurance policy gives coverage for risks to the property as a result of fire, theft or weather damage (III, 2010). This includes special types of insurance such as fire insurance, flood insurance, earthquake insurance, home, inland marine insurance. As opined by Prahl [14], during construction, the form provides all-risk coverage for fire and vandalism, as well as in transit coverage certain cases include added endorsement for a specific premium.

3 Research Methodology

Quantitative research that uses cross-sectional survey was adopted for the study to identify the level of awareness and adoption of the various insurance policies available to sustainable construction. Data for the study commenced with a review of literature on the subject matter to establish the level of awareness and adoption of various insurance policies on sustainable construction. It is worth noting that the literature review informed the design of the research used in the current investigation and self-administered questionnaire survey [3].

Using the survey method, questionnaires were self-administered to stakeholders that are directly involved with construction activities within the study area. These include mainly consultant Quantity Surveyors, Architects, Builders, Engineers, and Builders as well as Clients' representatives and insurance companies. Their list and register were obtained from their respective professional bodies that are shouldered with the responsibility of governing and controlling the profession and activities of their members. These include Nigerian Institute of Quantity Surveyors (NIQS) for Quantity surveyors, Architect Registration Council of Nigeria (ARCON), The Nigeria Institute of Builders (NIOB), Council of Registered Engineers of Nigeria (COREN), Financial Institution and Clients representation. The number of respondents in the study area is 336 as indicated in Table 1.

In calculating the sample size, the equation devised by Yamane [18] was adopted.

$N = n.$

$1 + N (e) 2.$

where "n" is the sample size for a group; "N" is that actual population of the group; and "e" is the level of significance (taking as 0.05).

Well-structured questionnaire with close-ended questions was used with the aim to examine the impact of various insurance policies on the performance of a project. Two sections were provided with the first planned to elicit information regarding personal data of the respondent and their organization. The second section was constructed to obtain information regarding the opinion on insurance policies which was measured on a 5-point Likert scale. Due to the inability to reach some of the listed professionals, financial institution and client representative a convenience sampling was adopted as the sampling technique for the study. A total of 118 questionnaires—representing the sample size—were administered from which 77 were retrieved representing about 65% response rate.

Table 1 Research population of respondent

S. No.	Firms	Population	Sample size
1	Quantity surveyors	57	24
2	Architect	27	16
3	Engineers	180	35
4	Builders	40	21
5	Financial institution	12	9
6	Clients representative	20	13
Total		336	118

Source NIQS, ARCON, NIOB, COREN (2019)

Cronbach's alpha (α) was computed to test the reliability of the results of the analysis and it indicates a value of 0.849 since the degree of reliability of the instrument is more perfect as the value tends towards 1.0 [12], it can be concluded that the instruments used for this research are significantly consistent and reliable. For the analysis, mean item score (MIS) and standard deviation were computed from the Likert scale and the values were employed to determine the significance and rank of the identified variables.

4   Data Presentation, Analysis and Discussion

4.1   Introduction

This session contains the analysis, presentation and discussion of data collected through questionnaire. This involved the use of frequencies, percentages and mean for presenting description finding of the survey. This technique was employed for analysing data related to the information of the respondents.

4.2   Characteristics of the Respondents

This comprises of data collected for the study on background information which comprises the organization type, academic qualifications, experience and professional status of respondents. The findings are presented in Table 2.

Table 2, shows the information of the respondents that sixty-five (77) questionnaires that were returned out of questionnaire sent out. 33.8% (26) of the respondents are from Public organisation and 66.2% (51) are from Private organisation. The table also shows that 14.2% of the respondents are Architect by profession, 28.6% Quantity surveyor, 14.2% Builder, 20.8% Engineer, 9.1% Insurance manager, 13.0% are from client organisation. It is also indicated from Table 1 that most of the respondents are M.Sc./M.Tech (Master of Technology) holder with 35.1%, B.Sc/B.Tech (Bachelor of Science/Bachelor of Technology) Holder representing 33.8%, HND (Higher National Diploma) 19.5%, PGD (Post Graduates Diploma) 7.8%, and Ph.D. (Doctor of Philosophy) holders representing, 3.9%.

Thus, results in Table 2 shows that most of the respondents are from NSE (Nigerian Society of engineers) 36.9% and the least CIIN (Chartered Institute of Insurers of Nigeria) and NIOB (Nigeria Institute of Builders) with 14.3% for each, respectively, NIQS (Nigerian Institute of Quantity Surveyors) 24.7%, NIA (Nigerian Institute of Architects) 15.6%. Also from the table, it can be deduced that regarding the years of working experience as presented on the table; 15.6% (12) were between 0 and 5 years, 32.5% (25) were between 6 and 10 years, 19.5% (15) were between 11 and 15 years, 11.7% (9) were between 16 and 20 years, 10.4% (8) has 21–30 years of

Table 2 Characteristics of respondents

Category	Information	Frequency	Percentage (%)
Respondent's Organisation	Private organisation	51	66.2
	Public organisation	26	33.8
	Total	77	100
Respondent's Profession	Architect	11	14.2
	Quantity surveyor	22	28.6
	Builder	11	14.2
	Engineer	16	20.8
	Insurer	7	9.1
	Client's representative	10	13.0
	Total	77	100
Academic qualification of Respondents	HND	15	19.5
	PGD	6	7.8
	B.Sc/B.Tech	26	33.8
	M.Sc/M.Tech	27	35.1
	Ph.D.	3	3.9
	Total	77	100
Respondent Professional body	NIA	12	15.6
	NIQS	19	24.7
	NIOB	11	14.3
	NSE	24	31.2
	CIIN	11	14.3
	Total	77	100
Years of experience	0–5	12	15.6
	6–10	25	32.5
	11–15	15	19.5
	16–20	9	11.7
	21–25	8	10.4
	26–30	8	10.4
	Total	77	100
Numbers of Insured projects handled	0–5	9	11.7
	6–10	12	15.6
	11–15	8	10.4
	16–20	15	19.5
	21–25	18	23.4
	26–30	15	19.5
	Total	77	100

Table 3 Level of awareness of insurance policies for sustainable construction

Insurance policies	Mean	S.D	Rank
Compulsory third-party motor vehicle insurance policy	4.23	0.765	1
Contractor's all risk policy (CAR)	3.98	0.944	2
Professional indemnity policy	3.94	0.706	3
Property insurance	3.94	0.882	4
Worker's compensation policy	3.75	1.075	5
Contractor plant and machinery policy	3.74	0.871	6
Public and product liability	3.45	0.733	7
Contract works insurance/industrial special risk policy	3.38	0.943	8
Latent defects policies	3.29	1.011	9

working experience. This result indicates that these respondents have come of age in active service in the Construction Industry and have acquired wealth of knowledge sufficient for any meaningful survey.

However, the table indicates that 11.7% of the respondent have handled project insured ranging from 0–5, 15.6% from 6–10, 10.4% from 11–15, 19.5% from 16–20, 23.4% from 21–25 and 19.5% from 26–30, which proofed that most firm or organisation ensures their project. The sum of 16–20, 21–25 and 26–30 convincible shows that projects are insured which, this has indicated the most organisation make use of insurance policies.

Table 3 shows the respondent awareness of the various policies rendered which clearly state that respondents generally agree with the various policies rendered on insurance in the construction industry. Compulsory third party motor vehicle insurance policy is the predominant insurance policy, this was ranked first with the MIS value of 4.23 and standard deviation (SD) of 0.765, Contractor's all risk ranked second with the MIS value of 3.98 and Standard deviation (SD) of 0.944, professional indemnity policy is ranked third with the MIS value of 3.94 and standard deviation (SD) of 0.706, Property insurance is ranked fourth with the MIS value of 3.94 and standard deviation (SD) of 0.882, Worker's Compensation Policy is ranked fifth with an MIS value of 3.75 with standard deviation (SD) of 1.075, Contractor Plant and Machinery Policy came as sixth in the group with MIS value of 3.73 with standard deviation (SD) of 0.871, Public and Product Liability with the MIS of 3.45 with standard deviation (SD) of 0.733 was ranked seventh, Contract Works insurance/industrial special risk policy with the 3.38 was ranked eighth with the standard deviation (SD) of 0.943, while Latent defect policies was the last with a mean of 3.29 with standard deviation (SD) of 1.011. As explained we can deduce that the respondents are more aware of Compulsory third-party motor vehicle insurance policy.

Table 4 indicates that the respondent awareness of the various policies rendered which clearly state that respondents generally agree with the various policies rendered

Table 4 Level of Adoption on the various policies in the construction industry

Insurance policies	Mean	S.D	Rank
Contractor's all-risk policy (CAR)	3.85	1.227	1
Professional indemnity policy	3.72	0.905	2
Property insurance	3.72	0.960	3
Compulsory third party motor vehicle insurance policy	3.6	1.101	4
Contractor plant and machinery policy	3.48	1.062	5
Worker's compensation policy	3.32	0.985	6
Public and product liability	2.86	1.087	7
Contract works insurance/industrial special risk policy	2.39	1.041	8
Latent defect policies	2.19	1.059	9

on insurance in the construction industry. Contractor's all-risk policy is the predominant ranked first with a mean of 3.85 and standard deviation of 1.227, professional indemnity policy ranked second with mean of 3.72 and SD of 0.905, Property insurance ranked third with mean of 3.72 and SD of 0.960, with the aid of standard deviation (S.D) the policies with same weighed mean 3.72 were rank using the one with lowest S.D to rank the one which comes first before the other.

Compulsory third party motor vehicle insurance ranked fourth with a mean of 3.60 and SD of 1.101, Contractor Plant and Machinery Policy was ranked fifth with a mean of 3.48 and SD of 1.062, Worker's Compensation Policy ranked sixth with a mean of 3.323 and SD 0.985, Public and Product Liability with a mean of 2.86 and SD 1.087 was ranked seventh, Contract Works insurance/industrial special risk policy with mean 2.39 was ranked eighth, while Latent defect policies are the last with a mean of 2.19 and SD of 1.059. However, it shows the least adoption of the insurance policy is Latent defect policies.

4.3 Discussion of Findings

This study is envisaged on evaluation the various insurance policies on construction from the analysis it was observed that respondent year of practice has a significant effect on their attitude toward the implementation of insurance policies, on construction projects, respondent with higher year of experience in practice have been involved in many insured projects and are aware of the various insurance policies available for construction project. Sequels from the analysis compulsory third party motor vehicle insurance policy is the highest known type of insurance policy known by construction professionals and Contractors' all risk (CAR) Policy is the highest effective in the types of insurance policies available for construction projects since all Contractors' All Risks (CAR) insurance is an all-inclusive insurance cover all physical loss or damage both to work itself (under construction or completed), and on plants and equipment's as well as materials which are in assertion with Rameezdeen et al. [16] which stated that Contractors' all risk (CAR) Policy is most reliable insurance policy used in the construction industry and the covers applies during the construction and maintenance period. However, from the results of the analysis which was also in consensus with what Suwisai and Clinton [17] said the contractor's all-risk insurance policy is an all-inclusive insurance cover used in construction.

Nevertheless, Professional indemnity insurance policy is also adopted and effective from the results on which closed to contractor all-risk insurance as opined by Baartz and Longley [2] as it names suggest, this insurance indemnifies an insured for amounts which the insured becomes legally liable to pay as a result of any actual or alleged negligent act, error or omission in the conduct of its business or profession. Cost and expenses incurred to investigate, defend or settle any claim are also included, sometimes in addition to policy limit. And with what to Looi [10] opined describing Professional insurance is a 'claim made' insurance, this means that the policy only responds to claims first made against the organization during the policy period, irrespective of when the act of negligence actually occurred.

5   Conclusion and Recommendation

In conclusion, Construction insurance plays an important role in transferring risks in the construction industry. On every project insurance is required, the employer should request from the contractor. There are different types of policies rendered by insurance for construction industry which cover from the building to the employer, workers on-site, materials, etc. The recognition of the awareness, adoption and effectiveness of each selected policies will help the construction industry and its various participants on how to manage themselves and their project through losses.

Therefore it is recommended that there should be awareness and seminar programs to re-orient all professionals and clients on various insurance policies available in the construction industry and also proper education on how to recover claims thus insurance organisation should be sincere with their clients by satisfying them during the period of uncertain occurrence.

References

1. Aniket D, Bhavin K (2015) Role of insurance as a risk management tool in construction projects. Int J Adv Res Eng Sci Manag 1–8:394–1766
2. Baartz E, Longley N (2003) Construction and infrastructure projects-risk management through insurance. Retrieved on 29 Oct 2018 from https://www.allens.com
3. Bryman A (2016) Social research methods, 5th edn. Oxford University Press, London
4. Caleb S, Scott M, Dylan et al (2018) Insurance retrieved on 3 Nov 2018 from www.investope dia.com/terms/i/insurance.asp

5. Emmett J, Therese M (eds) (2010) Fundamentals of risk and insurance (10th ed). A guide to insurance: what it does and how it works. R.R Donnelley & Sons, Inc. III Insurance Information Institute, New York
6. Hamzah AR, Wang C, Mohamad FS (2015) Implementation of risk management in Malaysian construction industry: case studies. J Constr Eng
7. JCT (2016) Standard form of building Contract, Private with Quantities: Joint contract Tribunal. RIBA Publications Ltd., London
8. Kavin P, Jitendra P (2019) Risk identification and applying risk management technique in construction project. Int J Res Appl Sci Eng Tech (IJRASET) 7(5):2321–9653
9. Liyadu Y (1985) On the responsiveness of insurance industry in Nigeria to social needs, paper presented to the public service lecture series Lagos
10. Looi M (2014) Construction insurance-important things you need to know. Retrieved 27 Aug 2018 from https://www.cila.co.uk
11. Lott RJ (2005) Controlling workers' compensation cost a risk management program. Retrieved 29 Jul 2018 from https://www.healthconsonsultantusa.com/Risk_mgmt_program_c ontrolling_worker_comp_cos.pdf
12. Moser CA, Kalton G (1999) *Survey Methods in Social Investigation*, (2nd ed.). Aldershot: Gower Publishing Company Ltd
13. Perera BAKS, Rathnayake RMCK, Rameezdeen R (2008) Use of insurance in managing construction risks: evaluation of contractors' all risks (CAR) insurance policy built-environment. Sri Lanka 08(2)
14. Prahl RJ (2004) Course of insurance basis. Retrieved 27 Jul 2018 from www.cwc.ca
15. Queen M, Satheesh Kumar S (2018) A study on insurance in construction industry. Int Res J Eng Technol (IRJET) 5(4):3991–3393
16. Rameezdeen R, Rathnayake R, Perera B (2008) Use of insurance in managing construction risks. Eval Contract All Risk (CAR) Insur Policy Built Environ J 2(8):1–50
17. Suwisai M, Clinton A (2015) Management of construction risk through contractor's all risk insurance policy. Retrieved 3 Nov 2018 from https://open.library.ubc.ca/cIRcle/collections/52660/items/1.0076338
18. Yamane T (1967) Statistics an Introductory Analysis (2nd ed.). New York: Harper ad Row

Chapter 14
To What Extent Is Biophilia Implemented in the Built Environment to Improve Health and Wellbeing?—State-of-the-Art Review and a Holistic Biophilic Design Framework

Carolyn Thomas and Yangang Xing

Abstract As human beings have detached themselves from natural environments by spending most of their time indoors, they have also distanced themselves from the positive experiences that nature provides. Sick building syndrome, nature deficit disorder amongst others, are examples of the impact of separating the built environment from nature. Biophilia is an innate affiliation to nature which stems from our evolutionary history, vital for sustaining health and wellbeing. Biophilic concepts have been explored from biophilic cities to biophilic hospitals. However, existing biophilic research is fragmented. In the last few decades, energy efficiency and carbon emissions have increased in importance for low environmental impact design, nonetheless, there is a need for more research in biophilic buildings which are beneficial to our health and wellbeing as well as causing less harm to the environment. This paper aims to investigate the application of biophilia in building design practices for improved health and wellbeing. Firstly, biophilic theoretical frameworks developed by leading biophilic experts have been examined and compared to health and wellness performance certifications such as WELL Building and Living Building Challenge (LBC) standards. Finally, a holistic biophilic framework inspired by Kellert and Calabrese has been elaborated to assess the biophilic features in the built environment. Multiple explorative case studies were employed for this paper, the findings revealed that the biophilic applications linked to direct experiences of nature were implemented inefficiently and lacked a holistic approach to improve health and wellbeing. The authors argue that biophilia needs to be included holistically to maximise the benefits of nature's experiences.

C. Thomas (✉) · Y. Xing
Nottingham Trent University, Nottingham, England, UK
e-mail: carolyn.thomas@ntu.ac.uk

Y. Xing
e-mail: yangang.xing@ntu.ac.uk

227
R. J. Howlett et al. (eds.), *Emerging Research in Sustainable Energy and Buildings for a Low-Carbon Future*, Advances in Sustainability Science and Technology,
https://doi.org/10.1007/978-981-15-8775-7_14

1   Introduction

Humans once lived in small communities, held small farms and acknowledged the dependencies on nature, aware of the importance of its cycles, seasons and weather for survival. Today, 55% of the world's population now resides in urban areas and is predicted to rise to 68% by 2050 [31]. People spend up to 90% of their time indoors [6, p. 2] and a further 6% in an enclosed vehicle [20]. Not only are humans the dominant species on earth but they are also distancing themselves from nature while corrupting the environment, but most importantly, impact the future generation's ability to survive. Humans have become an indoor species that favours artificial environments over natural ones.

Nature provides food, shelter, soil, water and air, vital to our survival, yet human beings are destroying these fundamental elements on a profound scale. Humans have altered the planet's physical, chemical and biological features on a geological level. As urban environments have become the daily setting for many, energy is primarily consumed in buildings and accounts for 40% of all energy use, and responsible for 36% of all CO_2 emissions [9]. Energy saved by design such as bioclimatic design and energy modelling, has improved buildings with considerable reductions in energy use, before adding photovoltaics, geothermal or wind. Although energy efficiency is a necessary move for buildings that cause less harm to the environment, they also need to consider the health and wellbeing of occupants.

In the last few decades, biophilia has been implemented to improve health and wellbeing, a technique that brings vegetation into cities, streets, and interiors. Biophilia is described as a preference for natural environments, as we have evolved for 99% of our species evolution outdoors, the human brain requires natural stimulations. As buildings are around for decades and urban environments in the daily setting for many, building design should promote health and wellbeing through interactions with nature. Urban environments are generally described as unhealthy which is reflected in the green building rating standards, such as BREEAM and LEED who have acknowledged the importance of implementing health and wellbeing measures.

This paper considers the field of biophilia, which has matured and broadened with a range of theoretical and conceptual frameworks, this will form the foundation for a holistic biophilic framework. The paper aims to identify the extent biophilia is implemented for improving health and wellbeing in the built environment.

1.1   Significance

Biophilia is "the urge to affiliate with other forms of life" [33, p. 85]. In the last few decades, biophilic design has received increased attention and significance in the built environment and is being implemented to improve health and wellbeing. Empirical studies confirm that biophilia improves concentration, decreases fatigue, improves mental wellbeing through the visual connections to nature. [13]. According to Louv, the absence of nature has contributed to decreased health and wellbeing and may be linked to increased obesity, depression, and attention disorders in children.

As a new biophilic framework to assess the experiences of nature is required to address the inadequacies in the built environment [8, 17, 37, 38]. Key concepts and theoretical backgrounds from the literature were employed to support the biophilia hypothesis. The research heavily relies on the practice of biophilia by Kellert and Calabrese for the design of a holistic biophilic framework.

1.2 Methodology Outline

A qualitative research design was employed for this paper, to understand a contemporary phenomenon, such as how biophilia is implemented in recently completed projects and why. Multiple case studies were selected to assess biophilic features, guided by the holistic biophilic framework. Four case studies were assessed for their biophilic applications, selected from the RIBA awards,

however, these are anonymised. Content analysis was used for the data collection, several sources and formats were employed:

- Project descriptions from the architect's websites
- RIBA website descriptions
- Observations from videos and images

The case studies were evaluated by building designers who have knowledge of biophilia and its applications. To evaluate the biophilic applications, case studies were analysed individually and collectively to look for trends and patterns, with results represented categorically through tables. Tables captured and summarised the findings for examination of similarities and differences.

2 A State-of-the-Art Review—Existing GBRT/Health and Wellness Performance-Based Standards and the Biophilia Hypothesis

2.1 The Biophilia Hypothesis and Multisensorial Experiences Provided by Nature

The term biophilia was initially devised by social psychologist Erich Fromm, who broadly defined it as "the passionate love of life and of all that is alive" [11, pp. 365–366]. Subsequently defined further and popularised by American biologist Edward O. Wilson, in his book Biophilia (1984, p. 85) as "the urge to affiliate with other forms of life". Furthermore, biophilia stems from the evolutionary history of the human species and is still vital to people's health and wellbeing in modern society [14, 15, 19, 34]. For 99% of our species history, humans developed to adapt in response to natural environments, there is a mismatch, the brains of early humans developed in different environments to today.

Humans left rural areas to reside in cities, which grew in numbers and size, people favour artificial environments over natural ones, with irreversible impact on the planet. Nonetheless, Wilson [35, pp. 31–41] explains that biophilia hasn't disappeared since the migration to cities and continues to play a vital role in the built environment. Kellert's work will be assessed further along with frameworks elaborated by experts in the field of biophilia. Biophilia is a much-needed innate affiliation to nature which we have distanced ourselves from.

Biophilia is a multisensorial experience, our physiognomy is designed to respond to natural environments. Non-dynamic environments shut down our senses, and lose contact with the world, our brains desire the variances, sounds, movements and scents provided by nature. Architects and scientists understand the importance space and place have psychologically and physically. However, previous studies have almost exclusively focused on the beneficial effects of visual connections to nature compared to all other senses, such as olfactory and acoustic.

The artificial settings imposed on individuals in modern society are mostly sensory deprived. Too often, buildings are bland environments, deprived of sensorial stimulation, resulting in places where fatigue and boredom set in. This has a negative impact on our psychological and physical wellbeing. Unfortunately, nature in the built environment is largely treated as a problem or an irrelevant matter, resulting in decreased interaction between people and nature. Additionally, sensory deprivation is distressing for the brain, decreasing its plasticity, the same thing all day long is harmful to our health and wellbeing [4]. Furthermore, the mood is defined and affected by what we do, see, hear and smell in that space [27]. Extended exposure to nature in healthcare facilities which implemented biophilia resulted in improved recovery rates, less pain relief administered, lower blood pressure, along with improved working conditions for staff [18, 22, 23, 30, 32]. The biophilia hypothesis supported by vast amounts of empirical studies, confirms that contact to natural environments offers a potent aesthetic stimulus. Direct contact with nature has the most profound impact on health and wellbeing, decreased stress, improved cognitive performance, along with improved mood and emotions.

Noise, unpleasant odours, artificial light and air conditioning cause stress making us sick, healthcare facilities and hospitals are rarely associated with improved comfort, health and wellbeing. Multisensorial encounters with the natural environment should be encouraged by incorporating indirect and direct contact with nature. This would promote sensorial stimulation and provide humans the much-needed exposure to nature to improve health and wellbeing. British Research Establishment (BRE) is also taking an active role on the impact the built environment has on health and wellbeing, Flavie Lowres, Associate Director of BRE states that "energy efficiency is now embedded in the construction thinking and processes…the focus is shifting more and more towards the health and wellbeing of the building occupants" (BRE 2018, p. 1). Biophilia is the missing link for true sustainability (BRE 2018), [36].

2.2 Green Building Rating Standards

Green building standards such as BREEAM's (Building Research Establishment Environmental Assessment Method, by BRE) new construction standard for non-domestic buildings include a category on health and wellbeing and accounts for 14% of the total credits. Credits are given for the provision of outdoor spaces, landscaped areas and biophilia to provide building users with direct experiences of nature. Daylighting and views onto nature, along with passive strategies are included [7, pp 72–126]. However, a recent study by Xue [36] argues that green building rating tools (GBRT) should shift from the energy-oriented approach to a human-centered one through a biophilic framework. Furthermore, health and wellness performance-based certification programs include biophilic design guidelines for improved health and wellbeing within topics such as air, water, nourishment, fitness, comfort and mind. However, a study by Obrecht [25] explains that many of the topics in GBRT and Health and Wellness performance-based standards relate to building management and services rather than the design of buildings. The importance of health and wellbeing was acknowledged in several government policies, green building standards and certification systems. For example, the UK National Planning Policy framework aims "to enable and support healthy lifestyles, especially where this would address identified local health and well-being needs" [24]. There is a lack of emphasis on health and wellbeing in green building design tools.

Different from energy-oriented GBRTs, there are two health and wellness performance-based certification schemes: WELL Building and Living Building Challenge (LBC) standards. The WELL Building Standard is developed by the International WELL Building Institute, (IWBE), WELL measures elements associated with building design that affects occupant health and wellbeing. Air, water, nourishment, light, movement, thermal comfort, sound, materials, mind and community, all topics are backed up by medical and scientific research. The Living Building Challenge (LBC) is developed by the International Living Future Institute, with several levels of certification, which measures building design elements related to place, water, energy, health & happiness, materials, equity and beauty.

Both schemes assimilated the application of biophilic design. As summarised in Table 1. Both WELL and LBC have included direct connections to nature such as

Table 1 Summary of biophilic measures in well and LBC

Certification schemes	Environmental category/imperative	Assessment criteria
WELL Building standard 2018 V2	M02 access to nature M07 restorative spaces M09 enhanced access to nature	Provide access to nature Nature incorporation Culture, place, flora, art, delight
LBC Living building challenge 2019 V4	11 access to nature 19 beauty and biophilia	Interior/exterior connection to nature Connect to place, climate, culture and community

access to nature, incorporation of nature, connection to place, culture and community. Both WELL and LBC prescribed outdoor biophilia and indoor biophilia for improving health and wellbeing, addressed direct connection to nature through natural lighting, views onto nature.

However, WELL is comprised of preconditions (which are required) and optimisations (which are recommended), M07 and M09 are within the optimisation category. In comparison to LBC, all imperatives are mandatory, furthermore, LBC provides a biophilic design guidebook based around Kellert's six elements. Environmental features, Natural shapes and forms, Natural patterns and processes, Light and space along with Evolved human-nature relationships. (Biophilic Design Guidebook, LBC, p. 11).

2.3 Biophilic Design Applications in the Built Environment and Associated Design Frameworks

In the last few decades, biophilia has received increased attention in the built environment [1, 10], to provide humans with vital exposure to nature. Biophilic urbanism has obtained increased consideration in academia and practice [1, 2, 10]. Biophiliccities.org founded by Tim Beatley aims to "advance the theory and practice planning for biophilic cities" [3, p. 1] and describes a holistic approach to biophilia to include conservation of wildlife along with providing humans the much-needed sensorial stimulation. An exemplary biophilic city is Singapore. The city has implemented an abundance of natural features and although the population has increased by 2 million over two decades the vegetation has increased from 36 to 47% [5, p. 1]. Park connectors allow people to walk, bike and jog between numerous areas of the city. Vegetation is used as a climate modifier, high-rise buildings, hospitals and schools use green facades, roofs to decrease the urban heat island effect whilst purifying air. Singapore's Koo Teck Puat Hospital KTPH received the biophilic award [12, pp. 5–7]. Climatic studies performed on the KTPH reveal temperatures were considerably lower in the afternoon due to shading from vegetation and evaporative cooling from water features [12, pp. 5–7]. Biophilia when implemented correctly can provide multifunctional solutions to common design problems in the built environment.

As a holistic approach to biophilic design is required, the biophilic application along with the biophilic principles as defined by Kellert and Calabrese [16], provides a suitable evaluation method for assessing biophilic features for improved health and wellbeing. According to Kellert and Calabrese [16] "Biophilic design seeks to create good habitat for people as a biological organism in the modern built environment that advances people's health, fitness and wellbeing". The following five fundamental principles of biophilic design were identified by Kellert and Calabrese (2015, pp. 6–7) as;

1. Biophilic design requires repeated and sustained engagement with nature;

2. Biophilic design focuses on human adaptations to the natural world that over evolutionary time have advanced people's health, fitness and wellbeing;
3. Biophilic design encourages an emotional attachment to settings and places;
4. Biophilic design promotes positive interactions between people and nature that encourage an expanded sense of relationship and responsibility for the human and natural communities;
5. Biophilic design encourages mutual reinforcing, interconnected, and integrated architectural solutions.

However, those principles can only be followed by the successful implementation of biophilia through a framework. As discussed previously, this set of principles is currently not employed in any health and wellness standards or GBRT. Several biophilic design frameworks have been created. Terrapin Bright Green created a framework (as in Appendix 1) which links biophilic features to health and wellbeing outcomes. Similarly, to Kellert and Calabrese's theoretical framework, Terrapin Bright Green's 14 biophilic patterns also distinguish between nature in the place, nature analogues and nature of the space. This framework is partially based on Kellert's biophilic theories and includes three categories of implementation. Furthermore, each is linked to aspects of health and wellbeing such as stress reduction, cognitive performance and emotion, mood and preference. The proposed framework could provide design strategies to address specific health and wellbeing outcomes for a range of environments. Finally, a conceptual framework focused on urban environments has been proposed by Beatley (Appendix 2) who coined the term biophilic city and linked the biophilia hypothesis to urban planning, to incorporate economy and improved health and wellbeing of population, along with recovery of urban landscapes (2011). He developed a biophilic pathway to urban resilience, the framework compared to Kellert and Calabrese, is aimed at the macro scale with a less detailed approach to its application. Based on previous research (Kellert and Calabrease 2015), [1, 2, 10], the authors have developed the following design framework which consists of three main experiences of nature and corresponding attributes, totalling to twenty-four biophilic features as in Fig. 1. The combination of the biophilic principles and application methods provides a complete framework for the evaluation or implementation of biophilia in the built environment.

2.4   Summary of the State-of-the-Art Review

It is observed that humans have distanced themselves from nature, spending most of our lives inside buildings and connecting with technology. The built environment has been linked to several illnesses and disorders, some more recognised than others: NDD and SBS, which further supports the idea that the lack of nature is affecting our health and wellbeing. Biophilia provides a holistic approach to sustainable design: conservation of natural environments; improved health and wellbeing; climate modification; energy efficiency. Its application has been defined by many researchers such

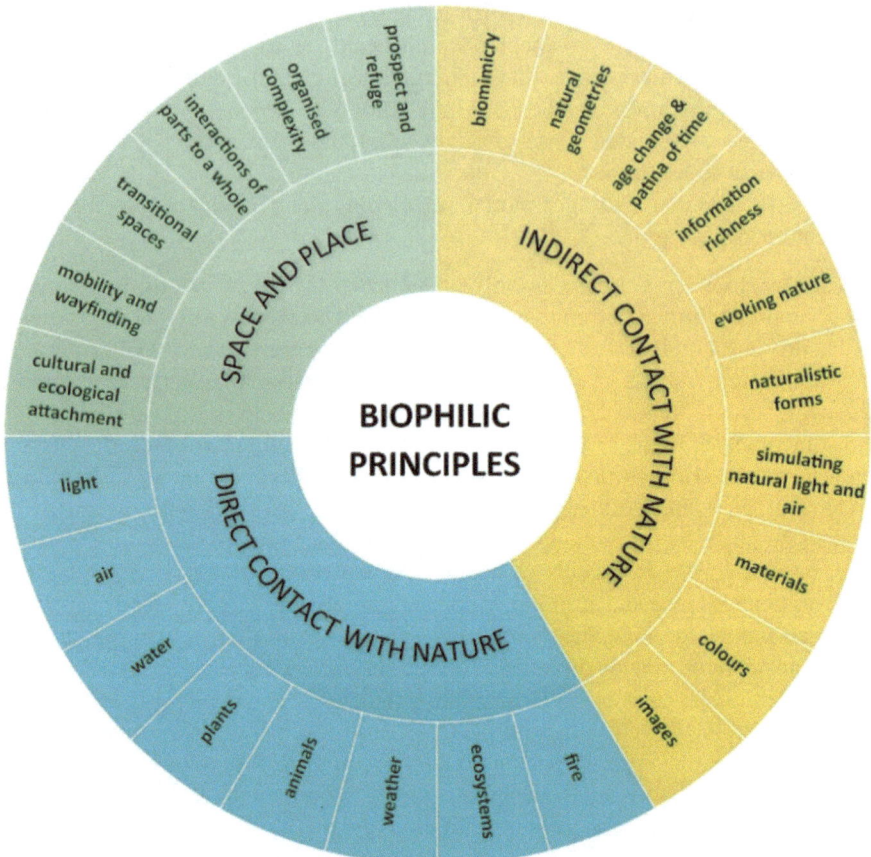

Fig. 1 Biophilic design framework. Adapted from Kellert and Calabrese (2015, pp. 6–20)

as Kellert and Calabrese's theoretical frameworks and Terrapin Bright's 14 biophilic patterns. Furthermore, empirical evidence supports the beneficial impact of biophilia on health and wellbeing, both psychologically and physically. Specifically, direct contact to nature through the application of light, air, water, plants and ecological landscapes.

Although biophilic features are widely implemented in urban environments, there are few which have adopted a holistic approach to sustainable design, in most countries and cities biophilia is not considered as an important feature for promoting health and wellbeing. Typically, buildings lack direct contact with nature; green views, plants and natural daylight or access to green areas are not a priority. Hospitals, educational buildings, office buildings and homes and generally devoid of vegetation and natural elements, despite a growing body of knowledge proving the beneficial impacts of experiences of nature on health and wellbeing. All main sustainability assessment standards such as BREEAM, WELL and LBC include natural ventilation, thermal comfort, lighting and ecological features, but lack the holistic approach to biophilia as defined by Kellert and others. Specifically, the principles, which as defined by Kellert and Calabrese, are fundamental.

3 Methodology

3.1 Developing a Framework for Assessing Biophilic Implementations

The aim of this paper is to investigate how and why biophilia is implemented in urban environments for improved health and wellbeing, by developing a holistic biophilic framework for its evaluation. Four case studies were selected, assessed for their design strategies and biophilic applications. The case studies were selected from the RIBA awards, "The RIBA National Awards are given to buildings across the UK recognised as significant contributions to architecture" [26]. The case studies are all projects which were recently completed and provide a good indication of the extent to which biophilia is implemented to promote health and wellbeing. The criteria for selection were large to medium-sized buildings in the UK, located in urban areas, from a variety of settings: educational, office spaces and dwellings.

As a new biophilic framework is necessary for the holistic application of biophilia, which is lacking in current GBRT and Health and Wellness standards, concepts and theories proposed by experts in the field of biophilia were assessed and compared. The state-of-the-art literature reviewed in Sect. 2 formed the basis for the elaboration of a holistic biophilic framework based on Kellert and Calabrese (see chart 1 "Practice of Biophilic Design). The framework is represented in the form of tables, one for the experiences and attributes and another for the principles (see Table 2).

- Stage 1: Kellert and Calabrese's experiences and application of biophilic design (2015, pp. 6–20).
- Stage 2: Kellert and Calabrese's biophilic principles (2015, pp. 6–7).

The data was collected from images, project descriptions from the architect's websites, RIBA website along with observations from videos and images to evaluate the application of biophilia. Case studies were analysed individually and collectively to look for trends and patterns, with results represented categorically through tables.

4 Results and Discussion

Two investigators evaluated the effectiveness of the holistic biophilic framework. A whose work currently focuses on urban community and productive renovation. Both

Table 2 Biophilic framework. Adapted from Kellert and Calabrese (2015, pp. 6–20)

Experience of direct contact with nature	A	B	C	D
1. Light: natural light, design strategies and materials				
2. Air natural ventilation, access to outside, operable windows				
3. Water: views of prominent water bodies, fountains, aquaria, wetlands				
4. Plants: abundant ecologically connected vegetation, flowering plants, local species				
5. Animals: design strategies, use of modem technologies to attract and view wildlife				
6. Weather: direct exposure to outside, porches, decks, balconies, gardens				
7. Natural landscapes and ecological features: interconnected plants, animals, water, soil				
8. Fire: fireplaces, creative use of light, colour, movement, and materials of varying heat conductance				
Experience of indirect contact with nature	A	B	C	D
1. Images of nature: images, paintings, sculptures, murals any other representational means				
2. Natural materials: wood, stone, wool, cotton, and leather				
3. Natural colours: earth tones from soil, plants, rocks, sunsets				
4. Simulating natural light and air; design strategies for lighting and ventilation, mimicking natural dynamic qualities				
5. Naturalistic shapes and forms patterns from nature				
6. Evoking nature: design principles from nature, biomorphic forms				
7. Information richness: diverse environments, options and opportunities				
8. Age change and the patina of time: naturally aging materials, weathering				
9. Natural geometries: mathematical opportunities from nature, self-repeating but varying patterns				
10. Biomimicry: forms and functions inspired from nature				
Experience of space and place	A	B	C	D
1. Prospect and refuge long views of surrounding settings, safety and security				
2. Organised complexity: orderly and organised complexity				
3. Integration of pans to whales: central focal points, linking of spaces, clear and discernible boundaries				
4. Transitional spaces: clear and discernible transitional spaces, hallways, porches				
5. Mobility and wayfinding: clear pathways, point of entry, exits				
6. Cultural and ecological attachment to place: culturally relevant design, local landscapes				
Biophilic principles	A	B	C	D
1. Repealed and sustained engagement with nature				
2. Advancing people's health, fitness and wellbeing				
3. Emotional attachment to settings and place				
4. Promotes interactions between people and nature, encourages a sense of relationship and responsibility for the human and natural communities				
5. Interconnected and integrated biophilic architectural solutions				

investigators have knowledge of biophilia and assessed the case studies for their biophilic features. All results from both investigators were used as this provided valuable information on the framework's effectiveness (see Appendix 3, 4, 5).

Four case studies were selected and assessed for their biophilic applications, guided by the framework derived from the literature. The results were checked against the five biophilic principles for a holistic approach to biophilic design. The case studies were analysed individually and collectively to look for trends and patterns, with results represented categorically with tables. This enabled to elaborate theories on the implementation of biophilia in urban environments. The results are summarised below.

4.1   Effectiveness of the Holistic Biophilic Framework in Assessing Biophilic Applications

Investigator A described the checklist as comprehensive and detailed. Most biophilic features that were implemented or absent were identified by both investigators. Further improvements to the framework have been identified and are as follows:

- Several biophilic design features were similar, causing some confusion to the category the feature belongs, further clarification is required either with detailed subcategories, or by combination of the features. Such as feature 4: Plants: abundant ecologically connected vegetation, flowering plants, local species, and Feature 7: Natural landscapes and ecological features: interconnected plants, animals, water, soil.
- Several biophilic features were unsuccessfully applied, causing some disagreements from the investigators. Especially when assessing the direct experiences of nature, such as plants and water features.
- The biophilic assessment checklist would benefit from including quantifying measures. This could be achieved with the addition of a column, to quantify each feature.
- Field measurements would have improved consistency in the results as some features were unnoticeable from content analysis, such as the creative uses of nature through artworks and imagery, colours and fabrics.

Furthermore, investigator A suggests images could be implemented to help identify each feature. The framework aimed to provide a holistic assessment of biophilic design in the built environment to improve health and wellbeing. As identified in the literature, a new biophilic assessment tool is required to address the deficiencies in the built environment [8, 17].

Further adjustments to the framework:

A focus group and a workshop, comprised of academics who were selected for their knowledge and expertise, discussed and assessed the clarity and usefulness of the framework, opinions and recommendations were implemented (see Appendix 6 for an amended version). Feature 6, 9, 10 from experience with indirect contact with nature, have been combined with feature 5. This category now represents shapes, forms, patterns, geometries and innovations from nature.

4.2   Holistic Biophilic Framework: Identified Trends and Patterns in Case Studies

The trends and patterns identified from the case studies findings suggest biophilia is implemented across all projects through various means. These have been summarised below and include both investigators findings:

- Experiences of direct and indirect contact with nature, along with experiences of space and place, have been implemented across all case studies.
- All three groups of experiences with nature have been partially implemented to advance people's health, wellbeing and fitness. Both investigators agree that none of the selected, recently completed projects, achieved all five biophilic principles to have a substantial impact on health and wellbeing.
- Across all case studies, experiences of direct contact with nature, through natural lighting and ventilation strategies, were implemented.
- Experiences of direct contact with nature which have the highest benefits for health and wellbeing when assessed against Terrapin Bright's framework were implemented ineffectively or not at all. Such as prolonged and repeated exposure to nature.
- The biophilic features which benefit health and wellbeing the most, such as direct contact with nature through vegetation, plants, animals, were lacking.
- Biophilia is not an integral part of current environmental assessment standards or health and wellness performance-based standards and certifications. Government incentives should be available to promote biophilic design, to include vegetation, green walls and roofs in urban areas, such as in Singapore.

Although biophilic features were implemented in all case studies, its application fails to meet the five principles, which are fundamental for improving health and wellbeing. This is largely due to implementing vegetation in transitional spaces such as halls, communal and intermediate areas. As previously stated in the literature, the results revealed the biophilic features implemented to improve health and wellbeing were achieved by including thermal comfort, daylighting and ventilation, reflecting the importance of energy efficiency measures, and health and wellbeing in the current standard assessment methods. However, in most case studies the experience of direct contact to nature such as plants: abundant ecologically connected vegetation, flowering plants and local species were lacking. Plants, green walls, lawns were disconnected. If biophilia is to be implemented for health and wellbeing, this should be reflected in the choice and selection of its application. Although building occupants had access to direct contact with nature, many spaces were devoid of biophilic features.

5 Conclusion

The findings from the holistic biophilic framework inspired by Kellert and Calabrese provided several key theories on the application of biophilic design: biophilia is acknowledged and implemented in urban environments but lacks a holistic approach. Biophilic features were applied in all case studies, however, its application is limited to transitional and intermediate spaces. Repeated and sustained engagement with nature is lacking in most case studies, nature is mostly implemented as an "added feature" rather than an integral part of the structure. This is further verified when compared to truly biophilic buildings from the literature such as in Singapore. None of the case studies selected achieved all five biophilic principles for its successful implementation, as defined by Kellert and Calabrese.

Biophilia is implemented in recent projects and contributes to improved health and wellbeing, however, its application lacks the holistic approach for its successful implementation as defined by researchers. Furthermore, its acknowledgment is growing and being recognised for its health and wellbeing benefits but also as a replacement for sustainable design. Additionally, its application needs further definition, biophilia should be consciously implemented with the knowledge that it improves health and wellbeing when implemented holistically, rather than just added as a last thought. Direct contact with nature is the most beneficial biophilic feature to improve health and wellbeing, however, natural ventilation and daylighting seem to be the main direct experiences applied in recently completed projects. The most beneficial features of biophilic design to improve health and wellbeing are implemented ineffectively due to site constraints, lack of acknowledgment, and the lack of incentives, especially when compared to biophilic cities such as Singapore.

5.1 Recommendations Identified from This Paper Are as Follows

- A quantitative approach for the evaluation of biophilia, using field measurements would provide a consistent method for the assessment of biophilic features.
- A holistic biophilic framework which includes the principles as defined in the literature should be included in environmental standards and health and wellness performance-based certification.
- Further research into the benefits of indirect contact with nature along with the effect of nature to other senses is required.

Acknowledgements We thank you for contributions from Xiaoying Ding for providing feedback on the biophilic design framework and case studies.

Appendix 1: Terrapin Bright Green's 14 Biophilic Patterns and Health and Wellbeing Outcomes

See Fig. 2.

14 PATTERNS		⋆	STRESS REDUCTION	COGNITIVE PERFORMANCE	EMOTION, MOOD & PREFERENCE
NATURE IN THE SPACE	Visual Connection with Nature	⋆	Lowered blood pressure and heart rate (Brown, Barton & Gladwell, 2013; van den Berg, Hartig, & Staats, 2007; Tsunetsugu & Miyazaki, 2005)	Improved mental engagement/ attentiveness (Biederman & Vessel, 2006)	Positively impacted attitude and overall happiness (Barton & Pretty, 2010)
	Non-Visual Connection with Nature		Reduced systolic blood pressure and stress hormones (Park, Tsunetsugu, Kasetani et al., 2009; Hartig, Evans, Jamner et al., 2003; Orsega-Smith, Mowen, Payne et al., 2004; Ulrich, Simons, Losito et al., 1991)	Positively impacted on cognitive performance (Mehta, Zhu & Cheema, 2012; Ljungberg, Neely, & Lundström, 2004)	Perceived improvements in mental health and tranquility (Li, Kobayashi, Inagaki et al., 2012; Jahncke, et al., 2011; Tsunetsugu, Park, & Miyazaki, 2010; Kim, Ren, & Fielding, 2007; Stigsdotter & Grahn, 2003)
	Non-Rhythmic Sensory Stimuli	⋆	Positively impacted on heart rate, systolic blood pressure and sympathetic nervous system activity (Li, 2009; Park et al, 2008; Kuhn et al., 2008; Beauchamp, et al., 2003; Ulrich et al., 1991)	Observed and quantified behavioral measures of attention and exploration (Windhager et al., 2011)	
	Thermal & Airflow Variability	⋆	Positively impacted comfort, well-being and productivity (Heerwagen, 2006; Tham & Willem, 2005; Wigö, 2005)	Positively impacted concentration (Hartig et al., 2003; Hartig et al., 1991; R. Kaplan & Kaplan, 1989)	Improved perception of temporal and spatial pleasure (alliesthesia) (Parkinson, de Dear & Candido, 2012; Zhang, Arens, Huizenga & Han, 2010; Arens, Zhang & Huizenga, 2006; Zhang, 2003; de Dear & Brager, 2002; Heschong, 1979)
	Presence of Water	⋆	Reduced stress, increased feelings of tranquility, lower heart rate and blood pressure (Alvarsson, Wiens, & Nilsson, 2010; Pheasant, Fisher, Watts et al., 2010; Biederman & Vessel, 2006)	Improved concentration and memory restoration (Alvarsson et al., 2010; Biederman & Vessel, 2006) Enhanced perception and psychological responsiveness (Alvarsson et al., 2010; Hunter et al., 2010)	Observed preferences and positive emotional responses (Windhager, 2011; Barton & Pretty, 2010; White, Smith, Humphryes et al., 2010; Karmanov & Hamel, 2008; Biederman & Vessel, 2006; Heerwagen & Orians, 1993; Ruso & Atzwanger, 2003; Ulrich, 1983)
	Dynamic & Diffuse Light	⋆	Positively impacted circadian system functioning (Figueiro, Brons, Plitnick et al., 2011; Beckett & Roden, 2009) Increased visual comfort (Elyezadi, 2012; Kim & Kim, 2007)		
	Connection with Natural Systems				Enhanced positive health responses; Shifted perception of environment (Kellert et al., 2008)
NATURAL ANALOGUES	Biomorphic Forms & Patterns	⋆			Observed view preference (Vessel, 2012; Joye, 2007)
	Material Connection with Nature			Decreased diastolic blood pressure (Tsunetsugu, Miyazaki & Sato, 2007) Improved creative performance (Lichterfeld et al., 2012)	Improved comfort (Tsunetsugu, Miyazaki & Sato 2007)
	Complexity & Order	⋆	Positively impacted perceptual and physiological stress responses (Salingaros, 2012; Joye, 2007; Taylor, 2006; S. Kaplan, 1988)		Observed view preference (Salingaros, 2012; Hägerhäll, Laike, Taylor et al., 2008; Hägerhäll, Purcell, & Taylor, 2004; Taylor, 2006)
NATURE OF THE SPACE	Prospect	⋆	Reduced stress (Grahn & Stigsdotter, 2010)	Reduced boredom, irritation, fatigue (Clearwater & Coss, 1991)	Improved comfort and perceived safety (Herzog & Bryce, 2007; Wang & Taylor, 2006; Petherick, 2000)
	Refuge	⋆		Improved concentration, attention and perception of safety (Grahn & Stigsdotter, 2010; Wang & Taylor, 2006; Wang & Taylor, 2006; Petherick, 2000; Ulrich et al., 1993)	
	Mystery	⋆			Induced strong pleasure response (Biederman, 2011; Salinpoor, Benovoy, Larcher et al., 2011; Ikemi, 2005; Blood & Zatorre, 2001)
	Risk/Peril	⋆			Resulted in strong dopamine or pleasure responses (Kohno et al., 2013; Wang & Tsien, 2011; Zald et al., 2008)

© 2014 Terrapin Bright Green / 14 Patterns of Biophilic Design

Fig. 2 Biophilic patterns and biological responses. Figure reproduced from (Terrapin Bright Green 2015, p. 1)

Appendix 2: Biophilic Pathways to Urban Resilience

See Fig. 3.

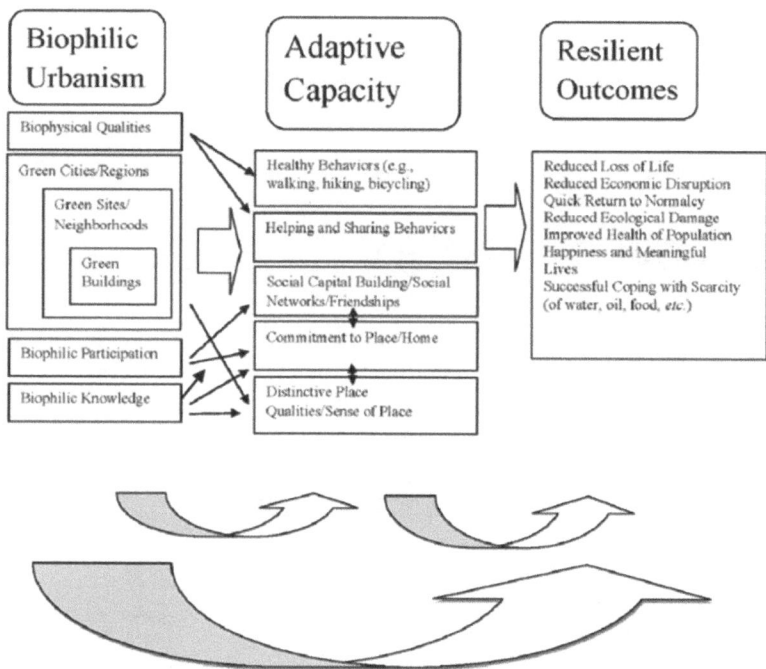

Fig. 3   Biophilic pathways to urban resilience. Figure reproduced from (Beatley & Newman 2013, p. 3333)

Appendix 3: Biophilic framework Results—Investigator A

Experience of direct contact with nature	A	B	C	D
1 Light: natural light design strategies & materials	✓		✓	✓
2 Air: natural ventilation, access to outside, operable windows	✓	✓	✓	✓
3 Water: views of prominent water bodies, fountains, aquaria, wetlands	✓			
4 Plants: abundant ecologically connected vegetation, flowering plants, local species	✓			✓
5 Animals: design strategies, use of modern technologies to attract and view wildlife				✓
6 Weather: direct exposure to outside, porches, decks, balconies, gardens	✓			✓
7 Natural landscapes and ecological features: interconnected plants, animals, water, soil	✓			✓
8 Fire: fireplaces, creative use of light, colour, movement, and materials of varying heat conductance	✓	✓		✓

Experience of indirect contact with nature	A	B	C	D
1 Images of nature: images, paintings, sculpture, murals other representational means		✓		
2 Natural materials: wood, stone, wool, cotton, and leather		✓		✓
3 Natural colours: earth tones from soil, plants, and rocks, sunsets		✓		
4 Simulating natural light and air: design strategies for lighting and ventilation, mimicking natural dynamic qualities		✓	✓	
5 Naturalistic shapes and forms: patterns from nature		✓		
6 Evoking nature: design principles from nature, biomorphic forms		✓		
7 Information richness: diverse environments, options and opportunities	✓	✓		✓
8 Age change and the patina of time: naturally aging materials, weathering				
9 Natural geometries: mathematical opportunities from nature, self-repeating but varying patterns.	✓			
10 Biomimicry: forms and functions inspired from nature				

Experience of space and place	A	B	C	D
1 Prospect and refuge: views of surrounding settings, safety and security				✓
2 Organised complexity: orderly and organised complexity	✓	✓	✓	✓
3 Integration of parts to a whole: central focal points, linking of spaces, clear and discernible boundaries		✓	✓	✓
4 Transitional spaces: clear and discernible transitional spaces, hallways, porches	✓	✓	✓	✓
5 Mobility and wayfinding: clear points for pathways, entries and exits.	✓	✓	✓	✓
6 Cultural and ecological attachment to place: culturally relevant design, local landscapes		✓	✓	✓

Biophilic principles	A	B	C	D
1 Repeated and sustained engagement with nature	✓			✓
2 Advancing people's health, fitness and wellbeing	✓	✓	✓	✓
3 Emotional attachment to settings and place	✓	✓	✓	✓
4 Promotes interactions between people and nature, encourages a sense of relationship and responsibility for the human and natural communities	✓			✓
5 Interconnected and integrated biophilic architectural solutions		✓		

Appendix 4: Biophilic Framework Results—Investigator B

Experience of direct contact with nature	A	B	C	D
1 Light: natural light design strategies & materials.	✓	✓	✓	✓
2 Air: natural ventilation, access to outside, operable windows.	✓	✓	✓	✓
3 Water: views of prominent water bodies, fountains, aquaria, wetlands…	✓	✓		
4 Plants: abundant ecologically connected vegetation, flowering plants, local species.				
5 Animals: design strategies, use of modern technologies to attract and view wildlife				✓
6 Weather: direct exposure to outside, porches, decks, balconies, gardens.	✓		✓	✓
7 Natural landscapes and ecological features: interconnected plants, animals, water, soil	✓			
8 Fire: fireplaces, creative use of light, colour, movement, and materials of varying heat conductance.	✓	✓		✓

Experience of indirect contact with nature	A	B	C	D
1 Images of nature: images, paintings, sculpture, murals other representational means		✓		✓
2 Natural materials: wood, stone, wool, cotton, and leather	✓	✓	✓	
3 Natural colours: earth tones from soil, plants, and rocks, sunsets		✓		✓
4 Simulating natural light and air: design strategies for lighting and ventilation, mimicking natural dynamic qualities	✓	✓	✓	✓
5 Naturalistic shapes and forms: patterns from nature		✓		✓
6 Evoking nature: design principles from nature, biomorphic forms		✓		
7 Information richness: diverse environments, options and opportunities	✓	✓	✓	✓
8 Age change and the patina of time: naturally aging materials, weathering		✓		✓
9 Natural geometries: mathematical opportunities from nature, self-repeating but varying patterns.		✓		
10 Biomimicry: forms and functions inspired from nature		✓		

Experience of space and place	A	B	C	D
1 Prospect and refuge: views of surrounding settings, safety and security			✓	
2 Organised complexity: orderly and organised complexity	✓	✓		
3 Integration of parts to a whole: central focal points, linking of spaces, clear and discernible boundaries	✓	✓	✓	✓
4 Transitional spaces: clear and discernible transitional spaces, hallways, porches	✓	✓	✓	✓
5 Mobility and wayfinding: clear points for pathways, entries and exits.	✓	✓	✓	✓
6 Cultural and ecological attachment to place: culturally relevant design, local landscapes	✓	✓	✓	✓

Biophilic principles	A	B	C	D
1 Repeated and sustained engagement with nature				
2 Advancing people's health, fitness and wellbeing	✓	✓	✓	✓
3 Emotional attachment to settings and place	✓	✓		
4 Promotes interactions between people and nature, encourages a sense of relationship and responsibility for the human and natural communities	✓		✓	✓
5 Interconnected and integrated biophilic architectural solutions		✓		✓

Appendix 5: Combined Results

See Table 3.

Table 3 Combined findings

Experience of direct contact with nature	A	B	C	D
1 Light: natural light design strategies & materials.	G	B	G	G
2 Air: natural ventilation, access to outside, operable windows.	G	B	G	G
3 Water: views of prominent water bodies, fountains, aquaria, wetlands…	G	B	W	W
4 Plants: abundant ecologically connected vegetation, flowering plants, local species.	Y	G	G	G
5 Animals: design strategies, use of modern technologies to attract and view wildlife	W	G	W	W
6 Weather: direct exposure to outside, porches, decks, balconies, gardens.	G	W	B	G
7 Natural landscapes and ecological features: interconnected plants, animals, water, soil	B	W	W	Y
8 Fire: fireplaces, creative use of light, colour, movement, and materials of varying heat conductance.	G	G	W	W

Experience of indirect contact with nature	A	B	C	D
1 Images of nature: images, paintings, sculpture, murals other representational means	G	G	G	W
2 Natural materials: wood, stone, wool, cotton, and leather.	B	G	G	Y
3 Natural colours: earth tones from soil, plants, and rocks, sunsets	W	G	G	W
4 Simulating natural light and air: design strategies for lighting and ventilation, mimicking natural dynamic qualities.	B	G	G	W
5 Naturalistic shapes and forms: patterns from nature.	G	B	W	W
6 Evoking nature: design principles from nature, biomorphic forms.	G	G	W	W
7 Information richness: diverse environments, options and opportunities.	G	G	W	W
8 Age change and the patina of time: naturally aging materials, weathering.	W	B	W	W
9 Natural geometries: mathematical opportunities from nature, self-repeating but varying patterns.	Y	B	W	W
10 Biomimicry: forms and functions inspired from nature	W	B	W	W

Experience of space and place	A	B	C	D
1 Prospect and refuge: views of surrounding settings, safety and security	W	B	B	Y
2 Organised complexity: orderly and organised complexity	G	G	B	Y
3 Integration of parts to a whole: central focal points, linking of spaces, clear and discernible boundaries.	B	W	W	W
4 Transitional spaces: clear and discernible transitional spaces, hallways, porches.	G	G	G	G
5 Mobility and wayfinding: clear points for pathways, entries and exits.	G	G	G	G
6 Cultural and ecological attachment to place: culturally relevant design, local landscapes.	G	W	W	W

Biophilic principles	A	B	C	D
1 Repeated and sustained engagement with nature	Y	W	W	W
2 Advancing people's health, fitness and wellbeing	G	W	W	G
3 Emotional attachment to settings and place	G	W	B	G
4 Interactions between people and nature, encourages a sense of relationship and responsibility for the human and natural communities	G	W	B	G
5 Interconnected and integrated biophilic architectural solutions	W	G	W	B

Legend:
- ▢ Agree feature implemented (green)
- ▢ Agree feature not implemented (white)
- ▢ XD Feature implemented (yellow)
- ▢ CT Feature implemented (blue)

Appendix 6: Amended Framework From Focus Group

Experience of direct contact with nature	A	B	C	D
1 Light: natural light, reflective materials and design strategies…				
2 Air: natural ventilation, access to outside, engineering strategies…				
3 Water: views of prominent water bodies, fountains, aquaria, constructed wetlands…				
4 Plants: abundant ecologically connected vegetation, flowering plants, local species…				
5 Animals: contact with local species/animal life, technologies to attract/view wildlife…				
6 Weather: exposure to outside conditions, porches, decks, balconies, gardens…				
7 Natural landscapes & ecosystems: biological diversity, ecological services, interactions and participation with nature…				
8 Fire: fireplaces, creative use of light, colour, movement, materials of varying heat conductance…				

Experience of indirect contact with nature	A	B	C	D
1 Images of nature: images, paintings, sculptures, murals, other representational means…				
2 Natural materials: wood, stone, wool, cotton, leather …				
3 Natural colours: earth tones from soil, plants, and rocks, sunsets…				
4 Simulating natural light and air: design strategies for lighting and ventilation which mimic natural dynamic qualities and variances…				
5 Nature inspired designs: patterns, geometries from nature, biomorphism, biomimicry…				
6 Information richness: information-rich and diverse environments, options & opportunities				
7 Age, change, and the patina of time: naturally aging materials, weathering, passage of time…				

Experience of space and place	A	B	C	D
1 Prospect and refuge: long views of surrounding settings, safety and security…				
2 Organised complexity: diverse and varied spaces, connection and coherence…				
3 Integration of parts to wholes: central focal points, linking of spaces, clear and discernible boundaries…				
4 Transitional spaces: clear and discernible transitional spaces, hallways, porches, patio…				
5 Mobility and wayfinding: clear pathways, points of entry, exits…				
6 Cultural and ecological attachment to place: culturally relevant design, local landscapes.				

Biophilic principles	A	B	C	D
1 Repeated and sustained engagement with nature.				
2 Advancing people's health, fitness and wellbeing.				
3 Emotional attachment to settings and place.				
4 Promotes interactions between people and nature, encourages a sense of relationship and responsibility for the human and natural communities.				
5 Mutual reinforcing, interconnected and integrated architectural solutions.				

References

1. Beatley T (2011) Biophilic cities: integrating nature into urban design and planning. Island Press, Washington
2. Beatley T, Newman P (2013) Biophilic cities are sustainable, resilient cities, vol 5, p 8. Available at: about:blank Accessed 19 Aug 2018, pp 3328–3345, p 3333
3. Beatley T (2018) Our mission. Available at http://biophiliccities.org/about/. Accessed 27 Aug 2018, p 1
4. Behling S (2016) Architecture and the science of the senses. Available at about:blank. Accessed 19 Aug 2018
5. Biophilic Cities (2018) Singapore. Available at about:blank. Accessed 27 Sept 2018, p 1
6. BREEAM (2018a) Assessing health and wellbeing in buildings. Available at about:blank. Accessed 12 Sept 2018, pp 2, 18
7. BREEAM (2018b) BREEAM UK new construction. Available at about:blank. Accessed 15 Sept 2018, pp 72–126
8. Browning WD et al (2014) 14 patterns of biophilic design, improving health and well-being in the built environment. Terrapin Bright Green, New York
9. European Commission (2018) Energy efficiency-building. Available at about:blank. Accessed 12 Oct 2018
10. Farr D (2011) Sustainable urbanism: urban design with nature. Wiley, Hoboken, NJ
11. Fromm E (1973) The anatomy of human destructiveness. Holt, Rinehart & Winston, New York, pp 365–366
12. Green Pulse (2018) Healing through nature Khoo Teck Puat Hospital. Available at about:blank Accessed 4 Sept 2018, pp 5–7
13. Kaplan R, Kaplan S (1989) The experience of nature: a psychological perspective. Cambridge University Press, Cambridge, UK
14. Kellert S (1997) Kinship to mastery: biophilia in human evolution and development. Island Press, Washington DC
15. Kellert S (2012) Birthright: people and nature in the modern world. Yale University Press, New Haven
16. Kellert S, Calabrese E (2015) Chart 1: Biophilic framework. Table 2: Biophilic framework. the practice of biophilic design. Available at about:blank. Accessed 23 Aug 2018, p. 3, 6, 6–7, 6–20
17. Kellert S et al (eds) (2008) Biophilic design: the theory, science, and practice of bringing buildings to life. Wiley, Hoboken, NJ
18. Kellert S, Heerwagen J (2007) Nature and healing: the science, theory, and promise of biophilic design. In: Guenther R, Vittori G (eds) Sustainable Healthcare Architecture. Wiley, Hoboken, NJ
19. Kellert S, Wilson EO (eds) (1993) The biophilia hypothesis. Island Press, Washington, DC
20. Klepeis NE (2001) *The national human activity pattern survey (NHAPS)*. Available at: about:blank. Accessed 9 Oct 2018
21. Living Building Challenge (2018) Biophilic design guidebook. Available from about:blank—the-guidebook. Accessed 8 Sept 2019, p 11
22. Louv R (2012) The nature principle: reconnecting with life in a virtual age. Algonquin Press, Chapel Hill
23. Marcus CM, Sachs NA (2014) Therapeutic landscapes: an evidence-based approach to designing healing gardens and restorative outdoor spaces. John Wiley, Hoboken, NJ
24. Ministry of Housing, Local Communities and Local Government (2018) The national planning policy framework. Available at about:blank. Accessed 15 Sept 2018, p 27
25. Obrecht TP et al (2019) Comparison of health and wellbeing aspects in building certification schemes. Available at about:blank. Accessed 15 Sept 2019, p 5
26. RIBA (2018) RIBA national award winners. Available at about:blank. Accessed 7 Sept 2018, p 1

27. Steinberg E (2015) Healing spaces: the science of place and wellbeing. Available at about:blank. Accessed 29 Aug 2018
28. Terrapin Bright Green (2015) Biophilic design patterns and biological responses. Available from about:blank. Accessed 23 July 2018, p 1
29. Thomas C (2019) Table 1: Summary of biophilic measures in WELL and LBC
30. Ulrich R (1993) Biophilia, biphobia and natural landscapes. In: Kellert S, Wilson E (eds) The biophilia hypothesis. Island Press, Washington DC, pp 2–137
31. UN Department of Economic and Social Affairs (2018) 68% of the world population projected to live in urban areas by 2050, says UN. Available at about:blank. Accessed 9 Oct 2018
32. Wells N, Rollings K (2012) The natural environment: influences on human health and function. In: Clayton S (ed) The Oxford handbook of environmental and conservation psychology. Oxford University Press, London
33. Wilson EO (1984) Biophilia. Harvard University Press, Cambridge, p 85
34. Wilson EO (1986) Biophilia: the human bond with other species. Harvard University Press, Cambridge
35. Wilson EO (1993) Biophilia and the conservation ethic. The biophilia hypothesis. Island Press, Washington, DC, pp, 31–41
36. Xue F et al (2019) Incorporating biophilia into green building rating tools for promoting health and wellbeing. Available at about:blank. Accessed 15 Sept 2019, p 1
37. Xing Y, Brewer M, El-Gharabawy H, Griffith G, Jones P (2018) Growing and testing mycelium bricks as building insulation materials. IOP Conference Series: Earth Environ Sci 121(2):022032
38. Xing Y, Jones P, Bosch M, Donnison I, Spear M, Ormondroyd G (2018) Exploring design principles of biological and living building envelopes: what can we learn from plant cell walls? Intell Build Int 10(2):78–102

Chapter 15
Thermal Conductivity Characterization of Industrial Small-Sized Building Materials: Experimental and Simulation Study

Mouatassim Charai⊙, Haitham Sghiouri, Ahmed Mezrhab, and Mustapha Karkri

Abstract A new experimental procedure for determining the thermal conductivity of small-sized building materials using the boxes method setup is described. The proposed approach does not require additional sensors. Indeed, the measurement is based on the permutation of sensors and then the interpretation of the steady-state heat balance of samples. For the characterization, local earthen blocks from eastern Morocco were developed. The measured thermal conductivity values were compared with those obtained by an accurate transient hot disk method. The comparison shows a good agreement and verifies the performance of the permutation approach. To evaluate the thermal performance of the developed building materials, annual simulations were performed for a typical earthen building located in two different climates in Morocco. In this paper, hot and cold semi-arid climates were chosen to study the impact of the characterized earthen walls on the annual energy consumption of the case study building. The results show that the earthen wall reduces the annual cooling loads by 21.2% and 18.1% for hot and cold semi-arid climates compared to the reference case, respectively. In addition, the studied earthen walls show a reduction in the maximum cooling peak load of up to 20% compared to the conventional walls. However, a slight difference in heating loads for buildings located in cold semi-arid climates was observed. The experimental tests demonstrated the performance of the proposed methodology, which could be used for laboratory testing and for updating building material databases. While the building performance analysis clearly proved that earth walls can play a great role in mitigating the cooling demand and improving the thermal comfort of buildings in summer.

M. Charai (✉) · H. Sghiouri · A. Mezrhab
Mechanics and Energy Laboratory, Mohammed First University, BV Mohammed VI BP 717, 60000 Oujda, Morocco
e-mail: mouatassim.charai@u-pec.fr

M. Charai · M. Karkri
CERTES, Université Paris-Est, 61 Avenue du Général de Gaulle, 94010 Créteil Cedex, France

© The Author(s), under exclusive license to Springer Nature Singapore Pte Ltd. 2021   249
R. J. Howlett et al. (eds.), *Emerging Research in Sustainable Energy and Buildings for a Low-Carbon Future*, Advances in Sustainability Science and Technology,
https://doi.org/10.1007/978-981-15-8775-7_15

Keywords Boxes method · Building materials · Building performance simulation · Energy consumption · Passive cooling · Thermal conductivity · Small sample

1 Introduction

Material selection in the building sector has become a more demanding criterion with regard to climate change and the energy consumption of buildings, which accounts for the largest percentage part of their overall life-cycle impact [1, 2]. In this context, and besides passive design measures, several codes and standards have recently been enforced to fulfil the need for low energy-intensity buildings and to strengthen the transition to sustainable constructions. Examples of such building standards and normative strategies to cope with the alarming global building energy situation are given in [3].

Building thermal regulations are one of the most energy-related standards practically used to improve the energy efficiency of buildings. These kinds of approaches interest in the thermal performance of materials constructing the external building envelope in order to design thermally compliant compositions. Taking as an example the Moroccan thermal regulation program for buildings, developed in 2014 by the Moroccan Agency for Energy Efficiency, the objective is to reduce by at least 39% the annual energy needs of new buildings through specifying a minimum performance threshold based on a thermal resistance estimation, which can be deducted directly from the thermal conductivity of used materials [4].

Therefore, the characterization of the thermal conductivity of materials is of critical concern in evaluating the thermal performance of buildings. There are many techniques available to measure the thermal conductivity of building components, which can be divided into two main categories: steady-state and transient methods. The first category determines directly the thermal conductivity of samples via the analytical resolution of Fourier's law, such as the Guarded Hot Plate (GHP) method by ISO 8302 [5] or EN 12,667 [6], while the second category of methods estimates the thermal conductivity property using relatively complex numerical models, such as the Hot Disk method by ISO 22,007–2 [7]. The transient methods provide rapid measurements based on the thermal time-dependent response of studied samples. Nevertheless, steady-state methods give much more accurate results than transient techniques due to the well-established theoretical model, and thus the measurement errors depend only on the experimental setup used. For this reason, several laboratories are investing in the development of new steady-state apparatuses to cover a wide range of building material samples. For instance, Buratti et al. [8], from the Laboratory of Building Physics at the University of Perugia, have designed a small box method apparatus with the aim of evaluating the thermal transmittance of small, homogenous and non-homogenous building components (masonry units, doors, windows and thermal bridges).

Indeed, the motive behind adapting large-scale experimental devices for small samples is to achieve successful identification even if there is a lack of raw materials and to avoid transport problems. Among methods of relatively large-scale samples is the boxes method [9], which requires samples with dimensions of $270 \times 270 \text{ mm}^2$ and a thickness range from 30 to 70 mm.

In this respect, this paper focuses on the adaptation of the boxes method for small samples in order to improve the boxes method's flexibility to cover samples of various sizes and forms. The objective of the proposed experimental methodology is to characterize small-sized building material specimens without using any additional sensor. A local building material from eastern Morocco was prepared and characterized using two methods: the proposed method and the transient hot disk method that represents, in this paper, the reference method for validation purposes. The test results were used as input data to run building energy simulations using Energy Plus software to evaluate the energy performance of the building material studied at the scale of a residential building.

2   Materials and Methods

2.1   *Measurement Device: Boxes Method Cell*

Figure 1 shows the experimental device used in this work. It is the EI702 cell, which represents the latest version of the boxes method apparatus that has been adopted in many research studies for the thermophysical characterization of building materials. A brief review of these studies is recently given by the authors in [10].

The EI702 cell is an all-in-one measurement device, which composed of two insulated boxes for determining the apparent thermal conductivity of materials as well as their thermal diffusivity property using two different techniques. The first box serves for the thermal conductivity measurement of samples through maintaining a steady-state unidirectional temperature gradient across the sample studied. While the second box is equipped with an incandescent lamp to allow the thermal diffusivity

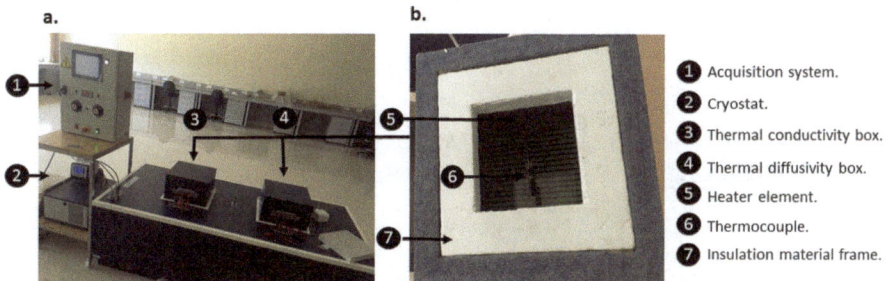

a.   b.

1 Acquisition system.
2 Cryostat.
3 Thermal conductivity box.
4 Thermal diffusivity box.
5 Heater element.
6 Thermocouple.
7 Insulation material frame.

Fig. 1   EI702 cell setup: **a** overall view, **b** interior view of boxes

characterization of samples via the well-known heat-pulse flash method [2]. For the thermal conductivity measurements, the instrument's uncertainty is less than 6%.

As shown in Fig. 1, the measurement apparatus consists of the following elements:

(1) An Acquisition system: it permits the users to fix the experimental inputs using an interactive touch-screen and allows them to display and record real-time temperature variations during the experiment.
(2) A Cryostat: it is used to maintain a constant low-temperature environment of the underside of the cell through a close-cycle system.
(3) A Thermal conductivity box: this box serves for the thermal conductivity measurement of the studied sample.
(4) A Thermal diffusivity box: this box serves for the thermal diffusivity measurement of the studied sample.
(5) A Heater element: it is the electrical heat source of the system to heat the upper side of studied samples during the thermal conductivity measurement.
(6) Probes: each box consists of two thermocouples type-K to measure the temperature variation on the upper and lower sides of the sample under study.

As above-mentioned, the boxes method requires samples with significant sizes. Therefore, in this paper, and in order to characterize small specimens, an insulation material frame will be used as a sample holder to cover the entire surface of the thermal conductivity box, as shown in Fig. 1b.

The mathematical model is described in the following section.

2.2 Theoretical Model

As indicated in the title, the paper focuses on the thermal conductivity box to examine a new experimental methodology for measuring the thermal conductivity of small samples without using any additional sensor. The sketch of the thermal conductivity box for a small-sized sample case is presented in Fig. 2a. Unlike conventional samples with a surface area of 270×270 mm^2 that cover the entire box, small samples are

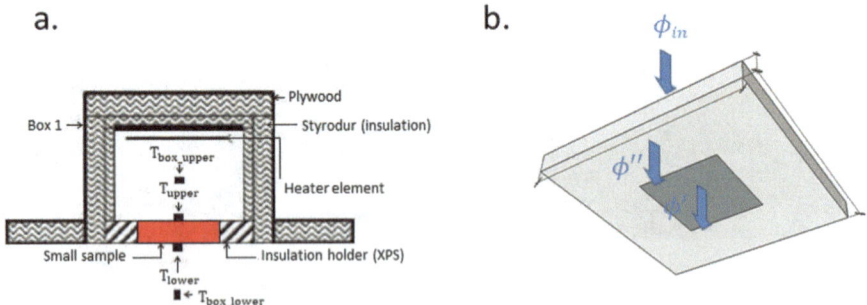

Fig. 2 Methodology illustration: **a** setup sketch, **b** sample heat balance

placed in the center of a frame made of insulation material (Fig. 2b), which serves as a holder sample to cover the box and separate the cold and hot environments. The upper side of the box represents the hot environment, which is equipped with an electrical heater element for heating the upper face of the sample with a uniform heat flow. On the other side of the sample, the ambiance is maintained at constant low temperature using a closed-cycle cryostat to ensure a steady unidirectional temperature gradient across the sample.

In general, when the steady-state is achieved, the relevant heat balance in the thermal conductivity box is expressed as follows:

$$\phi_{in} = \phi_{system} + \phi_{loss} = \frac{U^2}{R} \tag{1}$$

where ϕ_{in} is the heat flow generated by the heater element, U and R are the injected voltage and the heater element resistance, respectively, ϕ_{system} is the heat flow crossing the studied system (i.e. large sample or small sample with sample holder) and ϕ_{loss} is the heat loss through the box caused by the temperature difference between the upper side of the box T_{box_upper} and the room temperature T_{room}, which is described by the relationship:

$$\phi_{loss} = C(T_{box\_upper} - T_{room}) = C\Delta T_{loss} \tag{2}$$

Here, C is the heat loss coefficient of the box. This constant property is calculated theoretically according to the Carslaw and Jaeger formula [11] or experimentally by performing two thermal conductivity experiments on a well-known thermal conductivity material.

For the heat flow across the system, two main cases are to be considered according to the sample dimension: large samples and small samples with an insulation material frame. Therefore, the heat flow of the whole system can be expressed as follows:

$$\phi_{system} = \begin{cases} \phi_{sample} & \text{large sample} \\ \phi_{sample} + \phi_{frame} & \text{small sample} \end{cases} \tag{3}$$

where $\phi_{sample}, \phi_{frame}$ are the heat flow through the sample the insulation frame, respectively.

According to Fourier's law of heat conduction, it is well known that:

$$\phi_{sample} = \frac{\lambda S_s}{e_s}(T_{s\_upper} - T_{s\_lower}) = \frac{\lambda S_s}{e_s}\Delta T_s \tag{4}$$

$$\phi_{frame} = \frac{\lambda_f S_f}{e_f}(T_{f\_upper} - T_{f\_lower}) = \frac{\lambda_f S_f}{e_f}\Delta T_f \tag{5}$$

where λ is thermal conductivity of the sample, λ_f is the thermal conductivity of the insulation material frame, e, S, T_{upper}, T_{lower} are the thickness, surface, temperature of the upper and the lower sides of both solids, respectively.

Equations (2–5) may be substituted to Eq. (1) to express the thermal conductivity of studied samples as:

$$\lambda = \begin{cases} \frac{e_s}{S_s \Delta T_s}\left(\frac{U^2}{R} - C.\Delta T_{loss}\right) & \text{Large sample} \quad .a \\ \frac{e_s}{S_s \Delta T_s}\left(\frac{U^2}{R} - \left(\frac{S_f \Delta T_f}{e_f} + C.\Delta T_{loss}\right)\right) & \text{Small sample} \quad .b \end{cases} \qquad (6)$$

Equation 6 is the main formula for measuring the thermal conductivity using the boxes method. Unlike large samples, where the characterization is done by recording only the temperature variation of the upper and the lower faces of the sample, the characterization of small samples requires the addition of two thermocouples to monitor the temperature difference across the insulation material frame ΔT_f, and thus to evaluate the heat fraction dissipated by the frame.

In order to not use additional sensors and increase the flexibility of the boxes method, this paper characterized the thermal conductivity property of small-sized building material using a new, easy-to-use methodology presented in the following section.

2.3 Experimental Procedure

As indicated in the title, the paper focuses on the thermal conductivity box to examine a new experimental methodology for measuring the thermal conductivity of small samples without using any additional sensor.

In general, the thermal conductivity of large samples is characterized on the basis of Eq. 6a. The samples are placed in the thermal conductivity box and subjected to two different environments (cold and hot). The upper side of the sample is heated by the electrical heater element, while the upper side is kept at a constant low temperature via a controlled closed-cycle cryostat. In addition, the thermal conductivity chamber consists of four thermocouples, which are placed as illustrated in Fig. 2a. the thermocouples are used to monitor the temperatures required by the equation until reaching constant temperature signals. Two thermocouples are placed on either side of the sample to measure the temperature difference between the sample ΔT_s and the other two probes are placed on the upper and the lower side of the chamber. Another thermocouple is placed outside the chamber to measure the ambient temperature so that the fraction of heat loss can be deduced using the term $C.\Delta T_{loss}$. Finally, the thermal conductivity of large samples is simply calculated when equilibrium is reached.

On the other hand, the thermal conductivity of small samples can be characterized using Eq. 6b. As can be seen, the monitoring of the temperature of the upper and the lower faces of the sample holder is indispensable for the characterization. For this reason, some researchers added two supplementary sensors to take into account the heterogeneity of studied samples, as discussed in [12, 13]. Other researchers proposed to introduce a correction term to calculate the thermal conductivity of small samples [14]. However, the above-mentioned approaches have two main disadvantages. First, the addition of two sensors that increase the cost, calibration time, and experimental errors. Second, the correction term should be recalculated for each experiment when the insulation material or the sample dimensions are changed, which led to lengthen the duration of the experiment.

In this paper, the thermal conductivity of small samples was characterized using a switching approach. The proposed methodology consists of three steps:

- Step 1: placing thermocouples on the upper/lower sides of the sample until thermal equilibrium is reached.
- Step 2: Switching thermocouples from the sample to the insulation material frame.
- Step 3: recording the upper/lower temperature changes in the insulation material frame.

The permutation of sensors is performed after the observation of constant temperature signals of the upper/lower faces of the small sample studied. The thermal conductivity box should be made as quickly as possible to prevent any disturbance of the permanent regime of the EI702 cell. Then, the box is returned to its place to monitor the missing temperature difference of the upper/lower faces of the insulation frame. The thermal conductivity thus is calculated using Eq. 6b.

It should be noted that the EI702 cell is equipped with a real-time touch screen interface for the calibration of the thermocouples used. For more details on the calibration process, please refer to [10]. For validation purposes, the measured thermal conductivity values were compared to those obtained by the Hot Disk (TPS 2200) method. This method is based on a transient plan source technique that allows fast and accurate measurements according to ISO 22,007–2 [7]. The characterization requires two identical specimens (Fig. 3b). The TPS sensor is sandwiched between the samples to electrically heat them with constant electrical power. At the same time, it also monitors its temperature time-dependent response. Then, the thermal response is interpreted with a Thermal Constants Analyzer (TCA) to identify the thermal conductivity of studied samples using a parameter identification algorithm [15].

2.4 Used Material

In this work, a local earthen block from eastern Morocco was characterized. The samples were prepared in such a way as to have homogenous mixtures. For this purpose, the fine soil particles (<2 mm) were separated using a 2 mm sieve opening, as shown in Fig. 3.a. Furthermore, the particle size analysis showed that the soil used is a clayey soil with a clay fraction of 59%. The earthen blocks were prepared by hand mixing soil at constant water-to-clay ratio of 0.5. The mixture was then molded

a.   b.

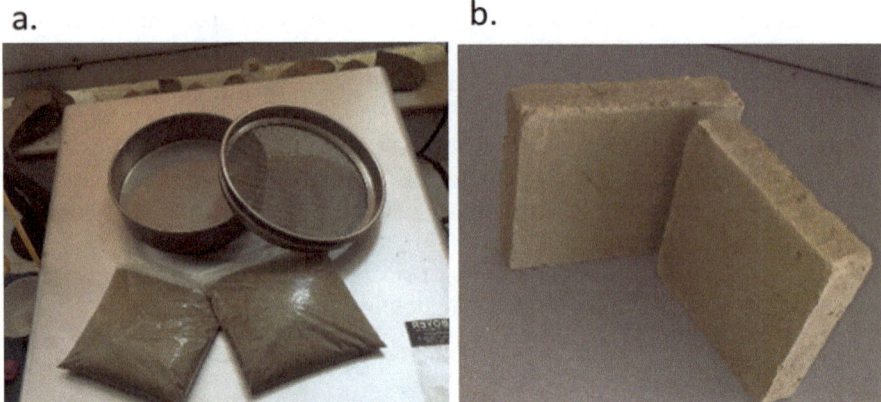

Fig. 3 Used soil: **a** sieving process, **b** final earthen blocks

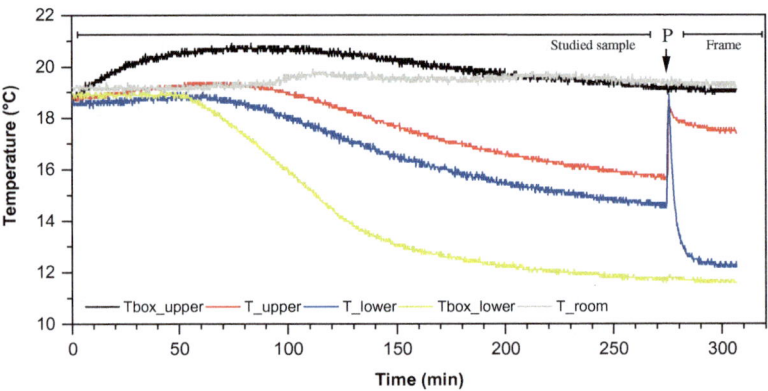

Fig. 4 Earth block thermogram

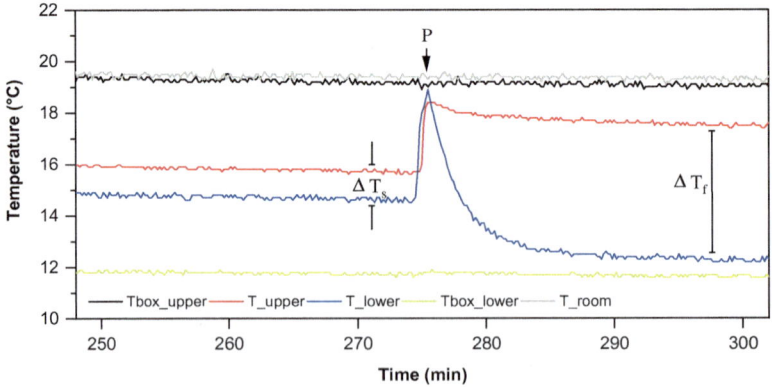

Fig. 5 Steady-state thermogram

and manually pressed into a square shape 15 cm × 15 cm. The samples were dried at room temperature for one month before the characterization.

3   Results and Discussion

3.1   Thermal Conductivity Measurement

Prior to the thermal conductivity measurements, all used temperature-sensing probes were calibrated using a touch-screen interface. It should be noted that this latest version of the boxes method (EI702 cell) is equipped with a real-time touch-screen interface for the calibration of used thermocouples. For more details about the EI702 calibration process, please refer to [10].

Three tests were conducted under the experimental conditions listed in Table 1.

Figure 4 shows a typical thermogram of the small sample studied using the proposed methodology. Once the steady-state of the small sample is established, the thermocouples on the cold and hot faces of the sample were quickly replaced on the insulation material frame. The permutation is clearly indicated by the appearance of the peaks seen in Fig. 4. The experience is left for a few minutes (approximately 30 min) to record the evolution of temperatures measured by different sensors for each face of the frame. This permutation step allows the evaluation of the heat dissipated by the frame and then deduced the thermal conductivity property of the small sample from Eq. 6a.

To be practical, the steady-state thermogram of the small sample was drawn in Fig. 5. It can be seen that the permutation peaks separate the constant signals of the small sample and the material frame from each other. As can be seen, the temperature difference between the thermal conductivity box and the room must be very small to limit losses and must be constant after the permutation to ensure the steady-state regime of the box. Finally, the thermal conductivity property of the small sample is directly deduced from the steady-state thermogram based on Eq. 6a.

The average measured value of the thermal conductivity is presented in Table 1. This value is compared to the results obtained by the reference (hot disk) method.

Table 1   Experimental conditions

Device	Loss coefficient (W/K)	Cryostat set point (C)	Voltage (V)	Resistance (Ω)
EI702	0.96	10	35	1800

Table 2   Measured thermal conductivity of studied material (W/mK)

Property	Proposed methodology	Hot disk method
Thermal conductivity	0.819 ± 0.004	0.807 ± 0.002

Results show a good agreement. The present approach was also tested to characterize different building materials (cement, plaster and XPS) with small dimensions. The finding results show the effectiveness of the proposed methodology (deviation lower than 3%).

The thermal conductivity is of great interest in building applications. In fact, knowledge of thermal conductivity of building materials permits the estimation of the annual energy consumption of buildings, with which the thermal performance of the building material is evaluated. In what follows, the potential use of the prepared local earthen blocks for residential buildings was discussed in order to promote the earthen construction in eastern Morocco.

3.2 Simulation: Case Study Building

Earthen construction is one of the great ecological alternatives to build low-environmental impact houses and to reduce the total carbon footprint of buildings. In this study, and for evaluating local earthen constructions in Morocco, a single-story building (Fig. 6) defined in [16] is used to present a typical residential building in Morocco. As shown in Fig. 7, the house is a one-floor building with a total area of 40 m^2. The ground floor contains two bedrooms and a living area. The building simulation was carried out under two different climatic conditions of Morocco (Oujda and Marrakech) using the EnergyPlus program, version 9.0 [17]. Here, Oujda city represents the cold semi-arid climate, while Marrakech represents the hot semi-arid climate. The climatic information of Oujda and Marrakech are listed in [18].

Table 3 lists the thermal properties of the external walls. The thermal properties of the other building envelope elements are summarized in Table 4. The thermal transport properties of the building components were chosen based on the Moroccan building database [19]. Two variances of the external building envelope were considered to evaluate the thermal performance of earth walls compared to conventional building materials, as shown in Table 3. The first variance represents the case study

a.   b.

Fig. 6 Case study building [16]

a.   b.   c.

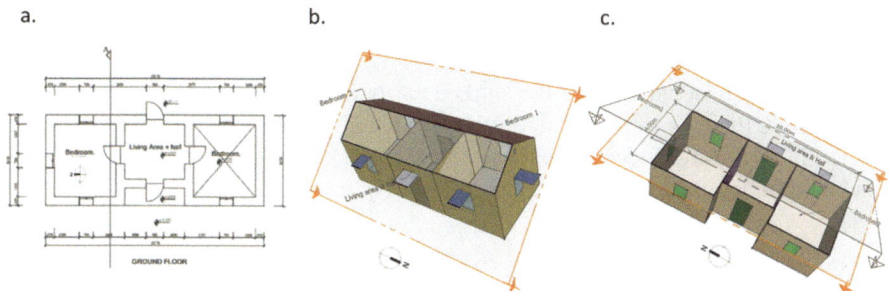

Fig. 7 Case study model: **a** 2D [16], **b-c** our sketch

Table 3 Thermophysimcal characteristics of external walls

	Thickness (cm)	λ (W/mK)	ρ (Kg/m³)	Cp (J/KgK)	R (m²K/W)
Earth wall: Earthen blocks 45 cm + earthen coating (U = 1.32 W/m²K)					
Earthen coating [a]	2	1.5	1500	2000	– –
Earthen block [b]	45	0.8	– –	– –	0.56
Earthen coating [a]	2	1.5	1500	2000	– –
Conventional wall: Concrete blocks 20 cm + cement coating (U = 1.64 W/m²K)					
Cement coating [a]	2	1	1700	1000	– –
Concrete block [c]	20	0.925	2100	920	– –
Cement coating [a]	2	1	1700	1000	– –

[a]Binayate database [19]; [b] Our work; [c] ASHRAE Handbook-Fundamentals [20]

Table 4 Material properties of the building elements

Element	Composition	Thickness (cm)	λ (W/mK)	ρ (Kg/m³)	Cp (J/KgK)	U-value (W/m²K)
Roof	Wooden stand/Clay tiles [a]	10	1	2000	800	– –
Floor	Soil [a]	30	1.5	1500	2000	– –
Doors	25 mm heavy wood [a]	2.5	0.29	1035	1600	– –
Windows	Simple glazing with wooden frame (SHGC [b] = 0.86)	0.6	0.9	– –	– –	4.07 [a]

[a]Binayate database [19]; [b] National Renewable Energy Laboratory [21]

building built with the prepared earthen blocks, while the second variance represents the reference case where external earthen walls are replaced by concrete blocks. As the thermal heat capacity of the studied earthen blocks is unknown, the earthen walls were defined by their thermal resistance using the no-mass material feature in EnergyPlus software. Indeed, the thermal resistance of a wall with a total thickness e and a thermal conductivity λ is given by the expression:

$$R_{th} = \frac{e}{\lambda} \tag{7}$$

The case study building is assumed to be occupied by three persons in accordance. With the annual occupancy schedule illustrated in Fig. 8a, and electrically lighted according to the lighting schedule shown in Fig. 8b. Lighting and electrical home appliance power densities were defined equal to 3 and 4 W/m², respectively. The building is naturally ventilated through openings and the passive night ventilation is considered during summer period, as indicated in Fig. 8c. The HVAC system was set to operate continuously at 20 °C for heating and 26 °C for cooling, according to the Moroccan thermal regulation code [4]. The input simulation data are listed in Table 5.

Table 5 Simulation parameters

Input	Value	Schedule
Heating Thermostat [a]	20 °C	ON if T_{in} < 20 °C
Cooling Thermostat [a]	26 °C	ON if T_{in} > 26 °C
Number of people	3 persons	Occupancy schedule (Fig. a)
Activity level [b]	70 W/m² (Standing/Relax)	Occupancy schedule (Fig. a)
Home appliance	4 W/m²	Occupancy schedule (Fig. a)
Lighting	3 W/m²	Lighting schedule (Fig. b)
Infiltration	0.5 ACH	Ventilation schedule including infiltration (Fig. c)

[a]Moroccan thermal regulation code [4]; [b] DOE Energy Plus [22]

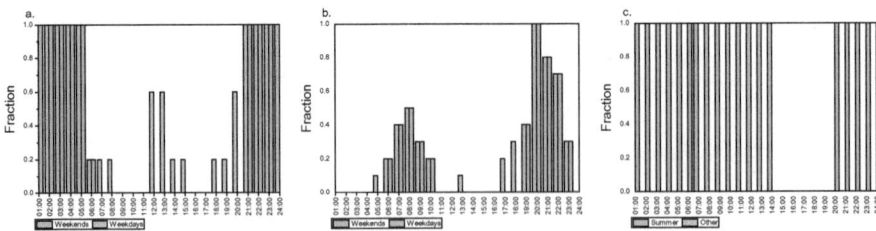

Fig. 8 Schedules: **a** occupancy, **b** artificial lighting, **c** ventilation

3.3   Simulation Outputs

Figure 9 shows the annual energy needs for the earthen walls and the reference case for Oujda and Marrakech. One can note that for both locations, the earthen walls are effective in reducing the total energy needs of residential buildings compared to conventional walls. Simulation results for these two locations show that the use of earthen walls has led to a reduction in heating and cooling needs, with a rate of 10.7% and 12.5% compared to the reference case for Oujda and Marrakech, respectively. These results demonstrate the potential use of earthen walls in semi-arid climate for reducing the annual needs of buildings.

To separately evaluate the impact of earthen walls on passive cooling and heating of the case study building, a monthly quantitative study of the heating and cooling loads were carried out. Figure 10 shows the monthly cooling needs of the case study building for Oujda and Marrakech. Cooling demand is predominating for Marrakech city with a peak monthly cooling demand of nearly 28.7 kWh/m^2 in July and a total yearly cooling demand of 96.9 kWh/m^2. However, the results show that the replacement of the reference case walls with the studied earthen walls has led to a significant reduction of about 23.4% of the maximum cooling peak and 21.2% of the total cooling needs compared to the reference case. In addition, the total cooling demand in Oujda was reduced from 59.4 to 48.6 kWh/m2. This amounted to a total reduction of about 18.1%. Furthermore, the use of earthen walls passively decreased the maximum cooling peak in July from 20.2 to 16 kWh/m^2, resulting in a reduction of about 20.6% compared to the reference case.

In terms of heating loads, Fig. 11 shows the profile of monthly needs for studied walls under Oujda and Marrakech climates. The finding results show that the use of

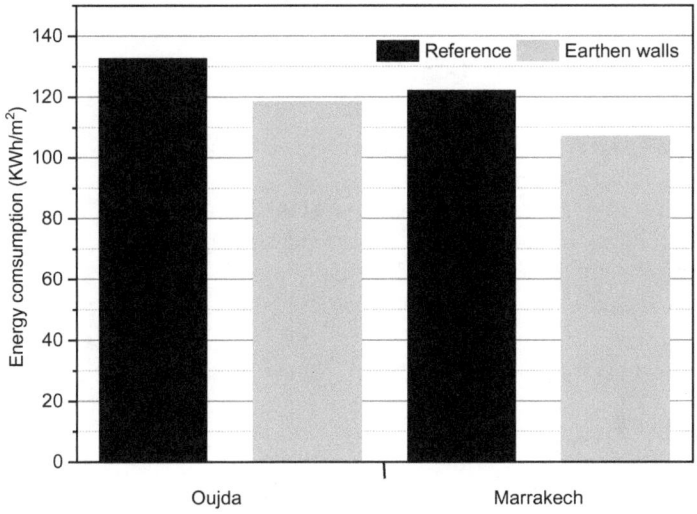

Fig. 9 Annual heat loads

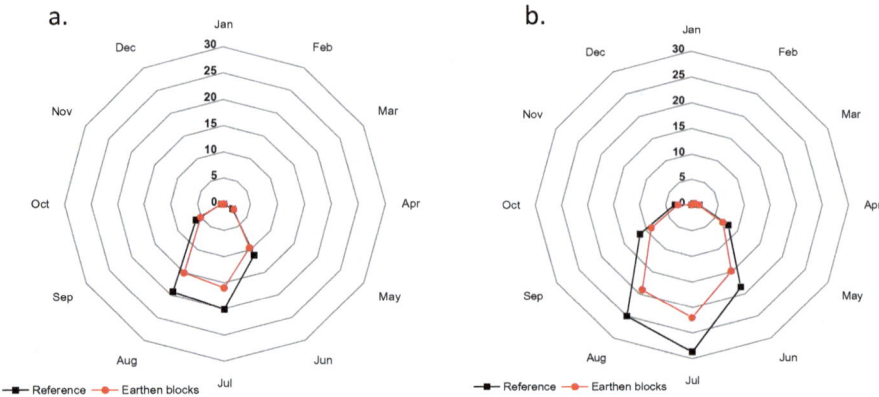

Fig. 10 Monthly cooling needs, **a** Oujda, **b** Marrakech

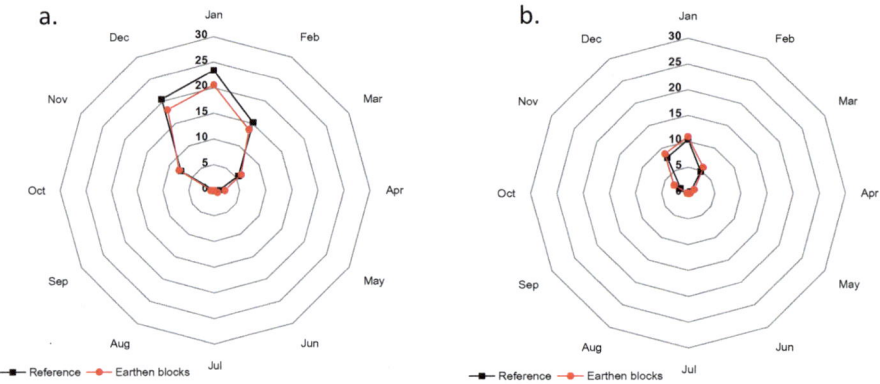

Fig. 11 Monthly heating needs: **a** Oujda, **b** Marrakech

earthen walls has a slight impact on the energy demand for heating as it reduced it by 4.7% in cold semi-arid climate compared to the reference case, while it does not change much for buildings located in hot semi-arid climate.

All these findings demonstrate the importance of earthen walls in passive cooling and reducing the annual energy needs of buildings in semi-arid climates, which is explained by the high thermal inertia of earthen building envelopes.

3.4 Discussion

The findings of the present work highlight the performance of a new easy-to-use methodology to measure the thermal conductivity of small building material samples using the boxes method, which is an industrial method commonly used for the characterization of samples with significant sizes. Also, in this work, local clay-based building material was developed and characterized. The experimental tests show that the thermal conductivity of studied earthen blocks is 0.819 W/mK. This thermal conductivity value was verified using the hot disk method.

Indeed, this new methodology could be used for measuring the thermal conductivity of industrial building materials of different sizes (large and small) and forms (solid, hollow, cubic or cylinder, etc.) without using additional sensors as discussed by Lachheb et al. [13], they characterized small building material composites using the boxes method. However, to take into account the whole system including the material frame, they used two additional sensors to measure the thermal conductivity of spent coffee plaster-based composites.

Concerning the experimental duration, the boxes method is a steady-state method that requires approximately 3 h to characterize conventional samples. For small samples and based on the proposed approach, the experimental duration increases by 30 min after the sensors' permutation in order to evaluate the steady-state temperature gradient across the insulating material frame.

In terms of performance, the studied earthen blocks show a good thermal resistance 1.32 m^2/W compared to concrete blocks 1.64 m^2/W, which can lead to a significant reduction in energy consumption of buildings. Simulation results indicate that earth walls significantly influence the energy consumption profile of buildings located in semi-arid climates and play an important role in passive cooling, highlighting the effectiveness of earth walls for improving the summer thermal comfort of buildings.

4   Conclusion and Perspective

This paper presents an experimental methodology to measure the thermal conductivity of building materials with different sizes via the boxes method set up in order to increase its flexibility in the laboratory testing. A local clay-based building material with small dimensions was prepared and characterized using the proposed methodology. The experimental tests show that the thermal conductivity of studied samples was 0.819 W/mK. The thermal conductivity characterization was also carried out using the hot disk method. It was found that the measured thermal conductivity values obtained by both methods show quite similar results (deviation smaller than 3%), highlighting the performance of the proposed methodology. By using this experimental approach, the boxes method could determine the thermal conductivity of specimens of different sizes (small and large) and forms (cubic, cylinder, etc.).

For performance analysis, numerical simulations were performed to study the impact of the prepared earthen block on the thermal performance of residential buildings in Morocco using Energy Plus software. The case study building was chosen to be located in two different semi-arid climates (hot and cold), represented respectively by the cities of Marrakech and Oujda. The simulation results show that the earthen wall reduces the annual cooling loads by 21.2% and 18.1% for hot and cold semi-arid climates compared to the reference case, respectively. In addition, the earthen walls studied show a reduction in the maximum cooling peak load of up to 20% compared to the conventional walls. However, a slight difference in heating loads for buildings located in cold semi-arid climates was observed. All these findings demonstrated the potential use of earth masonry for passive cooling and for constructing high-performant, low-environmental impact building envelopes in semi-arid climate.

The future work will be focused on the characterization of locally marketed and innovative building materials using the proposed methodology in order to develop local building material databases, which can be used in building retrofit and multi-objective studies.

Acknowledgements The authors would like to thank the "National Center for Scientific and Technical Research" (996183890) for funding this work through the PPR project "Promotion of solar energy and energy efficiency in the oriental region of Morocco".

References

1. Hyde R (2008) Bioclimatic housing: innovative designs for warm climates. Earthscan, London
2. Parker WJ, Jenkins RJ, Butler CP, Abbott GL (1961) Flash method of determining thermal diffusivity, heat capacity, and thermal conductivity. J Appl Phys 32:167–184. https://doi.org/10.1007/s00259-003-1399-3
3. Nag PK (2019) Energy Performance in Buildings: Standards and Codes. Office Buildings. Springer, pp 405–432
4. Thermal building regulations in Morocco,Moroccan Agency for Energy efficiency, AMEE, Morocco, available via: official website of AMEE. Accessed 12 Jun 2020
5. International Standard, ISO 8302 (1991) Thermal insulation—determination of steady-state thermal resistance and related properties
6. International Standard, ISO 8302 (1991) Thermal insulation—determination of steady-state thermal resistance and related properties
7. International Standard, ISO 22007–2 (2008) Plastics—determination of thermal conductivity and thermal diffusivity—Part 2: transient planeheat source (hot disc) method
8. Buratti C, Belloni E, Lunghi L, Barbanera MJ (2016) Thermal conductivity measurements by means of a new 'Small Hot-Box'apparatus: Manufacturing, calibration and preliminary experimental tests on different materials. Int J Thermophys 37(5):47. https://doi.org/10.1007/s10765-016-2052-2
9. Meukam P, Jannot Y, Noumowe A, Kofane TC (2004) Thermo physical characteristics of economical building materials. Constr Build Mater 18(6):437–443. https://doi.org/10.1016/j.conbuildmat.2004.03.010
10. Charai M, Sghiouri H, Mezrhab A, Karkri M (2020) New methodology for measuring the thermal conductivity of small samples using the boxes method with reduced sensors. Int J Thermophys 41(6):1–24. https://doi.org/10.1007/s10765-020-02649-0
11. Carslaw HS, Jaeger JC (1960) Conduction of heat in solids. Clarendon Press
12. Boumhaout M, Boukhattem L, Hamdi H, Benhamou B, Nouh FA (2014) Mesure De La Conductivite Thermique Des Materiaux De Construction De Differentes Tailles Par La Methode Des Boites 700:21–22
13. Lachheb A, Allouhi A, El Marhoune M, Saadani R, Kousksou T, Jamil A et al (2019) Thermal insulation improvement in construction materials by adding spent coffee grounds:

an experimental and simulation study 209:1411–1419. https://doi.org/10.1016/j.jclepro.2018.11.140

14. El Rhaffari Y, Boukalouch M, Khabbazi A, Samaouali A, Geraud Y (2010) Conductivité et diffusivité thermiques des matériaux poreux: application aux pierres du monument historique Chellah. Materiaux

15. Log T, Gustafsson SE (1995) Transient plane source (TPS) technique for measuring thermal transport properties of building materials. Fire Mater 19(1):43–49. https://doi.org/10.1002/fam.810190107

16. Obafemi AO, Kurt S (2016) Environmental impacts of adobe as a building material: the north cyprus traditional building case. Case Stud Constr Eng 4:32–41. https://doi.org/10.1016/j.cscm.2015.12.001

17. DOE EnergyPlus—Energy Simulation (2018). https://energyplus.net/. Accessed 12 Jun 2020

18. Sghiouri H, Charai M, Mezrhab A, Karkri M (2020) Comparison of passive cooling techniques in reducing overheating of clay-straw building in semi-arid climate. Building Simulation. Springer pp 65–88

19. AMEE. Binayate Perspective Library (2014)

20. Handbook Fundamentals, ASHRAE—American Society of Heating Vent (2017). Air-Conditioning Eng

21. Judkoff R, Neymark J, Polly B (2011) Building energy simulation test for existing homes (BESTEST-EX) (Presentation). NREL. https://doi.org/10.2172/1032671

22. DOE EnergyPlus—EnergyPlus Engineering Reference (2018). https://energyplus.net/. Accessed 12 Jun 2020

Part VI

Chapter 16
Coordinated Control Strategy to Improve Performance of PMSG Wind Power Systems

Youssef Errami, Abdellatif Obbadi, and Smail Sahnoun

Abstract This chapter is expected to help the reader understand the fundamentals of the control for the Wind Energy Conversion System (WECS) based Permanent Magnet Synchronous Generator (PMSG). At first, the modelling of the WECS is introduced, including the model of the Wind Turbine (WT), the dynamic models of the PMSG and the power converters. It can be helpful for the readers to know the operation of the PMSG wind system, which are the basics to design the control systems presented in the following chapter. Then, the controls of the Machine Side Converters (MSCs) and Grid Side Converter (GSC) are introduced. The proposed control methods were used to maximize the generated power from Wind Turbine Generators (WTGs), to keep a constant DC-bus voltage for the GSC and to control the powers fed to the grid.

1 Introduction

Because of the strong demand for clean and sustainable energy in the globe, wind generation is receiving considerable attention [1]. In addition, with the progress of wind resource, the focus is shifting toward the PMSG for its great advantages, such as simpler structure and reduced maintenance costs, higher efficiency, excellent power factor, lower losses and less maintenance [2, 3]. On the other hand, to control the WECS based PMSG and to achieve high performance and efficiency when connected to the power grid, power electronic converter systems are generally chosen as the interface between the WECS and the electric network. A PMSG control system includes a Machine Side Converters (MSCs) and Grid Side Converter (GSC) controllers. WT maximum power point tracking power curves are employed to set

Y. Errami (✉) · A. Obbadi · S. Sahnoun
Laboratory: Electronics, Instrumentation and Energy—Team: Exploitation and Processing of Renewable Energy, Department of Physics, Faculty of Science, University Chouaib Doukkali, Eljadida, Morocco
e-mail: errami.emi@gmail.com

© The Author(s), under exclusive license to Springer Nature Singapore Pte Ltd. 2021   269
R. J. Howlett et al. (eds.), *Emerging Research in Sustainable Energy and Buildings for a Low-Carbon Future*, Advances in Sustainability Science and Technology,
https://doi.org/10.1007/978-981-15-8775-7_16

reference for speed generators and power limiting techniques have been continuously examined. So, the MSC controls the PMSG active power to realize Maximum Power Point Tracking (MPPT), although the GSC maintains the dc-link voltage constant, synchronizes the ac power generated by the WECS with the electric network and it must have the aptitude of adjusting active and reactive power that the WECS exchange with the power grid in order to achieve a unity power factor of the system [4]. Numerous literature have discussed control approaches WECS based PMSG in a network context. The first class is to utilize the Vector Control approach which is normally adjusted for a particular functional point of the system. Gao et al. [5] have proposed comprehensive frequency regulation (CFR) scheme for rotor-speed-control-oriented wind power systems. Errami et al. [6] have discussed MPPT of WEGS based PMSG. Liu et al. [7] have proposed a super-capacitor energy storage unit connected parallel to the DC-link of the back-to-back converter to improve the high voltage ride through the performance of the WECS based PMSG. Another widely adopted strategy is to employ direct control techniques. In these approaches, the voltage vectors of power converters are chosen directly in compliance to the differences between the reference and actual value of torque or power. Jlassi et al. [8] have proposed a fault-tolerant PMSG using Direct Torque Control (DTC) and Direct Power Control (DPC) with two switching tables to increase the performance of the drive control. Zhang et al. [9] and Errami et al. [10] have discussed MPPT approach based DTC. Zhang et al. [11] have proposed a flux space vector-based DTC technique for PMSG employed in WECS. Zhao et al. [12] have presented the realization of MPPT technique without the loop power or speed regulator which is necessary for the control of decoupled current. The direct control techniques are simple to implement and they are characterized by fast dynamic response, no coordinate transformation and low parameter dependency. But they yield flux and torque ripples at variable switching frequency since predefined switching table and hysteresis regulator was employed.

This chapter presents Backstepping and Vector Control techniques for a grid-connected WECS topology based on PMSG and 3 Level Neutral Point Clamped (3L-NPC) (Fig. 1). Backstepping method is attractive essentially because of its ability to deal with nonlinear systems, uncertainties, disturbances, and errors of modelling. So, this work explores a nonlinear Backstepping approach to achieve power efficiency maximization. In addition, a pitch control scheme for WECS is presented to prevent wind turbine damage from excessive wind speed.

The rest part of the study is organized as follows. In Sect. 2, the modelling of the wind turbine systems and the PMSG are introduced. Backstepping and VC techniques for WECS are proposed in Sect. 3. Section 4 presents the simulation results to demonstrate the performance of the proposed approaches. Finally, the conclusion is made in Sect. 5.

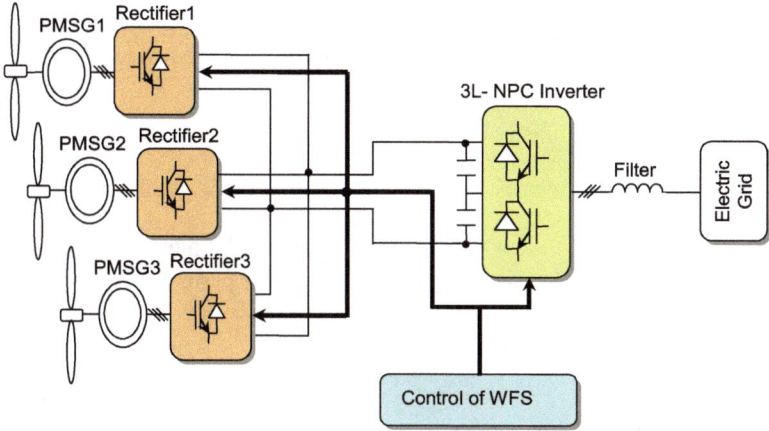

Fig. 1 Configuration of the system

2 System Model

2.1 Mathematical Model of the PMSG

The dynamic model of the PMSG in synchronous reference frame (dq) is done by [13, 14]:

$$\frac{di_q}{dt} = \frac{1}{L_s}(v_{gq} - R_s i_q - \omega_e L_s i_d - \omega_e \psi_0) \tag{1}$$

$$\frac{di_d}{dt} = \frac{1}{L_s}(v_{gd} - R_s i_d + \omega_e L_s i_q) \tag{2}$$

The expression for the electromagnetic torque can be described as:

$$T_e = \frac{3}{2} p_n [\psi_0 i_q] \tag{3}$$

The electrical angular velocity of the PMSG, ω_e is defined as:

$$\omega_e = p_n \omega_m \tag{4}$$

$$J\frac{d\omega_m}{dt} = T_e - T_{sm} - F\omega_m \tag{5}$$

where v_{gd}, v_{gq} are the d-axis and q-axis stator terminal voltages, respectively; i_q, i_d stator current in the dq frame; L_s is the inductances of the generator; R_s is the stator windings; ψ_0 is the permanent magnetic flux; p_n is the machine pole pairs; J is the total moment of inertia of the system (turbine-generator); F is the viscous friction coefficient and T_m is the mechanical torque developed by the turbine.

2.2 Wind Turbine

The wind turbine is used to convert the wind power to the mechanical one. So, the last is transformed into electrical power thanks to the permanent generators. The power generated by a wind turbine can be written as [10]:

$$P_{\text{Turbine}} = \frac{1}{2}\rho A C_P(\lambda, \beta) v^3 \tag{6}$$

where P_{Turbine} is the mechanical power of the turbine in watts, ρ is the air density (typically 1.225 kg/m^3), A is the swept area of the wind turbine blades (in m^2), C_P is the turbine power coefficient, v is the wind velocity (in m/s), β is the turbine blade pitch angle, and λ is the Tip Speed Ratio (TSR). Thus, if the air density, swept area and wind speed are constants, the output aerodynamic power is determined by the power performance coefficient of wind turbine system.

The wind turbine mechanical torque output T_{sm} given as:

$$T_{sm} = \frac{1}{2}\rho A C_P(\lambda, \beta) v^3 \frac{1}{\omega_m} \tag{7}$$

Besides, C_P is influenced by the tip-speed ratio λ which is defined as the ratio between the rotor blade tip and the speed of the wind, and is given by [15, 16]:

$$\lambda = \frac{\omega_m R}{v} \tag{8}$$

where ω_m and R are the rotor angular speed (in rad/s) and the radius of the swept area by turbine blades (in m), respectively. The computation of the power performance coefficient C_P requires the use of the information on blade geometry and blade element theory. As a result, these complex issues are usually empirically considered and a generic equation is used so as to model the power performance coefficient $C_P(\lambda, \beta)$ based on the modeling turbine system characteristics described in [17, 18] as:

$$C_P = \frac{1}{2}(\frac{116}{\lambda_i} - 0.4\beta - 5)e^{-(\frac{21}{\lambda_i})}$$
$$\frac{1}{\lambda_i} = \frac{1}{\lambda + 0.08\beta} - \frac{0.035}{\beta^3 + 1} \tag{9}$$

To progress the energy capture effectiveness from wind, an effective maximum MPPT strategy is important for WEC based PMSG. So, the blade pitch angle β

Fig. 16.2 Wind generator
power curves at various wind
speed

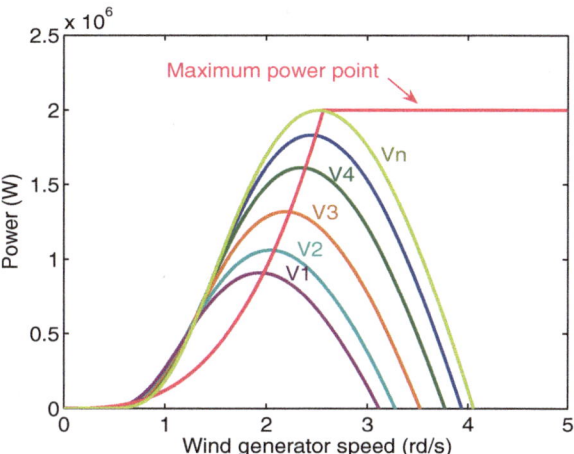

which is forever constant during MPPT approach and, there is an optimal value λ_{opt} at which the turbine converts the maximum power from wind. So, the optimal tip speed ratio λ_{opt} is employed to determine the optimal turbine rotational speed ω_{opt} as given by (10) [19].

$$\omega_{m-opt} = \frac{v\lambda_{opt}}{R} \tag{10}$$

The maximum power can be gained as:

$$P_{Turbine-max} = \frac{1}{2}\rho A \frac{R^3 C_{p\,max}\lambda_{opt}^3 \omega_{m-opt}^3}{C_{p\,max}}$$

where $C_p max$ is the maximum power coefficient and ω_{m_opt} is the reference velocity which is generated by a MPPT algorithm. Figure 16.2 depicts the typical wind turbine power curves for different wind velocities. So, the maximum power is extracted continuously from the wind because the system can operate at the peak of the $P(\omega_m)$ curve.

3   Proposed Control Approach

The proposed control approach comprises a Backstepping control for generators, pitch control for turbines and a Vector Control (VC) for grid side converter. The following sections describe different components of the proposed strategies in detail.

3.1 Backstepping Control and Pitch Angle Control System

The generator side converters are used as rectifiers and they are employed to keep the PMSG velocities at an optimal value obtained from the MPPT algorithm. The proposed control strategy for the generator side converter is based on SMC methodology. The adopted MPPT algorithm generates ω_{m_opt}, the references to speeds. Also, it is deduced from Eqs. (3) and (5) that the generator velocities can be controlled by regulating the q-axis stator current components (i_{qr}). According to the theory of Backstepping, the error of PMSG speed is selected as:

$$e_{B-\omega} = \omega_{m-opt} - \omega_m \tag{11}$$

As a result:

$$\frac{de_{B-\omega}}{dt} = \frac{d\omega_{m\_opt}}{dt} - \frac{d\omega_m}{dt} \tag{12}$$

Then, from (5), and (12), one can get:

$$\frac{de_{B-\omega}}{dt} = \frac{d\omega_{m\_opt}}{dt} - \frac{1}{J}(T_e - T_{sm} - F\omega_m) \tag{13}$$

Using (3), the time derivative $e_{B-\omega}$ can be calculated as:

$$\frac{de_{B-\omega}}{dt} = \frac{d\omega_{m\_opt}}{dt} - \frac{1}{J}\left[\frac{3}{2}p_n\left(\psi_0 i_q\right) - T_{sm} - F\omega_m\right] \tag{14}$$

To obtain the stabilizing function, consider the following function:

$$\Delta_{B-\omega} = \frac{1}{2}e_{B-\omega}^2 \tag{15}$$

The time derivative of $\Delta_{B-\omega}$ is given by:

$$\frac{d\Delta_{B-\omega}}{dt} = e_{B-\omega}\frac{de_{B-\omega}}{dt} = e_{B-\omega}\frac{d\omega_{m\_opt}}{dt} - \frac{e_{B-\omega}}{J}\left[\frac{3}{2}p_n\left(\psi_0 i_q\right) - T_{sm} - F\omega_m\right] \tag{16}$$

Besides, (16) can be written as:

$$\frac{d\Delta_{B-\omega}}{dt} = -\mu_{B-\omega}e_{B-\omega}^2 + \frac{e_{B-\omega}}{J}(J\mu_{B-\omega}e_{B-\omega} + T_{sm} - \frac{3}{2}p_n\psi_0 i_q + F\omega_m + J\frac{d\omega_{m\_opt}}{dt}) \tag{17}$$

where $\mu_{B-\omega}$ is the feedback constant of the closed-loop.

On the other hand, the q axis current component is identified as the virtual control variable in order to stabilize the PMSG speed and the sufficient condition for the global asymptotical stability is [20]:

$$\frac{d\Delta_{B-\omega}}{dt} \prec 0 \tag{18}$$

As a result, the PMSG speed convergence ω_{m-opt} is accomplished if one defines the following equations:

$$i_{q-ref} = \frac{2}{3 p_n \psi_0} (J \mu_{B-\omega} e_{B-\omega} + J \frac{d\omega_{m\_opt}}{dt} + F\omega_m + T_{sm}) \tag{19}$$

Also, to reduce the copper loss, the d axis current component is fixed to zero:

$$i_{d-ref} = 0 \tag{20}$$

To regulate the components of the current i_d and i_q to their references, we define the state errors as:

$$e_{B-d} = i_{d-ref} - i_d \tag{21}$$

$$e_{B-q} = i_{q-ref} - i_q \tag{22}$$

The time-derivate expression of e_{B-d} and ε_{B-q} are given by:

$$\frac{de_{B-d}}{dt} = \frac{di_{d-ref}}{dt} - \frac{di_d}{dt} = -\frac{1}{L_s}(v_{gd} + L_s\omega_e i_q - R_s i_d) \tag{23}$$

$$\frac{de_{B-q}}{dt} = \frac{di_{q-ref}}{dt} - \frac{di_q}{dt} = \frac{di_{q-ref}}{dt} - \frac{1}{L_s}(v_{gq} - \omega_e\psi_0 - R_s i_q - L_s\omega_e i_d) \tag{24}$$

Using Eqs. (23) and (24), we can write that:

$$e_{B-d}\frac{de_{B-d}}{dt} = e_{B-d}\left[-\frac{1}{L_s}(v_{gd} + L_s\omega_e i_q - R_s i_d)\right]$$

$$= -k_{B-d}e_{B-d}^2 + \frac{e_{B-d}}{L_s}\left[-v_{gd} - L_s\omega_e i_q + R_s i_d + L_s k_{B-d} e_{B-d}\right] \tag{25}$$

$$e_{B-q}\frac{de_{B-q}}{dt} = e_{B-q}\left[\frac{di_{q\text{-ref}}}{dt} - \frac{di_q}{dt}\right]$$

$$= -k_{B-q}e_{B-q}^2 + \frac{e_{B-q}}{L_s}\left[\begin{array}{c} L_s\dfrac{di_{q\text{-ref}}}{dt} - v_{gq} + \omega_e\psi_0 + R_s i_q \\ +L_s\omega_e i_d + ptk_{B-q}L_s e_{B-q} \end{array}\right] \quad (26)$$

where k_{B-d} and k_{B-q} are the feedback constants of the closed-loop. The new Lyapunov function can be chosen as:

$$\mathbb{R}_{B-q} = \frac{1}{2}e_{B-\omega}^2 + \frac{1}{2}e_q^2 + \frac{1}{2}e_d^2 \quad (27)$$

The derivate of the Lyapunov function (27) is:

$$\frac{d\mathbb{R}_{B-q}}{dt} = e_{B-\omega}\frac{de_{B-\omega}}{dt} + e_{B-d}\frac{de_{B-d}}{dt} + e_{B-q}\frac{de_{B-q}}{dt} \quad (28)$$

Using Eqs. (16), (25) and (26), we can rewrite (28) as:

$$\frac{d\mathbb{R}_{B-q}}{dt}$$

$$= -\mu_{B-\omega}e_{B-\omega}^2 + \frac{e_{B-\omega}}{J}(T_{sm} - \frac{3}{2}p_n\psi_0 i_q + F\omega_m + J\mu_{B-\omega}e_{B-\omega} + J\frac{d\omega_{m\_opt}}{dt})$$

$$-k_{B-d}e_{B-d}^2 + \frac{\varepsilon_{B-d}}{L_s}[-v_{gd} - L_s\omega_e i_q + R_s i_d + L_s k_{B-d}e_{B-d}]$$

$$-k_{B-q}e_{B-q}^2 + \frac{e_{B-q}}{L_s}\left[L_s\frac{di_{q-ref}}{dt} - v_{gq} + \omega_e\psi_0 + R_s i_q + L_s\omega_e i_d + k_{B-q}L_s e_{B-q}\right] \quad (29)$$

If the time derivative $\mathbb{R}_{B-q}$ is always negative, the global asymptotic stability in the currents is ensured. Then, using (29) backstepping control laws are chosen as follows:

$$v_{qr} = \omega_e\psi_0 + R_s i_q + L_s\omega_e i_d + L_s\frac{di_{q-ref}}{dt} + L_s k_{B-q}e_{B-q} \quad (30)$$

$$v_{dr} = L_s k_{B-d}e_{B-d} - L_s\omega_e i_q + R_s i_d \quad (31)$$

Also, substituting Eqs. (19), (30) and (31) into (29) gives:

$$\frac{d\mathbb{R}_{B-q}}{dt} = -\mu_{B-\omega}e_{B-\omega}^2 - k_{B-d}e_{B-d}^2 - k_{B-q}e_{B-q}^2 \leq 0 \quad (32)$$

Then Eq. (32) guarantees the asymptotic stability of the system and the regulation of the speed of the generators. The control diagram for each generator-side converter is depicted in Fig. 3. On the other hand, it is vital to limit the converted mechanical power during higher wind velocities and when the turbine output is above the nominal power. Figure 3 shows the block diagram of the pitch angle controller. The pitch

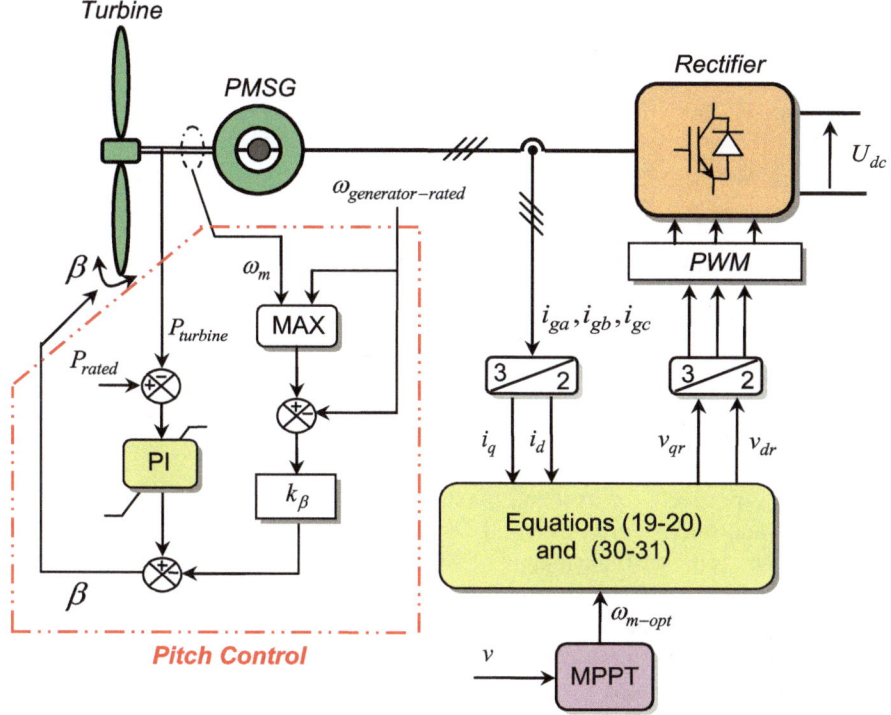

Fig. 3 Control of individual generator and pitch control

angle is kept constant at zero degrees until the velocity of the PMSG attains the rated speed. So, the pitch angle controller enters in operation and the angle of blades β, will increase to decrease the performance coefficient of power.

3.2   Grid Side Controller Methodology with VC

The 3L-NPC grid side converter is used to regulate the voltage of the DC bus capacitor, to provide grid synchronization and to regulate the reactive and active power flowing into the grid. Therefore, it can regulate the grid side power factor during wind variation [21, 22]. Hence, the Vector Control with PI control loops is used. On the other hand, in the rotating dq reference frame and if the grid voltage space vector is oriented on d-axis, the model for the grid side converter is shown in Eqs. (33) and (35):

$$L_f \frac{\mathrm{d}i_{d-f}}{\mathrm{d}t} = e_d - R_f i_{d-f} + \omega L_f i_{q-f} - V \qquad (33)$$

$$L_f \frac{di_{q-f}}{dt} = e_q - R_f i_{q-f} - \omega L_f i_{d-f} \tag{34}$$

Thus, the DC-link voltage deviation is obtained as:

$$C \frac{dU_{dc}}{dt} = \frac{3}{2} \frac{v_d}{U_{dc}} i_{d-f} - i_{inv} \tag{35}$$

Besides, the active power and reactive power can be expressed as [23]:

$$P = \frac{3}{2} V i_{d-f} \tag{36}$$

$$Q = \frac{3}{2} V i_{q-f} \tag{37}$$

where e_d, e_q are inverter d-axis and q-axis voltage components; i_{d-f}, i_{q-f} are d-axis current and q-axis current of grid; R_f is the filter resistance; L_f is the filter inductance; U_{dc} is the dc-link voltage; C is the dc-link capacitor. As a result, reactive and active power control can be achieved by controlling quadrature and direct grid current components, respectively. Figure 4 shows the control block diagram of the grid side converter. Double-loop structure is employed. The internal control loops regulate q-axis current and d-axis current. In addition, external voltage loop regulates the dc-link voltage via controlling the output power.

4   Simulation Result Analysis

In order to verify the efficacy of the proposed control, a simulation test system comprising of three PMSG-based WT and a multi-level NPC is built in Matlab/Simulink, as shown in Fig. 1.

The parameters of systems are listed in Appendix. Figures 5, 6, 7 and 8 show the performance of the proposed strategies. It is observed that, for each turbine, the rotor velocity vary according to the wind speed deviations. Also, if the wind velocity is less than the rated value v_n, then the coefficient of power is maintained at his maximums C_{p-max} and the pitch angle controller is deactivated Fig. 5b and Fig. 6a. The pitch control is not active until the wind velocity is up v_n and the power flowing into the grid reaches its maximum level. Figure 6c depicts the actual speed of PMSG1 and optimal speed. It can be observed that the proposed controller succeeded to track the optimal speed at different wind velocities. The active power flowing into the grid is shown in Fig. 7a. It is clear that active power increases with the increase in wind speed. The output voltage of the inverter is shown in Fig. 7b. Figure 7c shows the simulation result of DC-bus voltage that remains a constant value. Consequently, this proves the effectiveness of the implanted regulators. Figure 8 shows the variation and

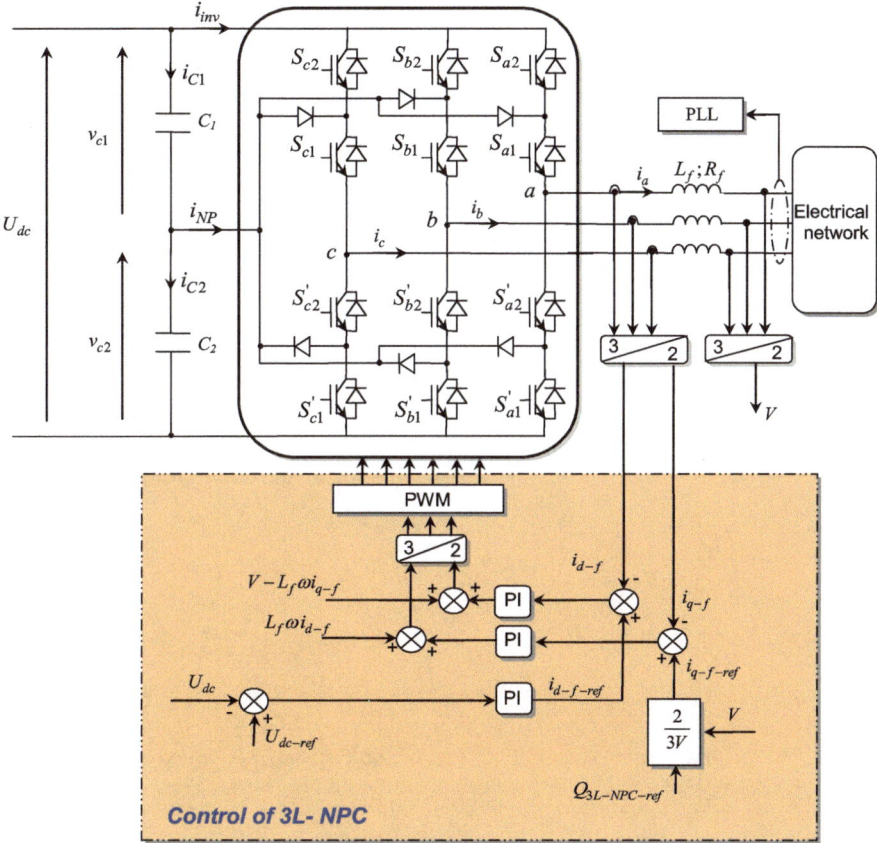

Fig. 4 Schematic of control strategy for inverter

a closer observation of three-phase voltage and current of GRID. The unity power factor operation is guaranteed by fixing the reactive power reference to zero.

5 Conclusion

In this chapter, the control strategies for the WECS based PMSG are introduced. Both the speed and power-control methods are employed to accommodate the requirement of the WECS. This chapter tries to help the readers to understand how to control the WECS based PMSG, which is the basis for the Backstepping and Vector Control techniques. The Backstepping approach is used to implement the MPPT. Control strategy based on Vector Control (VC) theory is applied for DC link voltage regulation and unity power factor control under varying wind conditions. The detailed derivation for the control laws has been provided and the conditions for the existence

Fig. 5 Performances of
WECS during wind speed
variations (1)

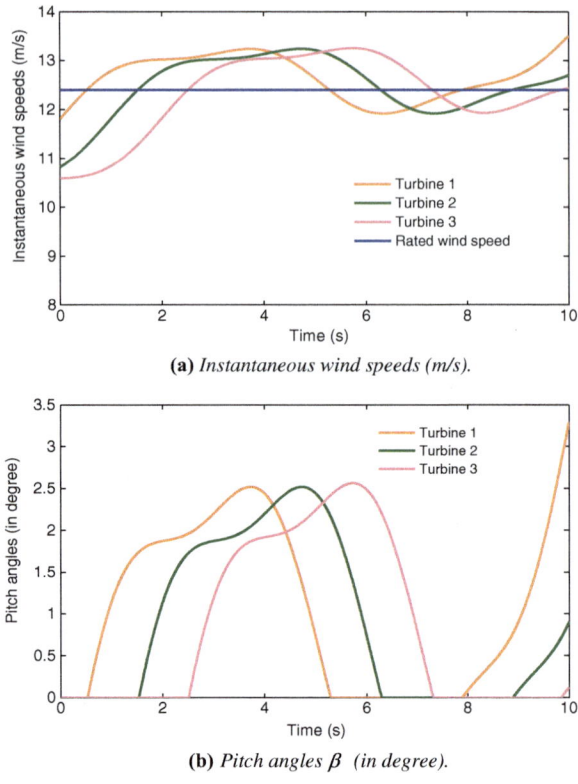

(a) *Instantaneous wind speeds (m/s).*

(b) *Pitch angles β (in degree).*

of the Backstepping approach are found by applying Lyapunov stability theory. The
simulation results show the effectiveness of the combining Backstepping algorithm
with VC technique.

Fig. 6 Performances of
system during wind speed
variations (2)

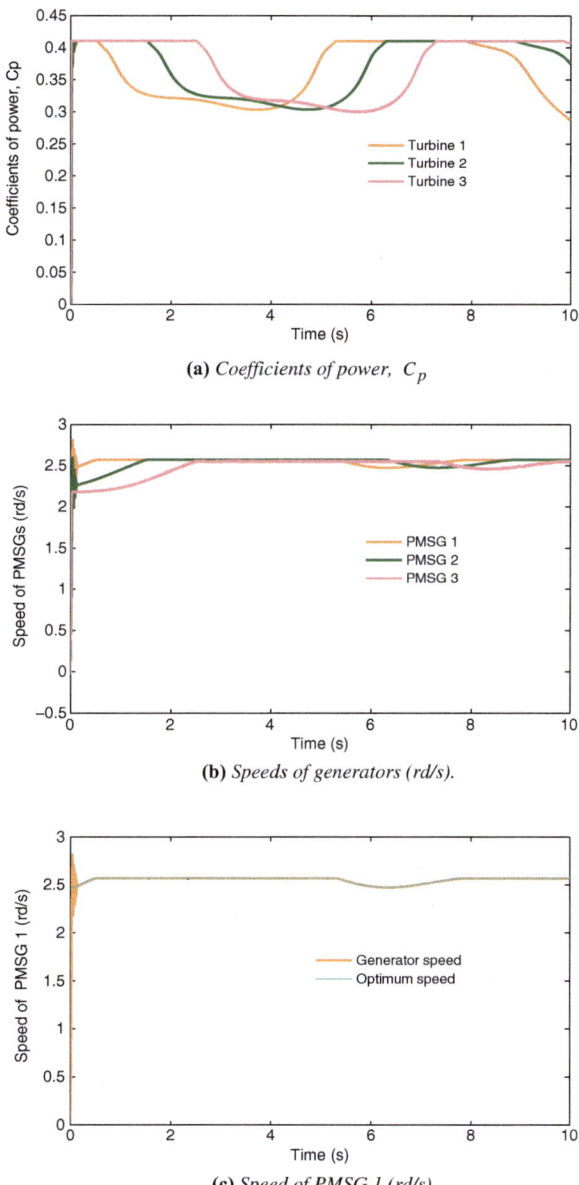

(a) *Coefficients of power,* C_p

(b) *Speeds of generators (rd/s).*

(c) *Speed of PMSG 1 (rd/s).*

Fig. 7 Performances of
system during wind speed
variations (3)

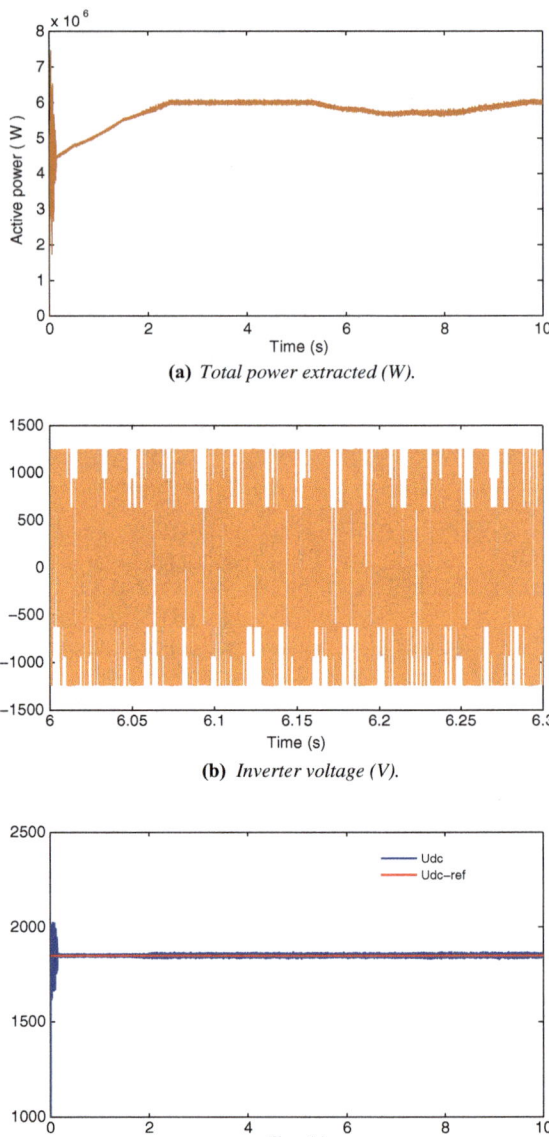

(**a**) *Total power extracted (W).*

(**b**) *Inverter voltage (V).*

(**c**) *DC link voltage (V)*

Fig. 8 Performances of system during wind speed variations (4)

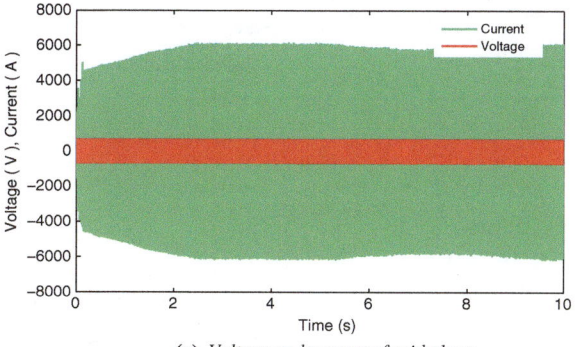

(a) *Voltage and current of grid phase.*

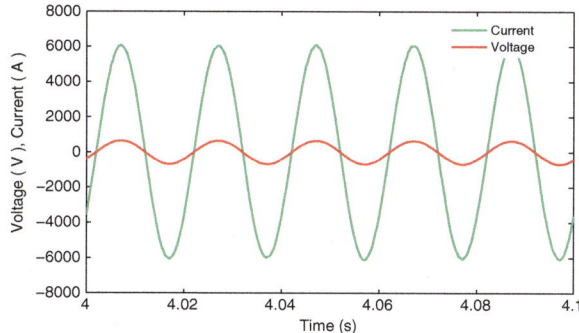

(b) *Voltage and current of grid phase (zoom).*

Appendix

See Tables 1 and 2.

Table 1 Parameters of the power synchronous generator

Parameter	Value
P_r rated power	2 (MW)
ω_m rated mechanical speed	2.57 (rad/s)
R stator resistance	0.008 (Ω)
L_s stator d-axis inductance	0.0003 (H)
ψ_f permanent magnet flux	3.86 (wb)
p_n pole pairs	60

Table 2 Parameters of the turbine

Parameter	Value
ρ the air density	1.08 kg/m^3
A area swept by blades	4775.94 m^2
v_n base wind speed	12.4 m/s

References

1. Effat Jahan Md, Rifat Hazari SM, Muyeen AU, Takahashi R, Tamura J (2019) Primary frequency regulation of the hybrid power system by deloaded PMSG-based offshore wind farm using centralised droop controller. J Eng 2019(18):4950–4954
2. Youssef Errami, Abdellatif Obbadi, Smail Sahnoun, Mohammed Ouassaid, Mohamed Maaroufi (2019) Performance evaluation of backstepping approach for wind power generation system-based permanent magnet synchronous generator and operating under non-ideal grid voltages. Int J Power Energy Convers 10(4). https://doi.org/10.1504/IJPEC.2019.10024034
3. Errami Y, Ouassaid M, Cherkaoui M, Maaroufi M (2015) Variable structure sliding mode control and direct torque control of wind power generation system based on the PM synchronous generator. J Electr Eng 66(3):121–131. https://doi.org/10.1515/jee-2015-0020
4. Errami Y, Ouassaid M, Maaroufi M (2015) Modelling and optimal power control for permanent magnet synchronous generator wind turbine system connected to utility grid with fault conditions. World J Model Simul 11(2):123–135
5. Gao DW, Ziping Wu, Yan W, Zhang H, Yan S, Wang X (2019) Comprehensive frequency regulation scheme for permanent magnet synchronous generator-based wind turbine generation system. IET Renew Power Gener 13(2):234–244
6. Errami Y, Ouassaid M, Maaroufi M (2013) Modeling and variable structure power control of PMSG based variable speed wind energy conversion system. J Optoelectron Adv Mater 15(11–12):1248–1255
7. Liu G, Hu J, Tian G, Xu L, Wang S (2019) Study on high voltage ride through control strategy of PMSG-based wind turbine generation system with SCESU. J Eng 2019(17):4257–4260
8. Jlassi I, Marques Cardoso AJ (2019) Fault-tolerant back-to-back converter for direct-drive pmsg wind turbines using direct torque and power control techniques. IEEE Trans Power Electron 34(11):11215–11227
9. Zhang Z, Zhao Y, Qiao W, Qu L (2014) Space-vector modulated sensorless direct-torque control for direct-drive PMSG wind turbines. IEEE Trans Ind Appl 50(4):2331–2341
10. Errami Y, Obbadi A, Ouassaid M, Maaroufi M (2018) Direct torque control strategy applied to the grid connected wind farm based on the PMSG and controlled with variable structure approach. Int J Power Energy Convers 9(1):58–88
11. Zhang Z, Zhao Y, Qiao W, Qu L (2014) A space-vector modulated sensorless direct-torque control for direct-drive PMSG wind turbines. IEEE Trans Ind Appl 50:2331–2341
12. Zhao Y, Wei C, Zhang Z, Qiao W (2013) A review on position/speed sensorless control for permanent-magnet synchronous machine-based wind energy conversion systems. IEEE J Emerg Sel Top Power Electron 1:203–216
13. Dao ND, Lee D-C, Lee S (2019) A simple and robust sensorless control based on stator current vector for PMSG wind power systems. IEEE Access 7:8070–8080
14. Errami Y, Ouassaid M, Maaroufi M (2014) Variable structure control for permanent magnet synchronous generator based wind energy conversion system operating under different grid conditions. In: 2014 second world conference on complex systems (WCCS), pp 340–345, 10–12 Nov. 2014. https://doi.org/10.1109/ICoCS.2014.7060996
15. Errami Y, Obbadi A, Sahnoun S, Ouassaid M, Maaroufi M (2019) Hybrid control strategy for wind turbine system driven permanent magnet synchronous generator. Prog Ind Ecol An Int J 13(2):124–143

16. Errami Y, Ouassaid M, Maaroufi M (2012) Control of grid connected PMSG based variable speed wind energy conversion system. Int Rev Model Simul (I.RE.MO.S), 5(2):655–664. ISSN: 1974–9821

17. Errami Y, Maaroufi M, Ouassaid M (2013) Maximum power point tracking of a wind power system based on the pmsg using sliding mode direct torque control. Renewable and sustainable energy conference (IRSEC), 2013 international, pp 228–233, 7–9 Mar 2013. https://doi.org/10.1109/IRSEC.2013.6529706

18. Errami Y, Benchagra M, Hillal M, Ouassaid M, Maaroufi M (2012) MPPT strategy and direct torque control of pmsg used for variable speed wind energy conversion system. Int Rev Model Simul (I.RE.MO.S) 5(2):887-898. ISSN: 1974–9821

19. Errami Y, Hilal M, Benchagra M, Ouassaid M, Maaroufi M (2012) Nonlinear control of MPPT and grid connected for variable speed wind energy conversion system based on the PMSG. J Theor Appl Inf Tech 39(2):205–217

20. Huang N, He J, Nabeel A, Demerdash O (2013) Sliding mode observer based position self-sensing control of a direct—drive PMSG wind turbine system fed by NPC converters. In: IEEE international electric machines drives conference (IEMDC), pp. 919–925

21. Errami Y, Maaroufi M, Cherkaoui M, Ouassaid M (2012) Maximum power point tracking strategy and direct torque control of permanent magnet synchronous generator wind farm. In: 2012 international conference on complex systems (ICCS), pp 1–6, 5–6 Nov 2012. https://doi.org/10.1109/ICoCS.2012.6458520

22. Errami Y, Maaroufi M, Ouassaid M (2012) Variable structure direct torque control and grid connected for wind energy conversion system based on the PMSG. In: 2012 International Conference on Complex Systems (ICCS), pp. 1–6, 5–6 Nov. 2012. https://doi.org/10.1109/ICoCS.2012.6458524

23. Errami Y, Ouassaid M, Maaroufi M (2013) Control scheme and power maximization of permanent magnet synchronous generator wind farm connected to the electric network. Int J Syst Control Commun (IJSCC), 5(3/4):214–230

Chapter 17
High-Temperature Heat Pumps for Sustainable Industry

Adrián Mota-Babiloni, Carlos Mateu-Royo, and Joaquín Navarro-Esbrí

Abstract High-temperature heat pumps (HTHPs) based on vapour compression technology are being considered an energy-efficient possibility for low-grade waste heat recovery in industrial applications. If the HTHP technology is extended, the industrial energy efficiency would be closer to the 2050 target of decarbonisation, and consequently, it would reduce the worldwide greenhouse gas emissions. This chapter gives a comprehensive overview of the current status and future possibilities of the HTHP technology based on the working principles, working fluids, configurations, existing prototypes, and the possibility of reversible operation. Many working fluid possibilities with low Global Warming Potential are available, but there is no perfect alternative, and the selection must consider many factors. The relatively higher temperature lifts make advanced configurations feasible. Existing HTHP prototypes cover a wide range of working fluids and configurations; nevertheless, they are going to be required for developing HTHPs capable of supply increasing temperatures from higher heat sources.

A. Mota-Babiloni (✉)
Postdoctoral Researcher, ISTENER Research Group, Mechanical Engineering and Construction Department, Universitat Jaume I, Av. de Vicent Sos Baynat, s/n, Castelló 12071, Spain
e-mail: mota@uji.es

C. Mateu-Royo
Researcher, ISTENER Research Group, Mechanical Engineering and Construction Department, Universitat Jaume I, Av. de Vicent Sos Baynat, s/n, Castelló 12071, Spain
e-mail: mateuc@uji.es

J. Navarro-Esbrí
Full Professor, ISTENER Research Group, Mechanical Engineering and Construction Department, Universitat Jaume I, Av. de Vicent Sos Baynat, s/n, Castelló 12071, Spain
e-mail: navarroj@uji.es

© The Author(s), under exclusive license to Springer Nature Singapore Pte Ltd. 2021   287
R. J. Howlett et al. (eds.), *Emerging Research in Sustainable Energy and Buildings for a Low-Carbon Future*, Advances in Sustainability Science and Technology,
https://doi.org/10.1007/978-981-15-8775-7_17

1 Introduction

In November 29th, 2019, the European Union parliament declared a global climate and environmental emergency. The target of reducing 40% carbon dioxide emissions by 2030 compared to 1990, as agreed in the Paris Agreement [14], is no longer viable, and it must be raised to 55%, as an intermediate step to achieve climate neutrality by 2050. Each law approved to protect the environment must be revised to make it compatible with keeping the global average surface temperature increase below 1.5° by 2100 compared to pre-industrial levels.

It is estimated that in 2017, the industrial processes and product use was responsible for the emission of 8% of the greenhouse gases in the EU, and fuel combustion and fugitive emissions from fuels, 54% [6]. According to the Environmental Protection Agency of the USA, 22% of the greenhouse emissions were caused by the industrial sector [5].

According to the estimations of [7], approximately 50% of global energy use would end up as waste heat in 2030, having the industry sector the second-highest portion of the theoretical waste heat recovery potential (approx. 28%). Data from 2015 illustrates that the potential waste heat in the EU has been estimated above 300 TWh/year, mostly available between 100 and 200 °C [13].

HTHPs becomes an option to upgrade low-temperature waste heat, being usable again for industrial or commercial purposes. If an HTHP lifts the waste heat temperature, there is a more significant potential for waste heat recovery. Up to date, different HTHPs technologies can be found: vapour compression, chemical, absorption, etc. [16]. However, vapour compression technology is the most widely considered [1], both from research and commercial point of view. This chapter explores the features of these HTHPs types, the current status of development and future possibilities.

2 Vapour Compression HTHPs

Vapour compression HTHPs operation principles are equivalent to other applications based on mechanical refrigeration and systems for residential, commercial and industrial purposes, commonly used for food conservation and freezing; heating, ventilation and air conditioning systems; stationary and mobile air conditioners; refrigerated transportation, cryogenics, etc. However, under this situation, the interest is based on the amount of heat that can be exchanged in the condenser (heat sink) given a quantity of heat available at the evaporator (heat source) and the energetic cost that represents from the compressor. Moreover, the range of temperatures observed in this application is significantly higher than seen in others, and therefore the components are subject to more restrictive conditions than usual. This fact must be considered during the design and construction of the systems. A schematic representation of the

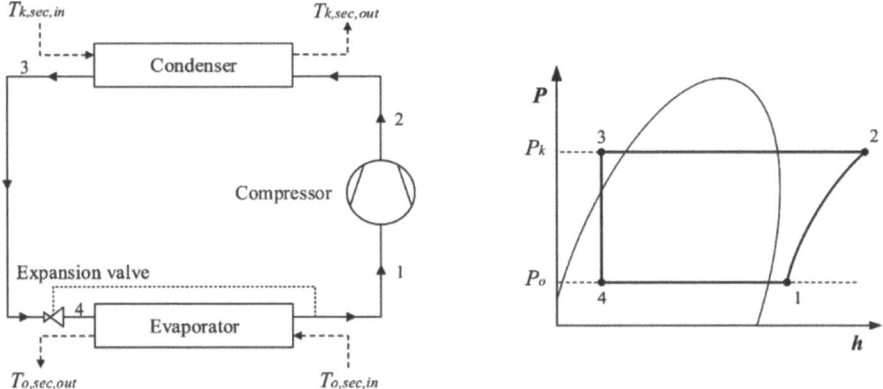

Fig. 1   Vapour compression HTHP, Ph diagram and Ts diagram

essential components common to all vapour compression HTHPs, and the pressure-enthalpy and temperature-entropy diagrams of its basic configuration are shown in Fig. 1.

3   Working Fluids

Typically, the most commonly used working fluid (also known as a refrigerant) for HTHPs is the HFC-245fa [1]. This fluid belongs to the third generation of refrigerants, as it is chlorine-free and does not decompose the atmospheric ozone. However, the 100-year time horizon global warming potential value is relatively high, 875, and can contribute to global warming if it is accidentally leaked to the atmosphere. Therefore, HFC-245fa must be substituted by environmentally friendly alternatives to transit to sustainable HTHPs in the industry.

In this way, HTHP sector can learn from other vapour compression application experience in the transition to the low GWP working fluids. There are different possibilities to classify the working fluids, but the most common today is the difference between natural and synthetic working fluids. Natural fluids are those that are directly available from nature, and no chemical transformation is needed to obtain it. Therefore, this option would be preferable to reduce the cost of acquisition of the fluid. However, synthetic fluids present characteristics that make them attractive under specific circumstances. Table 1 presents selected representative properties of the today's most promising low GWP refrigerants, compared with HFC-245fa, which is taken as a reference for calculating the benefits. Figure 2 shows the temperature-entropy diagram and the saturation curves of HFC-245fa and selected low GWP alternatives.

When considering the working fluids for HTHPs, different characteristics become relevant. Among them, the following ones can be highlighted: the critical temperature

Table 1 Main properties of working fluids for HTHPs

	M (g mol^{-1})	T_{crit} (°C)	P_{crit} (MPa)	ρ_V (kg m^{-3})^a	NBP (°C)	ODP	GWP$_{100}$	ASHRAE safety class
HFC-245fa	134.0	154.0	3.65	38.68	15.1	0	858	B1
HC-601	72.2	196.6	3.37	8.93	36.1	0	5	A3
HC-600	58.1	152.0	3.80	22.45	-0.5	0	4	A3
R514A	139.6	178.0	3.52	22.78	29.1	0	2	B1
HFO-1336mzz(Z)	164.1	171.4	2.90	24.07	33.4	0	2	A1
HCFO-1233zd(E)	130.5	166.5	3.62	30.66	18.3	0.00034	1	A1
HCFO-1224yd(Z)	148.5	155.5	3.33	40.18	14.6	0.00012	<1	A1

^{a}At saturated pressure of 75 °C

Fig. 2 T–s diagram of HFC-245fa and its low GWP alternatves

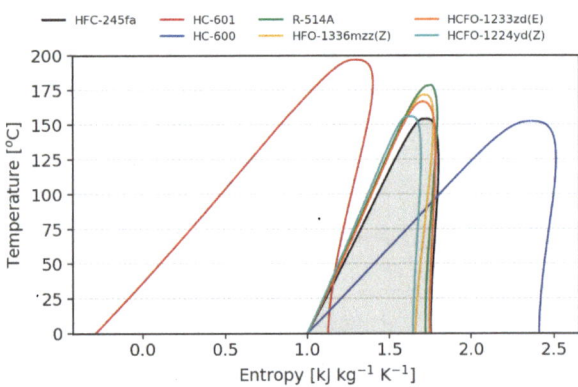

and pressure for defining the maximum condensation temperature without supercritical operation and the maximum design temperatures of the vapour compression pipelines and components, saturated vapour density and latent heat of vapourization, for defining the compressor size in new design installations or the heating capacity in drop-in or light retrofit replacements, the normal boiling point (NBP) to avoid air filtrations when the system is switched off and to define the minimum practical evaporation temperature.

Then, the ozone depletion potential (ODP) which must be below than certain level to respect the Montreal Protocol put into force, the Global Warming Potential (GWP) to define the direct emissions contribution, and finally, the ASHRAE Standard 34 safety classification, which indicates the flammability and toxicity levels of working fluids.

Other parameters from different natures, such as liquid density, vapour viscosity, specific heat capacity and thermal conductivity, material and lubricant oil compatibility, working fluid availability, and price, among many others, must be checked when selecting the refrigerant for the HTHP application.

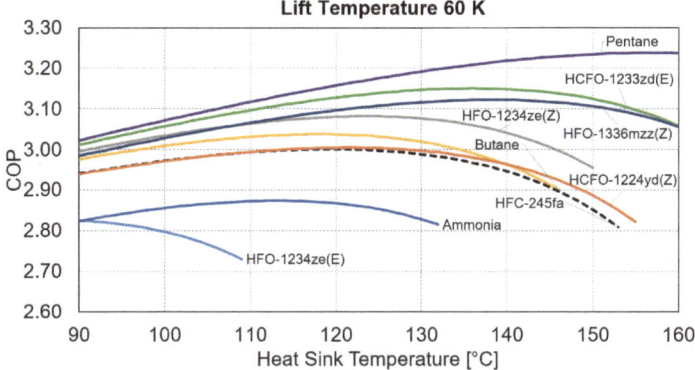

Fig. 3 Theoretical COP for different HTHP working fluids

For the basic cycle configuration, a theoretical simulation illustrates the COP for different heat sink temperatures, temperature lift of 60 K, isentropic efficiency of 0.8, subcooling degree of 5 K, superheating degree of 15 K (Fig. 3).

4 Advanced Configurations

Different variations from the basic vapour compression configuration can be proposed for various purposes: increasing the heating production, increasing the coefficient of performance, controlling the discharge temperature for the same heat source level, recovering heat at two different heat sink levels, among others. However, additional characteristics apart from thermodynamic and energetic parameters must be considered, as the initial cost, complexity, conditions control, etc. Schematic diagrams of these configurations are presented in Fig. 4.

A typical variation from the basic cycle is the internal heat exchanger (IHX) or liquid-to-suction heat exchanger configuration. Here, an additional heat exchanger is connected between to produce an additional subcooling effect at the liquid line while superheating the fluid at the suction line at the same time. In refrigeration cycles, it is commonly installed for increasing the cooling effect at the evaporator without introducing elements of mechanical subcooling. In heating instead, the interest is increasing the internal energy of the working fluid at the suction of the compressor to be then available at the discharge point, which is quite close to the entrance at the condenser [10].

Besides the inclusion of an IHX, a recent trend observed is the utilisation of ejector to improve the performance of the R-744 (CO_2) vapour compression refrigeration systems. The use of an ejector to avoid the thermodynamic losses at the expansion device in HTHPs has been theoretically studied by [3]. In addition to the conventional ejector outlet split heat pump cycle, the potential of a condenser outlet split ejector (CSHPC) enhanced was studied. The main differences between both cycles are that

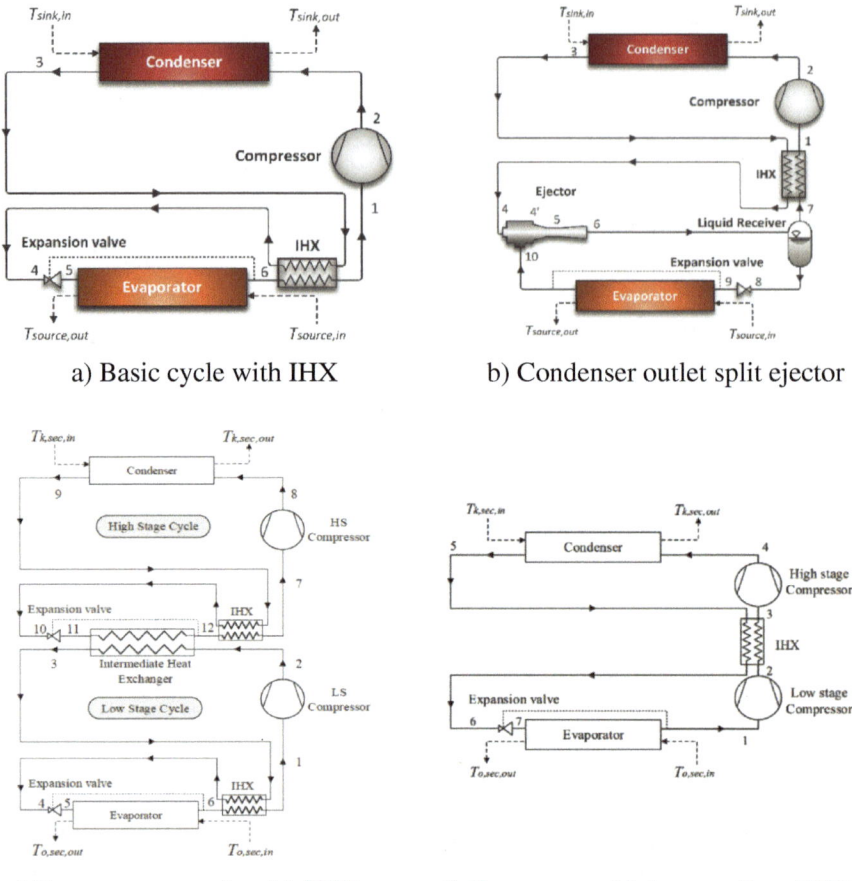

a) Basic cycle with IHX   b) Condenser outlet split ejector

c) Two-stage cascade with IHX   d) Two-stage with intermediate-IHX

Fig. 4 Advanced HTHP vapour compression configurations

the second one does not need a vapour–liquid separator, favouring the oil return to
the compressor, and that allows two evaporators, giving the possibility of recovering
heat from different sources. Compared with the basic cycle, the COP of CSHPC
could be improved up to 14.5 and 10.3%, the compressor size reduced to 26.8%, and
higher exergy efficiency in some main components.

Then, different vapour compression cycles can be connected by a heat exchanger to
cover high-temperature lifts without losing energy efficiency. This option is known as
cascade configuration, and the connection heat exchanger acts as a condenser for the
lower pressure circuit, and evaporator for the higher pressure circuit. Different stages
can be considered, even though two and three stages are the most commonly observed,
allowing the inclusion of the internal heat exchanger for each cycle. Moreover, it
offers great flexibility for the combination of working fluids to optimise the total COP

system, to restrict the installation of some circuits under particular safety conditions, or to reduce the refrigerant charge of one of the working fluids [12]. studied a wide variety of low GWP refrigerants for cascade refrigeration circuits optimising the IHX effectiveness and intermediate temperature to maximise the resulting COP. The resulting heating production for all options considered was comparable, but a proper selection of the working fluid pair can increase the COP between 10 and 17%. There is a trade-off between the minimum compressor size required and the final COP.

Finally, the compression can be split by injecting vapour or liquid in different ways, with similar purposes than the cascade vapour compression configuration. Among the different possibilities, Mateu-Royo et al. [9] proposed the two-stage cycle with intermediate-IHX, which was convenient only for higher temperature lifts.

This section focused only on introducing some promising options, but guidelines in working fluid selection presented in the previous section must be followed as well. The choice of working fluids plays an essential role in the selection of the optimal advanced HTHP configuration.

5 Experimental Prototypes

Today, there is a much wider variety in working fluids than configurations in the existing HTHP prototypes for which the results are available in the literature. While some of the most important investigations are presented in the following, a complete compilation of results can be found in Fig. 5.

Bamigbetan et al. [4] developed a hydrocarbon-based HTHP, and the selected working fluid was the R-600a (Isobutane). All the components except for the

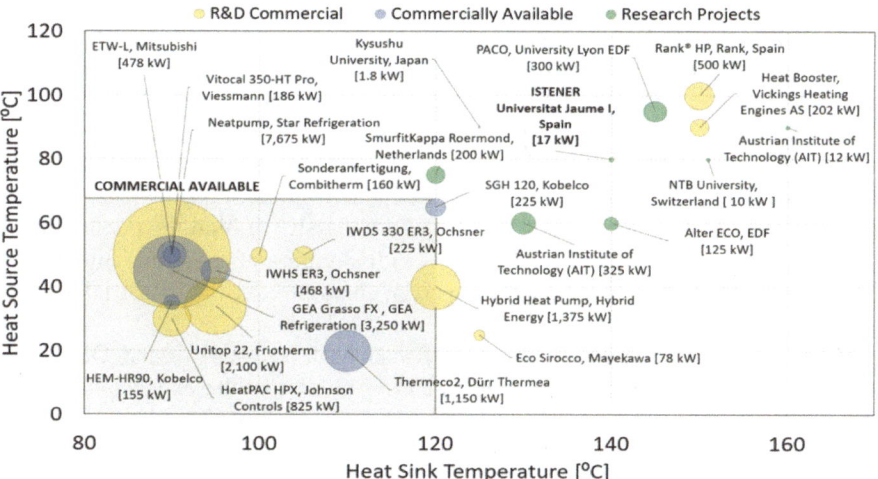

Fig. 5 Vapour compression HTHP prototypes

compressor were commercially available. Special attention was devoted to the modified compressor, based on a semi-hermetic, 4-cylinder piston propane model, to cover the higher temperatures and to ensure the proper lubrication. The compressor speed is varied from 30 to 50 Hz, the heat sink and source outlet temperatures between 115 and 117 °C, and between 85 and 86 °C. The average total compression and isentropic efficiencies are 74 and 83%, and the discharge temperature never reached 140 °C. This confirms the suitability of this prototype for reaching higher heat source temperatures.

A twin-screw compressor is a suitable solution in terms of compression ratio and volume flow rate for HTHPs using water as a working fluid. Wu [15] tested evaporating temperatures between 75 and 85 °C and condensing temperatures between 111 and 140 °C. The COP resulted from 1.85 to 6.10, and the highest heating capacity was 284.7 kW.

Arpagaus et al. [2] experimentally confirmed the benefit produced by the IHX in HTHP systems, considering the HCFO-1233zd(E) and HFO-1336mzz(Z) working fluid and variable-speed semi-hermetic piston compressor. The COP varied between 2.4 and 3.1, and heating capacity between 5.8 and 10 kW.

Mateu-Royo et al. [11] presented the experimental results of an HTHP with a modified scroll compressor and IHX. They tested this prototype using HFC-245fa at heat source temperatures between 60 and 80 °C, and heat sink temperatures between 90 and 140 °C, varying the heating capacity and COP between 10.9 and 17.5 kW and between 2.23 and 3.41, respectively. Low GWP alternatives can also be tested in this system with the potential of improving the HFC-245fa energetic results.

6 Reversibility

Finally, a future possibility to increase the economic feasibility of high-temperature heat pumps is introduced in this chapter. High-temperature heat pumps can only be operated when the heat is required. A storage tank for accumulating the excess heat (in the form of latent or sensible heat), but this can increase the cost significantly, and the velocity of charge and discharge of the tank depends on many factors. Besides, an innovative solution has been recently proposed, the reversible high-temperature heat pump/organic Rankine cycle [10]. If there is a necessity of high-temperature heat, the system can operate as an HTHP; if not, with a minimum additional investment, the system can act as an ORC generating clean electricity for the industrial plant or for injecting it to the net.

Most of the components of the HTHP can also be used for the ORC. The compressor can be adapted to operate as an expander, and the expansion valve is bypassed to circulate the fluid through a pump (Fig. 6). However, the overall efficiency of the system must be maximised, considering the built-in volume ratio of the compressor and expander, and therefore, a trade-off solution is required. Working as an HTHP mode, at the average evaporating temperature of 85 °C, a COP of 2.22 is

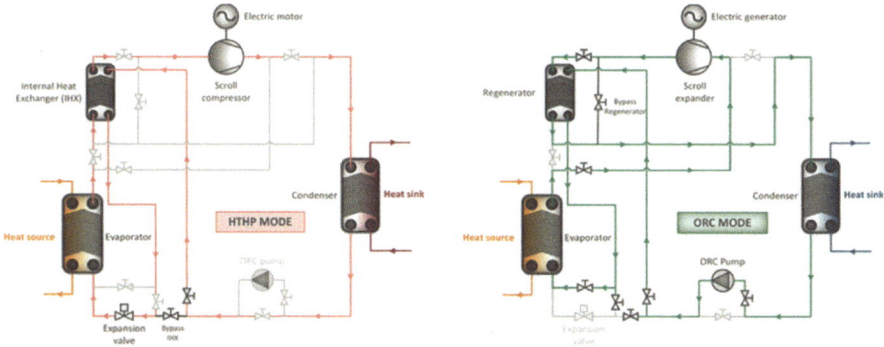

Fig. 6 Reversible HTHP-ORC operation modes

achieved condensing at 140 °C. For the same HTHP/ORC design, the ORC mode achieves a net electrical efficiency of 7.25% at a condensing temperature of 40 °C.

7 Conclusions

HTHPs for low-grade waste heat recovery has been developed during recent years to benefit the industrial energy efficiency, decarbonisation and reduce worldwide GHG emissions.

The criteria considered for refrigerant selection are sustainability, commercial availability, safety, and performance. HCFO-1233zd(E), HCFO-1224yd(Z), HFO-1336mzz(Z) and natural refrigerants are those refrigerants that can be considered as future low GWP alternatives for HTHPs.

Several studies confirm that IHX provides benefits to ensure the operation and increase the energy efficiency.Two-stage with IHX or cascade configurations are a future possibility for reaching higher performance than the basic configuration.

From an analysis of the existing literature, it can be concluded that the present boundaries are situated in a heat sink and heat source temperatures of 140 and 70 °C, respectively; several working fluids are being tested, and existing compressors are being adapted to the high-temperature situation, considering different compression technologies (piston, scroll, screw, …).

Given the potential of this technology, it is expected that research for HTHPs application in industry is going to increase, covering different working fluids, configurations, and compressor constructions, and higher heat source and sink temperatures than today to be able to compete with other waste heat recovery technologies.

Acknowledgements The authors acknowledge the Spanish Government for the financial support under projects RTC-2017-6511-3 and Adrián Mota-Babiloni through the postdoctoral grant FJCI-2016-28324. Furthermore, the authors acknowledge the Universitat Jaume I (Castelló de la Plana,

Spain) for the financial support under the projects UJI-B2018-24 and Carlos Mateu-Royo for the funding received through the PhD grant PREDOC/2017/41.

References

1. Arpagaus C, Bless F, Uhlmann M, Schiffmann J, Bertsch SS (2018) High temperature heat pumps: market overview, state of the art, research status, refrigerants, and application potentials. Energy 152

2. Arpagaus C, Kuster R, Prinzing M, Bless F, Uhlmann M, Büchel E, … Bertsch SS (2019) High temperature heat pump using HFO and HCFO refrigerants—system design and experimental results. In: V. Minea and IS Event Solutions (Eds.), Refrigeration science and technology proceedings. 25th IIR international congress of refrigeration, pp 4239–4247. https://doi.org/10.18462/iir.icr.2019.0242

3. Bai T, Yan G, Yu J (2019) Thermodynamic assessment of a condenser outlet split ejector-based high temperature heat pump cycle using various low GWP refrigerants. Energy 179:850–862. https://doi.org/10.1016/j.energy.2019.04.191

4. Bamigbetan O, Eikevik TM, Nekså P, Bantle M, Schlemminger C (2019) Experimental investigation of a prototype R-600 compressor for high temperature heat pump. Energy 169:730–738. https://doi.org/10.1016/J.ENERGY.2018.12.020

5. EPA (2015) Sources of greenhouse gas emissions. Climate change

6. EU (2018) Greenhouse gas emission statistics—emission inventories. Eurostat

7. Firth A, Zhang B, Yang A (2019) Quantification of global waste heat and its environmental effects. Appl Energy 235:1314–1334. https://doi.org/10.1016/j.apenergy.2018.10.102

8. Mateu-Royo C, Mota-Babiloni A, Navarro-Esbrí J, Peris B, Molés F, Amat-Albuixech M (2019) Multi-objective optimization of a novel reversible high-temperature heat pump-organic rankine cycle (HTHP-ORC) for industrial low-grade waste heat recovery. Energy Convers Manage 197:111908. https://doi.org/10.1016/J.ENCONMAN.2019.111908

9. Mateu-Royo C, Navarro-Esbrí J, Mota-Babiloni A, Amat-Albuixech M, Molés F (2018) Theoretical evaluation of different high-temperature heat pump configurations for low-grade waste heat recovery. Int J Refrig 90:229–237. https://doi.org/10.1016/j.ijrefrig.2018.04.017

10. Mateu-Royo C, Navarro-Esbrí J, Mota-Babiloni A, Amat-Albuixech M, Molés F (2019) Thermodynamic analysis of low GWP alternatives to HFC-245fa in high-temperature heat pumps: HCFO-1224yd(Z), HCFO-1233zd(E) and HFO-1336mzz(Z). Appl Therm Eng 152:762–777. https://doi.org/10.1016/j.applthermaleng.2019.02.047

11. Mateu-Royo C, Navarro-Esbrí J, Mota-Babiloni A, Molés F, Amat-Albuixech M (2019) Experimental exergy and energy analysis of a novel high-temperature heat pump with scroll compressor for waste heat recovery. Appl Energy 253, Article 113504. https://doi.org/10.1016/j.apenergy.2019.113504

12. Mota-Babiloni A, Mateu-Royo C, Navarro-Esbrí J, Molés F, Amat-Albuixech M, Barragán-Cervera Á (2018) Optimisation of high-temperature heat pump cascades with internal heat exchangers using refrigerants with low global warming potential. Energy 165:1248–1258. https://doi.org/10.1016/j.energy.2018.09.188

13. Papapetrou M, Kosmadakis G, Cipollina A, La Commare U, Micale G (2018) Industrial waste heat: estimation of the technically available resource in the EU per industrial sector, temperature level and country. Appl Therm Eng 138:207–216. https://doi.org/10.1016/J.APPLTHERMALENG.2018.04.043

14. United Nations (2015) Paris climate change conference-November 2015, COP 21. Adoption of The Paris Agreement. https://doi.org/FCCC/CP/2015/L.9/Rev.1ss

15. Wu D, Jiang J, Hu B, Wang RZ (2019) Experimental investigation on the performance of a very high temperature heat pump with water refrigerant. Energy 116427. https://doi.org/10.1016/j. energy.2019.116427
16. Xu ZY, Wang RZ, Yang C (2019) Perspectives for low-temperature waste heat recovery. Energy 176:1037–1043. https://doi.org/10.1016/j.energy.2019.04.001

Chapter 18
Replacing Fossil Fuels by on-Site Sources of Energy in a Residential Building in Chalus, Iran

Milad Moradibistouni, Brenda Vale, and Abbas Mahravan

Abstract The study of energy and its relationship to the environment is now critical due to its effect on the quality of present and future life. The need to reduce greenhouse gases to combat climate change implies an increasing use of non-renewable sources of energy. This article aims to evaluate the possibility of replacing fossil fuels by renewable sources of energy in a residential building in Chalus, an Iranian city with a high potential for using on-site sources of energy. This study first investigates the annual operating energy used by households in Chalus. To do this, the type of appliances and average time of their use, including heaters and air conditioning units, is extracted from official reports together with a local field study. The ability of renewable on-site sources of energy to supply this load is calculated by considering the specific characteristics of the region. The result shows these sources have the potential to provide approximately 98% of the annual energy the household consumes.

Keywords Renewable sources of energy · Domestic energy use · Sustainability

1 Introduction

After the energy crisis in the 1970s, attention was turned to the idea of the house generating all the energy it needs from its site in both the UK [1, 2] and North America [3, 4]. The focus then was on whether sun, wind, and to some extent biomass could provide all the energy the house and its occupants required. This article returns to

M. Moradibistouni (✉) · B. Vale
Victoria University of Wellington, Wellington, New Zealand
e-mail: Milad.bistouni@vuw.ac.nz

B. Vale
e-mail: Brenda.vale@outlook.co.nz

A. Mahravan
Razi University, Kermanshah, Iran
e-mail: A.mahravan@razi.ac.ir

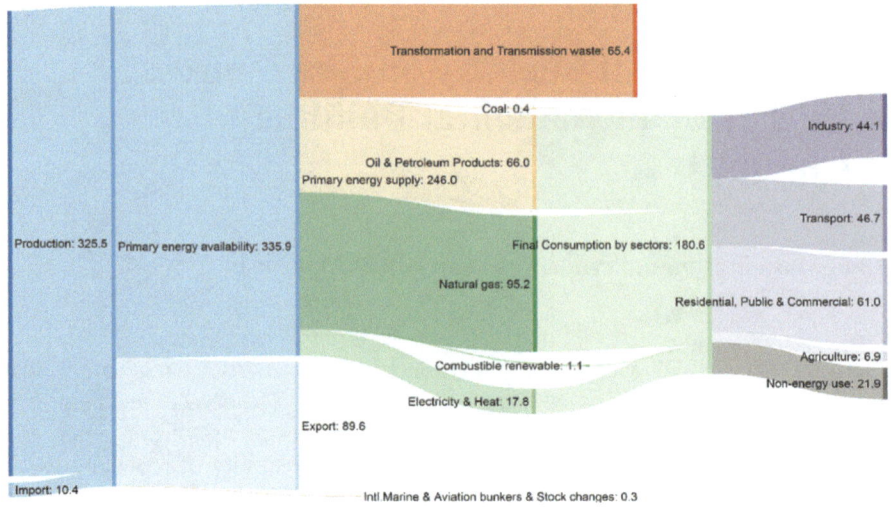

Fig. 1 Iran's energy flow in 2014 (MMBOE). *Source* adapted from [5]

this idea of the house being able to collect the energy it needs from its site but in the very different context of northern Iran.

In 2014, more than 89% of the energy entering Iran's network was generated from natural gas, liquid fuels, and coal, which all are non-renewable sources of energy. In the same year, over 1% of Iran's energy demand was generated by renewable sources (Fig. 1). Although electricity provided approximately 9% of the country's energy, more than 94% of this was generated from fossil fuels with 26% of all primary energy wasted during conversion, transfer, and distribution [5]. This report [5] also shows that in 2014, the residential, commercial, and general (RCG) sector was the biggest in terms of energy demand, accounting for approximately 34% of all energy use.

The RCG sector was not only the biggest user of energy in Iran but also the third-highest producer of greenhouse gasses (GHG), after power plants and transport (Fig. 2). The RCG sector was responsible for 23% of all GHG emissions. Figure 2 also indicates that more than 99% of all GHG in Iran came from natural gas and liquid fuels, which are non-renewable fossil fuels.

Figures 1 and 2 show the dependency of the residential sector on non-renewable energy sources and particularly natural gas. These facts show the necessity of investigating the replacement of fossil fuels by renewable sources of energy in the residential sector.

1.1 Research Aim

This article aims to evaluate the possibility of substituting the fossil fuels which serve a typical 150 m^2 meter residential building in Chalus with sustainable and renewable

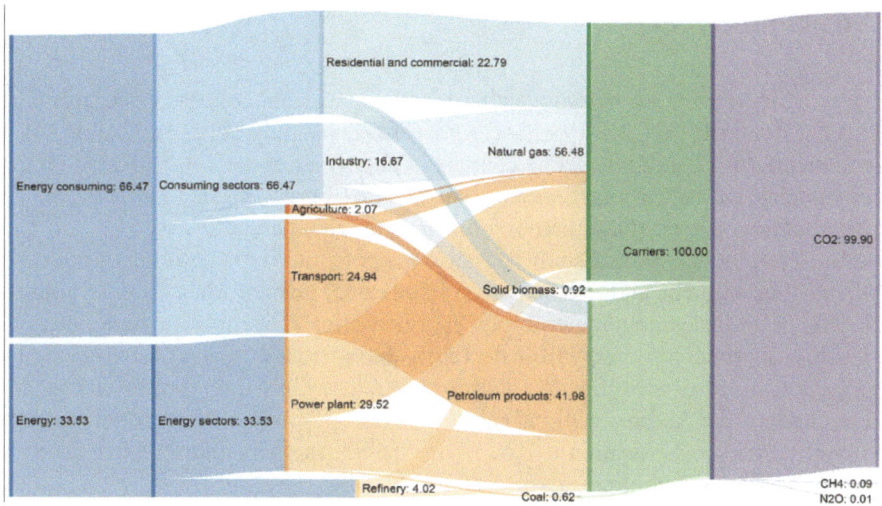

Fig. 2 GHG emission flow chart by Iranian energy sector (2014). *Source* adapted from [5]

sources, which can be harvested from the house plot. This requires calculating the annual energy consumption of the house, based on average consumption of all such houses, and the potential energy that can be generated by renewable sources on the site. The values are then compared to see if it is possible to replace non-renewable sources of energy in this region.

2   Background

In recent years, how to design and construct sustainable houses has been a significant subject of investigation. Using conventional sources of energy more efficiently and replacing conventional sources of energy with renewable have emerged as very important factors affecting the sustainability of housing in addition to its being environmentally friendly, affordable, and with a minimum of waste generated [6, 7]. However, the majority of such studies stress the importance of replacing fossil fuels with sustainable ones, without analysing whether this is possible [8–10]. Other studies have looked at the potential of only one renewable source of energy. For example, Monforti and Belis [11] looked at the potential and consequences of using biomass energy in the residential sector, and Velkin et al. [12] investigated the possibility of using solar energy in residential buildings in the Ural region in Russia. The same holds true for other renewable sources of energy such as wind [13] and wave energy [14]. However, less has been written on integrating all available renewable sources of energy in a specific region so as to compare this with the annual energy use of a residential building in that region. This is the gap this article aims to fill.

3   Case Study Site

Chalus, a city with mild, humid weather located in Mazandaran Province in Iran at 36° 39′ 17″ North, 51° 25′ 19″ East, forms the case study (Fig. 3). Chalus has a high potential for using renewable sources of energy sources as it is located on the southern edge of the Caspian Sea with a high ratio of green space to urban area.

Mazandaran is one of the three regions in Iran which are allowed to harvest and use forest products [15], and so there is also the potential for generating bio-energy, although this would be at the scale of the region rather than the house site. Moreover, data from the Mazandaran meteorological organisation [16] show the high potential of Chalus for producing energy from the renewable sources of sun and wind (Table 1).

Figure 4 presents the climatic conditions of Chalus based on data from Mazandaran meteorological organisation [16] plotted on a psychometric chart, where each red circle represents one month. The figure helps in understanding when there is need for heating and cooling.

Figure 4 shows humidity in Chalus that always exceeds the comfort zone, which reveals the importance of natural and mechanical ventilation. The temperature is only in the comfort period for the four months of May, June, August, and September. In the rest of the year, there is a need for heating except for July, where the temperature exceeds the upper end of the comfort level and could potentially be translated as a minimal need for cooling. However, the field study results [17] show that most

Fig. 3   Map of Iran showing Mazandaran Province and Chalus *Source* author

Table 1 Chalus annual weather condition (Nowshahr station)

	Spring	Summer	Autumn	Winter	Annual
Average temperature (centigrade)	18.3	25.5	15.3	8.8	17.0
Average relative humidity (percent)	80	80	82	83	81
Total rainfall (mm)	57	174	780	251	1262
Total sunny hours (h	579	617	387	325	1908
Maximum wind speed (m/s)	27	20	19	22	27

Source [16]

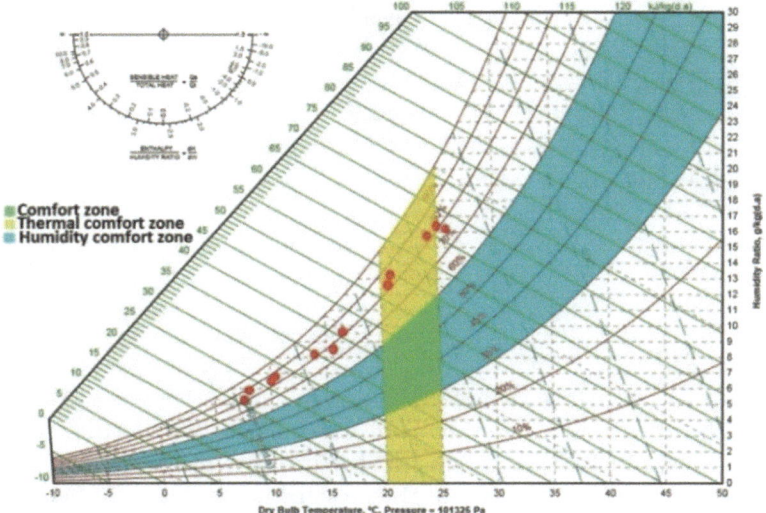

Fig. 4 Chalus climatic conditions plotted on a psychometric chart. Each red circle represents one month of a year. *Source* author

households used cooling appliances in summer due to the high level of humidity which affects the feeling of comfort.

4   Method

This study aims to evaluate the potential of on-site renewable sources of energy to see if these can generate the required annual energy demand of a typical house in Chalus. The Statistical Yearbook of Mazandaran shows that the average area of a residential building in the region consented in 2010 was 143 m^2 [18]. Also based on the Municipality of Chalus report [19], the maximum allowed density in the city is a house occupying 40–60% of the plot area based on different variables. Consequently, the area of the house in this study is assumed to be 150 m² and the lot area 300 m².

The house is a one-storey building and has two bedrooms, a dining and living area, which will be called the living room, a kitchen, a toilet, a shower, and a shared space between the two, called the cold bath. The building is a timber-framed house on a suspended timber floor, with a pitched roof covered with clay tiles. This is the typical method of construction in Chalus.

First, the annual operating energy consumption of the 150 m² residential building in Chalus was calculated using Eq. (1). For this purpose, energy using domestic appliances was classified into five categories of cooling, heating (including water heating), kitchen appliances, lighting, and others, as these are the main users of household energy [20–24]. Then, using data collected from the Iran National Statistics Organizations (INSO) report [25], types of appliance used by the majority of households in each category and the average time the majority of families used them were determined. This report was the latest version of its kind available. Using this data, the annual energy used by the household was calculated.

Equation 1: Annual energy consumption

$$E_{(kWh/year)} = P_{(W)} \times t_{(h/day)} \times 365(\text{number of days during a year})/1000_{(W/kW)} \tag{1}$$

E: energy (kWh/year)/P: wattage of the appliance (W)/t: number of hours of use per day/Source: [26]

However, data from INSO report [25], although highly reliable, mix data from all different regions and cities, and there is no detailed data breakdown for each city. Given the climate of Iran and the cultural habits of its citizens vary widely using averages could be misleading. To see if data extracted from the INSO report [25] are applicable to Chalus, a field study was conducted [17]. This investigated over 50 households in Chalus through a questionnaire-based survey. Households received the data extracted from the INSO report [25] and were asked to rate the consistency of these data with the types of appliances they used and the time they used them. They could add appliances and times to the data as needed. Where the survey results showed a considerable variation with data from the INSO report [25], data from the survey were used. In fact, most of the data from the report were confirmed by the survey respondents, and amendments were only made in the two categories of cooling and lighting.

In the next step, the potential yields of appropriate renewable sources of energy in Chalus, classified into three categories of biomass, solar, and wind energy, were studied. Chalus has the potential for generating energy from the sea, but this would be at a scale larger than that of the house. There is also the potential for geothermal in Iran, but there is no uncovered source in Chalus [27]. Finally, the energy each source could produce considering the site characteristics was calculated using the equations in Table 2 to see if the renewable sources could provide enough energy for the 150 m² house. Biomass was considered separately as this would be generated at the regional scale. Another consideration was the type and efficiency of technologies

Table 2 Calculation of the potential of renewable sources of energy

Field	Equation number	Equation	Source
Solar energy output from the PV system	Equation (3)	$E = A\,r\,H\,PR$	[28, 29]
Potential wind energy calculation	Equation (4)	$P_{\max} = \frac{16}{27}\frac{\rho}{2}v^2\frac{\pi D^2}{4}$	[30]
ρ Calculation	Equation (5)	$\rho = \frac{P}{RdT}$	[31]

used for generating energy from these sources. In this study, the chosen technologies were those which were widely available based on an online search.

5 Data

5.1 Calculating the House Energy Consumption

This study focused on the operating energy of the residential building ignoring embodied energy at this stage. This operating energy includes all the energy consumed by the main appliances in the household classified into five categories of heating (including water heating), cooling, kitchen appliances, lighting, and others. To calculate the operating energy needed, data from INSO report [25] and the field study [17] were used.

Table 3 lists the different types of heating appliances and their time of use, showing over 79% of Iranians use gas heaters as their main source of heating. Table 3 also shows more than 67% of gas heater users turned their heaters on for between 1501 and 2160 h during winter, which gives an average usage of 1830 h. The field study [17] regarding the type and average use of heating appliances confirmed the data extracted from INSO report [25]. This means gas heaters form the primary heating source in Chalus. It is worth noting the typical house has three heaters, one in the living room and one in each of two bedrooms, and each is estimated to be used for an average 1830 h a year.

Table 3 Percentage and monthly use of heating appliances in Iran

Type of appliance	%	Time of use for each appliance (%)			
		500 h or less	50–1000 h	1001–1500 h	1501–2160 h
Gas heater	79.4	4.7	11.0	16.4	67.9
Radiant heater	7.7	0.1	3.2	6.5	90.2
Oil heater	3.1	22.2	21.8	15.3	40.7
Central heating	2.8	3.7	7.9	13.6	73.4
Others	7.0				

Source [25]

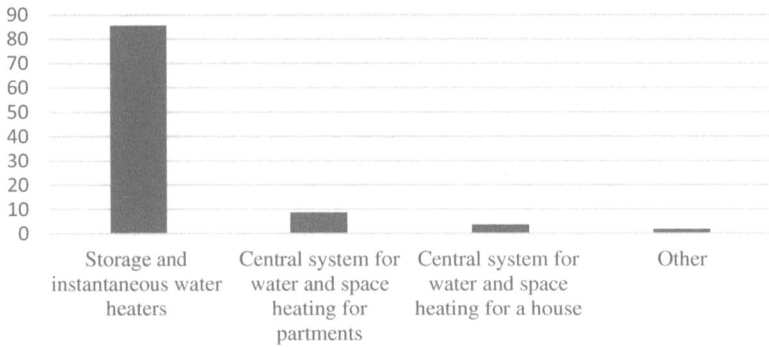

Fig. 5 Appliances for hot water production. *Source* [25]

Hot water. Hot water is used in a residential building for showering, washing dishes, and laundering. According to the INSO report [25], over 85% of Iranians use storage or instantaneous water heaters (although these are different types of water heater, they are combined in the statistics) (Fig. 5). For all water heater types, more than 90% of households used natural gas as their fuel [25]. The field study [17] also showed all respondents used fossil fuel (mostly natural gas) water heaters. The INSO report [25] only mentioned the number of days households used their water heaters giving no detailed breakdown of the information. The field study [25] showed the time each household used their water heaters varied considerably, making it hard to find a reliable average. As a result, similar studies were investigated to find a reliable figure. Yousefi [32] studied the economic and environmental costs of replacing gas and electric water heaters with solar water heaters in Iran. By considering relevant standards, Yousefi [32] calculated each Iranian needs 60 l of hot water a day, with the output water heated to 60 °C. Having this information, the energy each household needs for hot water for each person can be calculated using Eq. (2).

Equation 2: Energy required for hot water

$$Q = VMC\Delta T \tag{2}$$

Q: energy required for hot water (KJ)/V: volume of water/M: the specific weight of water = 1 kg/L/C: specific heat of water = 4.18 (kJ/kg C)/ΔT: difference in temperatures of input and output water/Source: [33, 34].

To use Eq. (2), it is necessary to know the input temperature to the water heater. In this study, it is assumed the input water temperature is equal to annual average ambient temperature, which in Chalus is 17 °C (Table 1). Using Eq. (2), a household of four in Chalus needs 4374 kWh of energy a year for its hot water.

Cooling. Based on Fig. 4, the temperature of Chalus only exceeds the comfort level for cooling in July. However, the field study [17] showed due to high average relative humidity in the region of more than 81% and its effect on feeling comfortable,

Table 4 Percentage and monthly use of cooling appliances (h)

Type of appliance	%	Time of use for each appliance (%)			
		500 h or less	50–1000 h	1001–1500 h	1501–2160 h
Water cooler	57.3	34.5	29.1	20.4	16.0
Water cooler and fan	12.7	NA	NA	NA	NA
Heat pump and fan	9.4	NA	NA	NA	NA
Fan	9.0	NA	NA	NA	NA
Heat pump	8.8	34.0	23.6	13.5	29.0
Others	1.7				

Source [25]

most households in Chalus use some form of cooling during the summer and some even in rest of the year.

At the national scale, Table 4 shows that the main cooling appliance used by more than 57% of Iranians was a water cooler. However, the field study [17] showed that most households in Chalus use a heat pump for cooling their houses. So in this study, the main cooling appliance was assumed to be a single heat pump. Also, Table 4 shows that 34% of heat pump users nationally use their heat pump for 500 h or less during summer. This level of use was confirmed by the field study [17].

Lighting The type and number of lights can be remarkably different. The INSO report [25] shows that 10–50 W compact fluorescent lamps (CFLs), incandescent lamps (from 40 W to over 150 W), and 18–65 W fluorescent lamps are the most used lightning types in Iran, used respectively by 60%, 23%, and 13% of households. The field study found the same results. In this study, therefore, three types of CFL (23 W), fluorescent (36 W), and incandescent lamps (100 W) are assumed as the main sources of lighting. Based on observations, there are 13 lights in the house:

- CFL lamps (one in each of the toilet, the shower room, the kitchen, the living room, and one in each bedroom),
- incandescent lamps (one in each of two bedrooms, one in the living room, and one in the cold bath),
- three fluorescent lamps (one in the kitchen and two in the living room).

Aside from the type and number of lights, another important factor is the time these lights are on. Based on Table 5, more than 40% of respondents used CFL and fluorescent lamps for 4–8 h a day (average 6 h) and over 60% used their incandescent lamps for 1–4 h a day (average 2.5 h). However, the field study produced different results, possibly due to the climate and having more cloudy hours in comparison with the national average. Households in Chalus used their lights approximately 10% more than the national averages. As a result, in this study, CFL and fluorescent lamps are used for 6.5 h a day and incandescent lamps for 2.75 h a day.

Kitchen. A variety of appliances are found in Iranian kitchens. Table 6 sets out the three large appliances used by more than 50% of respondents in the INSO report [25] and their average monthly use. For calculating the time a refrigerator works at

Table 5 Daily use of different lighting types

Season	Type	Time of use for each type (%)			
		1–4 h/day	4–8 h/day	8–12 h/day	More than 12 h/day
Summer	Incandescent	67.0	24.0	7.0	2.0
	CFL	36.5	45.0	13.5	6.0
	Fluorescent	30.3	49.5	14.7	5.5
Winter	Incandescent	60.0	25.8	11.7	2.5
	CFL	26.0	41.5	24.6	8.0
	Fluorescent	20.6	47.1	24.2	9.1
Year	Incandescent	63.5	24.8	9.3	2.2
	CFL	31.2	43.2	19.1	7.0
	Fluorescent	25.1	48.2	19.4	7.3

Source [25]

Table 6 Monthly use of appliances in the kitchen (used by more than 50% of Iranians)

Appliance type	%	Time of use for each appliance (%)			
		Less than 10 h per month	10–50 h per month	50–110 h per month	111–200+ hours per month
Oven	61.1	1.3	20.1	53.7	24.9
Washing Machine	79.4	85.3	13.4	1.2	0.1
Refrigerator	54.1	240 h per month			

Source [25]

maximum power, the time that it is connected to the electricity supply (24 h a day for approximately 30 days a month) should be divided by three [35].

From Table 6, the most popular monthly use times are 50–110 h for an oven and less than 10 h for a washing machine. The field study respondents agreed their large appliances were the oven, washing machine, and refrigerator. Some respondents mentioned other appliances such as a meat grinder and food processor. However, neither of these appliances was found in the whole field study sample. The respondents also confirmed the times for oven and refrigerator use extracted from the INSO report [25], except for the washing machine, which the majority of respondents used for 10 h a month. As a result, uses of 75, 10, and 240 h a month were assumed for the oven, washing machine, and refrigerator, respectively.

Other appliances. When it comes to small appliances because their incidence is very variable, this study only counted those used by more than 50% of respondents in the national survey. The INSO report [25] showed the most commonly occurring small appliances were a vacuum cleaner, TV, and iron, which were used by more than 86%, 73%, and 71% of respondents, respectively. Table 7 shows the monthly

Table 7 Monthly use of small appliances (used by more than 50% of Iranians)

Appliance	%	Time of use for each appliance (%)			
		Less than 10 h/month	10–50 h/month	50–110 h/month	111–200+ hours/month
Vacuum cleaner	86.5	90.5	9.2	0.2	0.1
TV	73.6	2.6	8.9	29.1	59.4
Iron	71.0	96.8	3.0	0.1	0.1

Source [25]

use of these appliances. Taking the field study and Table 7 together, the iron and vacuum cleaner are used for 10 h/month and the TV for 155.5 h/month.

5.2 On-Site Production of Energy

In this section, the potential production of energy using the natural sources of biomass, solar, and wind is calculated by considering the site characteristics. Although biomass is not strictly an on-site source, it is plentiful in the area and the raw biomass could be converted to useful energy on-site.

Biomass energy. Bio-energy, which can be extracted from all organic material such as trees, plants, and animal waste, is basically the accumulation of the sun's energy through photosynthesis, and its further incorporation into livestock, and all types of organic wastes, including wastes from the food industry [36]. One of the most important aspects of biomass over other types of renewable sources of energy is the fact that bio-energy production is not affected by weather conditions and can be continuous. In Iran, annual biomass potential has been evaluated to be approximately 331 million tonnes, which can potentially provide 218 TW of electricity annually [37].

Based on Amini [15], Mazandaran, Gilan, and Golestan are three provinces in Iran which are allowed to harvest and use forest products, due to their environmental conditions and ratio of green space to urban area. Moreover, a Chalus city council report [35] shows, in addition to the tourist industry, agriculture and animal husbandry are the most prominent industries in Chalus. This suggests Chalus has great potential for using biomass energy. Pazouki et al. [38] evaluated the available potential of biomass in Chalus to be approximately 4 TWh a year.

There are different methods of converting natural materials to energy such as thermochemical and biochemical conversion, but this study only looks at biochemical conversion [37–40]. Biochemical processes use enzymes, bacteria, or other microorganisms to break down biomass into fuel such as biogas or bioethanol using anaerobic fermentation. This can produce methane from landfills or ethanol from fermentation

Table 8 Biogas potential of different materials

Raw material	Water content (%)	Dry material gas production rate (m³/kg)	The raw material needed to produce 1 m³ biogas (kg)	
			Dry	Fresh
Pig manure	82	0.3	4.0	22.2
Cow manure	83	0.2	5.3	31.0
Chicken manure	70	0.3	4.0	13.3
Human manure	80	0.3	3.3	16.7
Rice straw	15	0.3	3.8	4.5
Wheat straw	15	0.3	3.7	4.4
Corn stalks	18	0.3	3.5	4.2

Source [42]

[41]. Since an assumption about conversion has to be made, in this study it was decided to extract bio-energy using a 10 m³ biogas system (PX-HBS-10 M³). Table 8 shows the quantity of materials that could be put into a digester to produce 1 m³ of biogas.

In this study, rice straw was proposed as the main fuel for the biogas system due to its availability in the region and its low water content, meaning less fresh material has to be fed into the digester. As 1 m³ biogas can generate 2. 5kWh electricity [43, 44], feeding 38 kg dry rice straw a day into the biogas system, which is completely possible, could produce 10 m³ biogas, which equates to 25 kWh a day or 9125 kWh a year. A digester to produce this much gas would be a three-metre high cylinder with a diameter of over 2 m. This would be placed underground and would occupy approximately 12 m² of the plot. There is also a need for space to store the rice straw to feed into the digester. This area would depend on how often the straw was delivered to the site. The 150 m² plot is probably large enough to contain the digester and straw storage area, but such a system may be more appropriate for rural rather than urban areas. The other and possibly more acceptable option in urban areas is to have larger centralised biogas systems.

Solar energy. Solar energy is one of the most important sources of renewable energy in the world [45]. "…about 0.1% of this energy [solar energy intercepted by the earth], when converted at an efficiency of 10% would generate four times the world's total generating capacity of about 3000 GW" [46]. This study looks at two types of solar technologies, photovoltaic systems and solar water heaters, both of which can be used at the domestic scale.

Photovoltaic systems. Photovoltaic systems are designed to generate electricity directly from solar radiation. Use of these systems which are noiseless and practical has grown remarkably in recent years [47]. The energy that can be generated in different regions depends on various factors, including average solar radiation and panel specification. There are many different formulae for calculating the energy that could be generated from photovoltaic panels. This study uses Eq. (3) which is

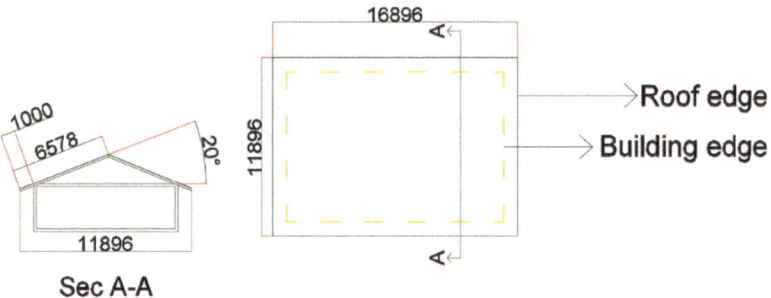

Fig. 6 House roof section

accepted by most researchers [28, 29]. The photovoltaic panels to be used in this study are JKM360M-72-V from Jinko Solar. These are 1.95 × 0.992 × .04 m panels with an efficiency of 18.57% [48].

Equation : Solar energy output from *the* PV system

$$E = A*r*H*PR \tag{3}$$

E: annual solar energy output of the photovoltaic system (kWh)/A: total solar panel Area (m²)/r: solar panel yield (%) or the efficiency of the panel/H: average solar radiation.yr-1/PR, the coefficient for losses (range between 0.5 and 0.9), which is considered 0.7 in this study/Source: [28, 29].

The total solar panel area is limited by the need to use the south-facing side of the roof. Due to the high annual rainfall in Chalus, roofs in the region normally have a slope of 20–40 °C for drainage [49]. For the same reason, all buildings normally have some type of overhang to protect the wall. In this study, the slope of the roof is 20°, and the roof overhangs the wall by 1 m measured down the roof slope (Fig. 6).

The area of the south-facing half of the roof is thus 110.5 m² (17 × 6.5), and the area of each PV panel is 1.94 m² (1.95 × 0.992). This means 56 photovoltaic panels could be installed covering 108 m² of the south-facing roof. However, some space should be kept to ensure a flow of air around the panels for maximum efficiency of generation and for the solar water heater. As a result, 28 grid link photovoltaic panels (54.32 m²) covering 50% of the maximum available space are assumed. The average solar radiation in Chalus is approximately 1387 kWh/m² [50], and using Eq. (3) the total annual solar energy which can be harvested in Chalus is 9493 kWh. In order to store surplus energy rather than selling it back to the grid, it would be possible to add storage to the system through installing a Tesla Powerwall battery. This battery, which can be installed on floor or walls and also inside or outside the house, has a useable capacity of 13.5 kWh and dimensions of 1150 mm by 753 mm by 147 mm [51].

Solar water heating. Solar water heaters are designed to provide hot water at around 60 °C for residential use either using pumped or natural circulation of water, and systems usually include a collector, storage tank, and connecting pipes [52].

Most researchers and producers of solar hot water systems claim such systems are able to produce 70% of the hot water a household needs [32–53]. In this study, an A591270/2Tg Rheem premier active solar water heater is assumed [54]. A household of four in Chalus needs 4374 kWh energy annually for their hot water. Assuming the solar water heater produces 70% of this, the fossil fuels that can be replaced by installing a solar water heater equate to approximately 3062 kWh/year.

Wind energy. "The kinetic energy of the air flow [the wind] provides the motive force that turns the wind turbine blades that, via a drive shaft, provide the mechanical energy to power the generator in the wind turbine" [55]. There are many ways of calculating the potential energy that could be generated by the wind, but one of the most accepted methods is the Betz Equation (Eq. 4).

Equation 4: Potential wind energy calculation

$$P_{max} = \frac{16}{27} \frac{\rho}{2} v^2 \frac{\pi D^2}{4} \text{[Watt]} \tag{4}$$

P: potential wind energy calculation/ρ: air density (kg/m3)/V: average wind speed/D: turbine rooter diameter/ρ (kg/m^3) can be calculated using Eq. (5)./Source: [30]

Equation 5: ρ Calculation

$$\rho = \frac{P}{Rd\ T} \tag{5}$$

T: average monthly air temperature (K)/P: average monthly air pressure (Pa)/Rd: gas constant for dry air, which its value is 287 J/kg K/Source: [31].

In order to calculate the potential of wind energy in Chalus, it is necessary to determine the average wind speed, average monthly air temperature, and average monthly air pressure of the site. Data collected from the Mazandaran meteorological organisation (Table 1) reveal the average annual air temperature on the site is 17 °C or 290.15 K, and the average monthly air pressure is 101,325 Pascals as the site is located at sea level. The average wind speed (V) on the site is 16 m/s, and it never goes below 5 m/s, which is the minimum speed needed for a wind turbine to work [56]. The next step is to determine the turbine rotor diameter for calculating the maximum energy which can be generated in Chalus. Two small AEOLOS-H500W wind turbines were chosen for the on-site installation for this study. This machine has a rotor diameter of 2.7 m and would be mounted on a 12 m high tower (6 m free-standing and 6 m guyed) in the garden of the house. Each turbine needs approximately 18 m^2 for assembly and the ends of the guys enclose an area of 36 m^2 on the site. Using Eqs. 4 and 5, each AEOLOS turbine can potentially generate over 470 kWh/year or 940 kWh a year.

Table 9 Appliances and their use of energy

Appliances		N	DU	MU	AU	PR	AEU	P	SS
Heating	Heater	3	–	–	1830	1500	8235	47	12,609
	Water heater						4374	25	
Cooling	Heat pump	1		–	500	2300	11,500	7	1150
Lighting	Incandescent	4	2.7	–	4015	100	401	6	984
	CFL	6	6.5	–	14,235	23	327		
	Fluorescent	3	6.5	–	7117	360	256		
Kitchen	Oven	1	–	75	900	1400	1260	11	1932
	Washing Machine	1	–	10	120	2000	240		
	Refrigerator	1	–	240	2880	150	432		
Other	Vacuum cleaner	1	–	10	120	1600	192	4	676
	TV	1	–	155	1866	188	351		
	Iron	1	–	10	120	1100	132		
	Total						17,351	100	17,351

Number N, daily use (h) DU, monthly use (h) MU, annual use (h) AU, power rating (W) PR, annual energy use (kWh) AEU, portion (%) P, sub-section total SS

6 Results

6.1 The Annual Energy Use of the House

Table 9 summarises the average time and energy use of all appliances, with average wattages extracted from an Iranian energy efficiency organisation report [57]. The annual operating energy for the 150 m^2 house in Chalus is then calculated using Eq. (1). Although most ovens and heaters in Iran are powered by natural gas, where appropriate this use is expressed in kWh.

Table 9 shows the annual energy demand of the 150 m^2 residential building in Chalus is 17,351 kWh. Based on Table 9, heating is the biggest user of energy, responsible for 72% of the total (12,609 kWh/y), which breaks down into 47% and 25% for space and water heating, respectively. The next biggest user of energy is the kitchen category being responsible for 11% of annual energy (1932 kWh/y). Cooling, lighting, and other appliances are respectively responsible for 7%, 6%, and 4% of annual energy use.

6.2 Annual Energy Production of on-Site Renewable Sources

Table 10 tabulates the calculated potential of different renewable sources of energy for the typical house in Chalus.

Table 10 Potential of renewable sources of energy in Chalus

Source of energy	Annual potential (kWh)	(%)
Biomass energy	9125	37
Solar energy (photovoltaics)	9493	38
Solar energy (solar water heater)	3062	15
Wind energy	940	4
Total	22,620	100

Table 10 shows that renewable sources of energy can potentially generate 22,620 kWh energy a year which is more than needed for the 150 m² residential building. However, this number assumes all systems works to 100% of their capacity for 365 days a year and without any energy loss during production and transfer. There will also be days when different systems are out of service due to technical servicing or environmental conditions. To account for this, the calculated energy is reduced by 25% to give a more reliable evaluation.

As a result, the energy that could be generated in Chalus using renewable sources of energy would be 16,965 kWh/year. Compared with annual energy need of the household of 17,351 kWh, on-site renewable sources of energy could potentially produce approximately 98% of all energy the building needs over a year based on current usage and appliance efficiencies.

7 Limitations of the Study

This study has made a number of assumptions leading to the need for further considerations. Most households in Chalus make use of more appliances than listed in Table 9. This study is only based on averages.

This study calculated the potential of renewable sources of energy in energy production based on the assumption that all systems work with an efficiency of 100% for 365 days a year. Although the calculated values were reduced by 25% to account for systems not operating all the time, this is a big assumption and systems like PV panels may operate for some years without interruption.

Next, the space these systems need for their operation on the site has been based on the systems selected. Whether having such systems on-site is acceptable to people is another area of study. The cost of buying and installing the systems is also an issue, and whether the savings in buying fossil fuel energy will offset the costs of having and operating the renewable systems is an area for further investigation. It might be better to first invest money in saving energy by insulating the building or buying low energy use appliances, rather than investing in renewable energy systems.

Lastly, at present, there is no infrastructure in Chalus for installing and servicing renewable energy systems. Although once installed PV panels require little maintenance, other systems like a biogas digester require regular feeding, and wind turbines also need regular maintenance. Having local firms who can do this work is an important aspect of small-scale renewable energy systems, outside the scope of this study.

8 Discussion and Conclusion

Because of the need to reduce fossil fuel use in the face of climate change, this study has tried to evaluate the possibility of replacing fossil fuels with renewably generated energy for a stand-alone residential building. While it is important for centralised energy-generating sources to switch to renewable energy sources, this study has returned to the idea of seeing whether the energy for a single house could be generated from its plot. A typical $150\,m^2$ house for four people, located in Chalus, was chosen as the case study. Chalus was selected as an Iranian city with a high potential for using different types of renewable energy.

This study focused on operating energy. Based on existing national data and field studies, it was possible to show an average house in Chalus has an annual energy consumption of 17,351 kWh. Of this, heating is responsible for over 72% of all energy used in the house (12,609 kWh/year). The breakdown of the remainder is 1932, 1150, 984, and 676 kWh for large appliances, cooling, lighting, and other appliances, respectively. Looking at the available on-site sources of biomass, solar, and wind showed a potential for generating 22,620 kWh annually, of which 56% was from solar, 40% from biomass, and only 4% from wind. Allowing a 25% reduction for uncertainty in the generation and the need to have systems off-line for servicing reduced the total 16,965 kWh/year or 98% of current estimated demand. Effectively, it would be possible to provide sufficient energy from renewable sources with no changes in behaviour or equipment. Given only half the roof was covered with PV panels, having more of these and only relying on solar energy with no wind or biomass might be the way forward in further investigations.

However, lessons learned when use of on-site renewable energy was first mooted in the 1970s showed it was much cheaper to save energy than to generate it. So although it is theoretically possible to supply the existing house with renewable energy, a first step should be to save energy before investing in energy-generating equipment. As heating (space and water heating) combined account for over 70% of all energy, insulating to save energy seems a key action. However, the first piece of renewable energy-generating equipment to install might be a solar water heater as these are cheaper than PVs and could produce a 3062 kWh/year reduction in overall annual energy use without any change to the building. Given the small proportion of energy produced by the small wind turbines and their visual intrusion and space they need, they are probably not worth having for a typical house on a small plot.

The same is true of the biogas generator which is more acceptable at the scale of a farm or small neighbourhood. This means the energy use of the house needs to be reduced from 17,351 to 12,555 kWh which can be generated from solar energy alone. The next step would be to look at the embodied energy of the required solar systems and their replacement cycles as part of the life cycle of the house. This needs to be compared with energy bound up in centralised systems for renewable energy generation. In fact, these need to be calculated as part of the grid linked PVs which only achieve their desired output on an annual basis, with more energy generated in summer than in winter.

What this study shows is that there is a high potential for replacing residential fossil fuel use by renewable energy in the city of Chalus, with a consequent important reduction in greenhouse gases. Although it is better first to invest in reducing energy, there is merit in investing in renewable generation at the house scale as a way of showing people a vision of a future world without fossil fuels.

References

1. Vale B, Vale R (1975) The autonomous house: design and planning for self-sufficiency. Thames and Hudson, London
2. Harper P, Boyle G (1976) Radical technology—food and shelter, tools and materials, energy and communications, autonomy and community. Wildwood House, London
3. Todd NJ, Todd J (1984) Bioshelters, ocean arks, city farming: ecology as the basis of design. Sierra Club, San Francisco, California
4. Olkowski B, Olkowski H, Van der Ryn S (1979) The integral urban house: self-reliant living in the city. New Society Publishers, Gabriola Island
5. Amini F, Gahramani N, Saberfattahi L, Soleimanpour P, Tavanpour M (2015) Iran and world energy facts and figures. Ministry of Energy, Tehran, Iran
6. Fujihira K (2019) Comprehensive strategy for sustainable housing design. In: Cakmakli A (ed) Different strategies of housing design. IntechOpen, London
7. Golubchikov O, Badyina A (2012) Sustainable housing for sustainable cities, 1st edn. United Nations Human Settlements Programme, Nairobi
8. Shahzad U (2015) The need for renewable energy sources. Int J Inf Technol Electri Eng 4(4):16–19
9. Owusu1 PA, Sarkodie SA (2016) A review of renewable energy sources, sustainability issues and climate change mitigation. Cogent Eng 3(1) NP
10. Altintas K, Turk T, TuVayvay O (2016) Renewable energy for a sustainable future. Marmara J Pure and Appl Sci 1:7–13
11. Monforti-Ferrario F, Belis C (2018) Sustainable use of biomass in residential sector—A report pr pared in support of the European Union Strategy for the Danube Region (EUSDR), EUR 29542 EN. Publications Office of the European Union, Luxembourg
12. Velkin VL, Shcheklein SE, Danilove V (2017) The use of solar energy for residential buildings in the capital city. In: IOP conference series: earth and environmental science 72. Ekaterinburg, Russian Federation
13. Ingeli V, Jankovichová E, Tien MN, Cekon M (2014) Integration of small wind energy source for optimization of energy efficiency in residential building. Adv Mater Res 1041 Brno
14. Franzitta V, Catrini P, Curto D (2017) Wave energy assessment along sicilian coastline. Based on DEIM point absorber. Energies 10(376)

15. Amini F, Gahramani N, Saberfattahi L, Soleimanpour P, Tavanpour M, Farmad M, Khodi M (2015) Energy balance sheet of Iran Ministry of Energy. Electricity and Energy Dept. Office of Planning for Electricity and Energy, Tehran, Iran

16. Mazandaran Meteorological Organization Annual Chalus climatic charactristics, http://www.mazmet.ir. Last Accessed 12 June 2017

17. Moradibistouni M (2014) Analytical study of the optimization of energy consumption in residential buildings with design approach of a residential unit in consistent with the temperate and humid climate, architecture. Azad University of Nour, Mazandaran, Iran

18. Country Planning and Budget Organization (2014) Statistical yearbook of Mazandaran. Statistical Center of Iran, Iran

19. Municipality of Chalus http://chalouscity.ir/homepage.aspx?site=DouranPortal&tabid=1&lang=fa-IR. Last Accessed 30 Sept 2017

20. US Department of Energy (2014) Energy savers: tips on saving money & energy at home, Office of Energy Efficiency and Renewable Energy, United States of America

21. Koller C, Junge R, Schuetze T, Jacques Talmon-gros M (2018) Energy toolbox: framework for the development of a tool for the primary design of zero-emission buildings in European and Asian cities. Sustainability 9(2244)

22. Global Energy Assessment (2012) Toward a sustainable future. Cambridge University Press, Cambridge, UK

23. Biswas WK (2014) Carbon footprint and embodied energy consumption assessment of building construction works in Western Australia. Int J Sustain Built Environ 3(2):179–186

24. Malama A, Makashini L, Abanda H, Ng'ombe A (2015) A comparative analysis of energy usage and energy efficiency behavior in low- and high-income households: the case of Kitwe. Zambia Res 4(4):871–902

25. Iran National Statistic Organizations(INSO) (2010) Review of the results of energy consumption in the household sector in urban areas' survey. Iran National Statistic Organizations, Tehran, Iran

26. Grisso R (2014) Estimating appliance and home electronic energy use energy series. Virginia Cooperative Extension, United States of America

27. Bahrami M, Abbaszadeh P (2013) An overview of renewable energies in Iran. Renew Sustain Energy Rev 24:198–208

28. Swain A (2017) Solar energy generation potential on national highways. Int Res J Eng Technol (IRJET) 04(09):462–470

29. Hossain MS, Li B (2016) Renovation of NZCB in a poor solar irradiation zone: an investigative case study of residential buildings in Chongqing urban areas. Int J Energy Environ 7(1):49–61

30. Carriveau R (2011) Fundamental and advanced topics in wind power. InTech, London

31. Babayani D, Khaleghi M, Tashakor S, Hashemi-Tilehnoee M (2016) Evaluating wind energy potential in Gorgan-Iran using two methods of Weibull distribution function. Int J Renew Energy Develop 5(1):43–48

32. Yousefi H, Norolahi Y, Toghyani S (2016) Economical and environmental cost analysing of replacing gas and electric water heaters with the solar water heaters. Environ Sci Technol 18(3):26–35

33. Danca M-F (2018) A simplified superheating Rankine pump with possible application in irrigation. arXiv preprint

34. Zhou Z, Li C, Wang L, Ebert A (2016) A: Performance analysis of a collective solar domestic water-heating system in the temperate zone of Yunnan Province, China. J Eng Sci Technol Rev 9(3):60–65

35. Tavanir Energy use in hose holds. http://news.tavanir.org.ir. Last Accessed 07 June 2017

36. McKendry P (2002) Energy production from biomass (Part 1): overview of biomass. Bioresour Technol 83:37–46

37. Norolahi AG, Kheirooz M, Ansari B (2013) Evaluating the potential ability of biomass sources (animal waste) to generate biogas. In: The third Iranian conference on renewable energies and distributed generation, Isfahan, Itan

38. Chalus City Council Chalus in one look. http://chalusshora.ir. Last Accessed 16 June 2018

39. Pazouki M, Adl M, Pazouki M (2014) Potential uses of biomass energy in the Persian Gulf region. Iran
40. Sharma S, Meena R, Sharma A, Kumar Goyal P (2014) Biomass conversion technologies for renewable energy and fuels: a review note. IOSR J Mech Civil Eng 11(2):28–35
41. Cross B (1995) The world directory of renewable energy suppliers and services. James & James Science Publishers, London
42. Puxin Biogas Plants (2010) Puxin family size and medium size biogas systems in P.B. Plants. Puxin Biogas Plants, Norway
43. Rashidi Z, Karbassi A, Ataei A, Ifaei P, Samiee-Zafarghandi R, Mohammadizadeh MJ (2012) Power plant design using gas produced by waste leachate treatment plant. Int J Environ Res 6(4):875–882
44. Jørgensen PJ (2009) Biogas—green energy: process, design, energy supply, environment. Researcher for a Day, Tjele
45. Alrikabi NK (2014) Renewable energy types. J Clean Energy Technol 2(1):61–64
46. Thirugnanasambandam M, Iniyan S, Goic R (2010) A review of solar thermal technologies. Renew Sustain Energy Rev 14(1):312–322
47. Halog A, YU M (2015) Solar photovoltaic development in australia—a life cycle sustainability assessment study. Sustainability 7(2):1213–1247
48. Jinko Solar: Solar Module. https://www.jinkosolar.com. Last Accessed 20 Feb 2018
49. Varjavi MS (2015) Identification of the Mazandaran indigenous architecture concepts by studying the style of the wooden structures in the northern parts of the country: with sustainable architecture approach. Appl Environ Biol Sci 5(108):825–832
50. Fathi S, Mirabdolah Lavasani A (2017) A review of renewable and sustainable energy potential and assessment of solar projects in Iran
51. Tesla: Powerwall. https://www.tesla.com/en_NZ/powerwall. Last Accessed 12 Feb 2020
52. Kalogirou S (2009) Thermal performance, economic and environmental life cycle analysis of thermosiphon solar water heaters. Sol Energy 83(1):39–48
53. Parliamentary Commissioner (2012) New Zealand Office of the Parliamentary Commissioner for the environment: evaluating solar water heating: sun renewable energy and climate change. Parliamentary Commissioner, New Zealand
54. Rheem (2009) Technical specifications on Rheem water heating products. Rheem, New Zealand
55. Gielen D (2012) Wind power, renewable energy technologies: cost analysis series. Int Renew Energy Agency (IRENA)
56. Mirzaei G (2013) Practicability to Install Wind Turbines in Mazandaran Province. J Basic Appl Sci Res 3(4):654–662
57. Iran Energy Efficiency Organisation (2017) Energy consumption of electrical devices in household. Renewable sources of energy and electricity harvesting department Iran

Part VII

Chapter 19
Simulation of an Adsorption Machine with Auxiliary Heater for CO_2-Neutral Air-Conditioning of Electric Utility Vehicles

Lukas Wildner, Michael S. J. Walter, and Stefan Weiherer

Abstract Air-conditioning of vehicles is important to improve driver comfort and motivate users to invest. The heating and cooling processes required for this are very energy-intensive processes. In conventional battery-operated commercial vehicles, the energy required to operate the air-conditioning system is used from the battery. This reduces the range of the vehicles by up to 30%. One of the greatest challenges in making electric commercial vehicles usable across the board is to increase their driving range. To reduce energy losses through air-conditioning, it is preferable to develop a technology that is independent of the battery. There are a number of options for controlling the temperature of a moving vehicle, but only a limited number that is CO_2-neutral. In this paper, we focus on adsorption chiller technology in combination with an auxiliary heater based on bioethanol. To understand the advantage of an adsorption machine, a simulation model can provide useful data on scaling and ease of use and thus be the basis for design and assembly of a prototype system. Therefore, a mathematical model of the adsorption technology is combined with the known dimensional parameters of electric vehicles, and the results are presented in form of a simulation model.

Keywords Adsorption · Simulation · Air-conditioning · Electric utility vehicle · Fuzzy controller · Auxiliary heater

1 Motivation

In order to solve the major challenge for the market penetration of battery-operated electric utility vehicles (BEUVs), their range needs to be increased and charging time shortened. In addition to the vehicle power train, common heating and cooling processes are very energy-intensive and reduce the driving range. Usually, this

L. Wildner (✉) · M. S. J. Walter · S. Weiherer
Ansbach University of Applied Sciences, Residenzstr. 8, 91522 Ansbach, Germany
e-mail: lukaswildner@posteo.de

© The Author(s), under exclusive license to Springer Nature Singapore Pte Ltd. 2021
R. J. Howlett et al. (eds.), *Emerging Research in Sustainable Energy and Buildings for a Low-Carbon Future*, Advances in Sustainability Science and Technology,
https://doi.org/10.1007/978-981-15-8775-7_19

Fig. 1 Research vehicle
Renault ZOE R240

problem is solved by increasing the battery capacity, which leads to a significant increase in weight and thus reduces the payload of a BEUV (Fig. 1). 4

A much more efficient way of solving this problem would be to supplement vehicle air-conditioning with an additional CO_2-neutral system. Conventional auxiliary heaters are already established on the market and can easily be converted to bioethanol. These can bring a significant improvement in winter season. A usable solution for ensuring cooling in summer season can be the adsorption technology. Adsorption tanks can be operated reversibly and climate neutrally with water and heat. Building on our previous work, we now want to check whether the adsorption technology is suitable to complement the validated auxiliary heater in terms of performance and geometric as well as application constraints.[1]

This paper discusses how to create a simulation model that represents the most important technological mechanism as accurately as required to design the main components of the heating and cooling system [1–5].

2   State of the Art

The air-conditioning of a BEUV is usually provided by air-conditioning systems such as used in combustion vehicles. These technologies require a large amount of energy from the battery. Heat pumps, for example, have high energy efficiency but consume electrical energy for operation. The most sold BEUV (Renault ZOE) is equipped in standard configuration with a heat pump system for heating and cooling. In order to save as much energy from the battery as possible to ensure more driving distance, it is recommended to use a more efficient technology. Additional heaters can replace the air-conditioning of BEUVs in the cold season. The heat is supplied directly by the combustion of fuel. As heat generation is largely independent in energy supply, the use of auxiliary heaters has little effect on the battery. In recent years, we implemented and verified a bioethanol-powered auxiliary heater in the

Renault ZOE research vehicle (Fig. 1) and verified its performance and application during cold seasons [1, 3, 4].

In warm seasons, an intelligent method is needed to replace the previous compression cooling systems, which show a comparably low efficiency related to the system's weight. Adsorption cooling, however, shows potential for efficient use in electric utility vehicles. Abdullah et al. [5] point out that adsorption cooling is increasingly used when further improvements are developed. The problem, however, is the lack of systems in the required performance class for use in automotive applications. Adsorption processes have been mathematically investigated and simulated several times. The thermodynamic process of adsorption and desorption of water is extremely complex. The integration of the kinetics into a higher-level system is usually only carried out in a simplified form. In addition, studies are known in which no simulation model is determined, but a test facility for integration into a higher-level system exists. However, no simulation model of an adsorption chiller with additional heating is known that was developed for electric vehicles. This work covers both the simulative design and the integration of such a system into electric commercial vehicles [3–9].

2.1 Adsorption

Adsorption describes the accumulation of a substance, e.g., water, on another substance, e.g., a porous rock, without any change of substance occurring. The porous volcanic rock focused for the adsorption machine is generally called zeolite, which translated from ancient Greek means 'boiling stone'. The material has a decisive property which makes two thermodynamic uses possible. Zeolite is extremely porous and thus has an impressively large surface in relation to its mass. In contact with simple water (Fig. 2), a process takes place that can be symbolized using the example of a sponge. The water molecules are attracted to the rock and bound to the surface. This process takes place very quickly, especially in a vacuum, which extracts energy from the original space of the water. Cold is produced. On the other hand, the water molecules in the porous corridors of the rock structure collide with each other and pass the rock, thus releasing energy to the zeolite. Heat is generated [7, 9, 10].

The adsorption process can be operated reversibly by supplying heat. Heat energy is supplied to the volcanic rock, which is loaded with water, in order to increase the kinetic energy of the water. The gaseous water can thus be expelled from the rock. Attention must be paid to the effect that no other gas, such as carbon dioxide or nitrogen oxides, occupies the empty pores of the zeolite. The so-called desorption process for the regeneration of the sorption medium thus restores the initial state and enables reversible use (Fig. 3) [7, 9, 10].

Regeneration of the adsorption machine requires not only the reversal of the process sequence but also time in which no cooling capacity can be delivered. This

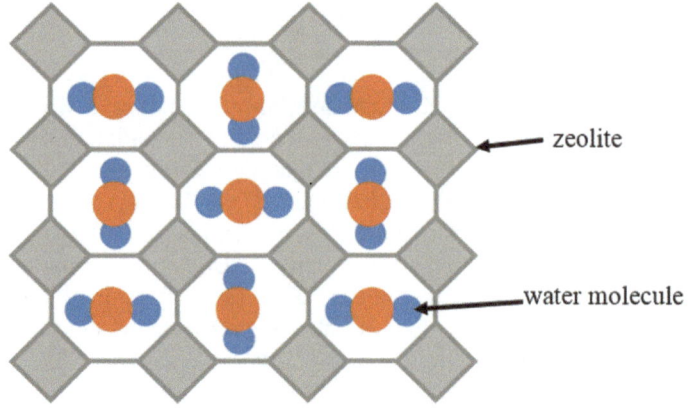

Fig. 2 Zeolite structure with incorporated water molecules

Fig. 3 Functional diagram
of an adsorption machine

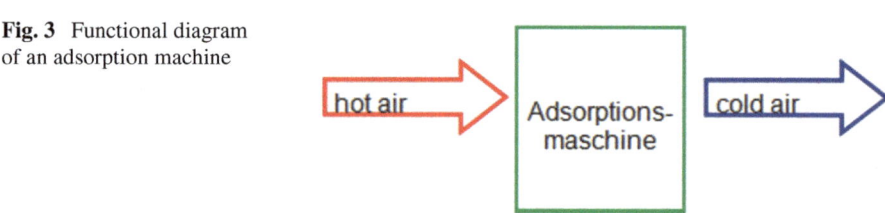

dead time of adsorption/desorption cannot be avoided. The duration of the process depends on the temperature, pressure and humidity of the desorption area [7, 9, 10].

2.2 Adsorption Chiller for Vehicles

Adsorption technology is known for some time, but has not yet been integrated into vehicles. Although there are mobile refrigerator systems, they work cyclically and maintain the cold generated for a longer period of time through high-quality insulation. The use in a continuous cooling process is quite rare. In order to investigate the advantages of adsorption cooling for vehicles in continuous operation, a research team built a test bench that was to be operated with the engine exhaust heat of a truck. In the work of Vasta et al. [3], an adsorption machine was thus developed for cooling the driver's cab. It showed an efficient cooling performance that is quite capable of ensuring the comfort temperature of 18 °C. As the temperature of combustion exhaust gases can also be reached with a bioethanol-fired auxiliary heater, an energetic use is to be expected.[3, 7]

2.3 Simulation of Adsorption Process

For the description of the adsorption process in different applications, a mathematical model with the previously determined values was used to simulate the multitude of properties as comprehensively as possible. Bathen and Breitbach [7] provided mathematical descriptions for different process options. However, the model used is generally referred to as the linear-driving-force model (LDF). The load of the sorption material with water vapor is calculated depending on the sorption material and the geometry of the sorption chamber. It simplifies laboratory testing without losing too much process data and demonstrates energy and environmental benefits of adsorption technology [7].

$$\mathrm{d}q/\mathrm{d}t = k_{\mathrm{eff}} \cdot {}^{A_P}/_{\rho} \cdot \left(q - q_{gl}\right) \tag{1}$$

$$k_{\mathrm{eff}} = 15 \cdot {}^{D_{\mathrm{ges}}}/_{r_P^2} \tag{2}$$

$\frac{\mathrm{d}q}{\mathrm{d}t}$ denotes the solids loading per time, including the adsorptive quantity in the pore volume. q_{gl} is being subtracted as potential for the solids loading in the boundary film. A_P stands for the particle surface and ρ for the density. k_{eff} represents the effective mass transfer coefficient. This gets calculated with the total diffusion resistance D_{ges} and the relative pore r_P diameter of the adsorptivity. With the values for the loading of the sorption chamber determined in this way, the potential energy transfer can be deduced from the generally known steam table and the prevailing temperature. All loss mechanisms of the material or energy transport are of course associated with an energy loss, which needs to be considered individually in the calculations [4, 7, 9].

3 Realization of the Model Construction

The thermodynamic mechanisms of a BEUV, an adsorption machine and an auxiliary heater are analyzed and integrated into a simulation model. In our previous work, we already integrated an auxiliary heater into a Renault ZOE R240 (2015). Based on this work, the geometry of the electric vehicle could be taken into account as boundary conditions in the simulation model as well as the specific thermodynamic properties of the applied bioethanol-fueled auxiliary heater. The system covers both summer and winter operating conditions. In winter, the vehicle is to be supplied with thermal energy (hot air) and in summer with cooling energy (cold air). Care should be taken to avoid excessive operation of the auxiliary heater, especially in summer. This can save fuel and counteract unnecessary heating of the vehicle and the environment. If operation takes place in winter (Fig. 4, 1. case), the auxiliary heater is used as a heat source to directly heat the driver's cabin. In summer, instead (Fig. 4, 2. case), the heat sink of the adsorption machine is used to cool the driver's cabin, while the heat

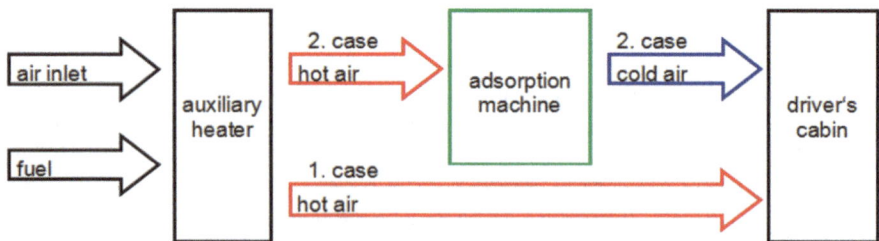

Fig. 4 Functional scheme of air-conditioning in two use cases: 1. Case: winter and 2. Case: summer

of the auxiliary heater is used to reload the zeolite inside the adsorption machine. In order to guarantee a continuous cooling capacity, an expansion of the adsorption machine is necessary. It is recommended to divide the sorption storage into at least two parts which can be operated independently of each other (Figs. 5 and 6). Thus, the first storage chamber can be regenerated by an adequate heat input, while the second storage chamber provides heat sink by adding water (Fig. 5).

If, after a certain time, sorption chamber 2 is saturated with water and the cooling capacity falls below a predefined limit value and sorption chamber 1 is therefore sufficiently regenerated, the process can be reversed in step 2 (Fig. 6). The chamber 2 is then regenerated by heat input, while the chamber 1 supplies cooling capacity by adding water. The water required for operation remains in the system in a cyclic process. There is no immediate need to provide further tank systems for water supply.

Fig. 5 Scheme of cyclic cold generation with adsorption machine (Step 1)

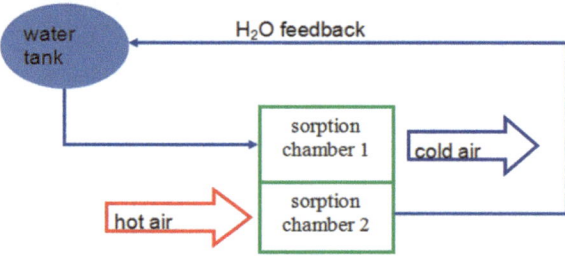

Fig. 6 Scheme of cyclic cold generation with adsorption machine (Step 2)

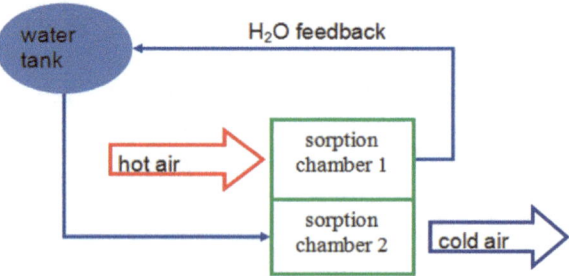

The adsorption machine is divided into two independent chambers. While one storage chamber is filled with water and thus produces cold, the second storage chamber can be discharged with the heat from the auxiliary heater. This avoids a dead time in which no cooling capacity can be taken from the storage tanks. For the performance calculation and the integration of the adsorption technology into the calculation model, ten kilograms of zeolite 13X are assumed, from which a constant cooling performance of approximately 1.2 watts is assumed. The process runs under pressure between 0.1 and 30 mbar, expecting a maximum adsorption volume of 0.34103 (kg og water per kg og zeolite).[10] Based on the inconsistent charge states of the sorption chambers, their interchanging operation is defined by the filling level of the water tank and a timer for the desorption phase [3, 7, 9, 10].

3.1 Controlling System

Since the influence of the instantaneous temperature represents an increase in complexity of the model structure that is difficult to comprehend, its influence on the process is usually simplified. The expansion of the complexity also shows a more exact description, but is not in equilibrium with the additional computing effort and the gain in controllability. Accordingly, the influence of these is also excluded in this paper, and a simplified model, the so-called LDF (e.g., 1–2), is used. The description is limited to the thermodynamic model of fluid phase adsorption within a homogeny transport system layer model. The adsorption chiller is powered by water and a bioethanol-fueled auxiliary heater. The controllability of the system is supported by a specially developed fuzzy controller. Since the chiller is built as a storage, the vehicle can be heated or cooled before the driver arrives, even without operating the heater. Thus, the vehicle is already preheated when the driver arrives [7, 11].

To summarize the functionality of the auxiliary heater, the driver's cab and the adsorption machine, their thermodynamic description is integrated into the simulation model and established in a control loop with MATLAB/Simulink (Fig. 7).

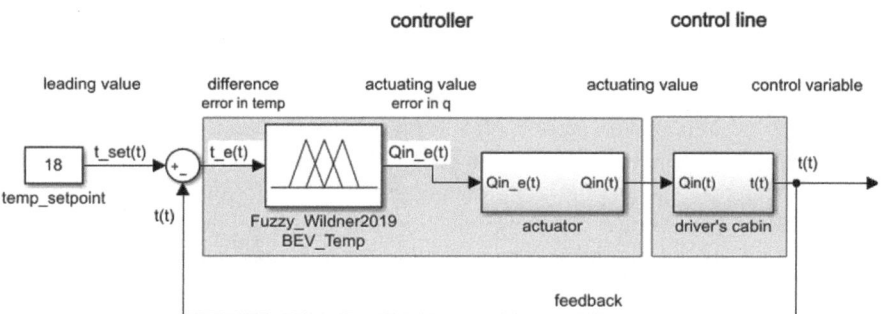

Fig. 7 Control loop of the simulation model

The driver's cab is the controlled section in which the effect of the energy input or output is described and returns the current value of the temperature in the vehicle as a control variable $t(t)$. The current value of the temperature is returned as feedback and placed in front of the control loop. The difference between the set temperature t_set(t) and the actual temperature t_e(t) is entered into a fuzzy controller as the error in temperature. There an actuating value Qin_e(t) is generated by 'IF THEN' rules, which generates the power level of the air conditioning.

With the operability of an ordinary air-conditioning system as a model, 21 rules for a fuzzy controller could be set up and be successfully integrated. The controller works with seven functions for the error of the current temperature and three functions for the desired blower level (based on the individual comfort requirements of the driver). The resulting power level is assigned with five functions. These functions indicate a range of values that determine the sensitivity of the fuzzy controller. The more functions, the more sensitive the controller and the more complex the calculation for the output value is. On basis of a mathematical model for the auxiliary heater and the adsorption machine, the expected energy supply is calculated and transferred to the controlled system as an actuating value. The energy output of the actuator in the form of cold or heat acts on the driver's cab. In this version of the control loop, external disturbance variables, such as strong fluctuations in the outside temperature or the opening of doors, are not taken into account [1, 7, 8, 11].

4 Conclusions

A simulation model of an adsorption machine and an external heating unit for BEUV was created and tested with the standard operating profile. The results indicate that the considered concept on a system on heating and cooling of a BEUV may be very compatible with the given requirements concerning performance and usability. The expected output in the previously calculated one- to two-kilowatt range is adequate to achieve sufficient cooling capacity for individual vehicles within the developed control loop. The simulation was started with two sorption containers that were completely filled with water, i.e., emptied according to the sorption chambers state of charge (short: SOC), and were initially in the state of desorption. After the water of a sorption chamber was expelled, i.e., after the SOC is full, the cooling process could start with a delay of 50s. The loading and unloading time of the sorption chambers depends on the intensity of heat energy supplied (by the auxiliary heater) and the heat transfer from the external heat source and the sorption chambers. The data of the auxiliary heater considered and heat exchangers allow a uniformly intense desorption of the sorption chambers as the energy input is mainly limited by the heat exchanger surface. In the simulation, three cycles of both sorption chambers were sufficient to cover the cooling demand. The target temperature of the driver's cab is reached with an acceptable time delay of about seven minutes, whereby the operation time is only just over six minutes (Fig. 8).

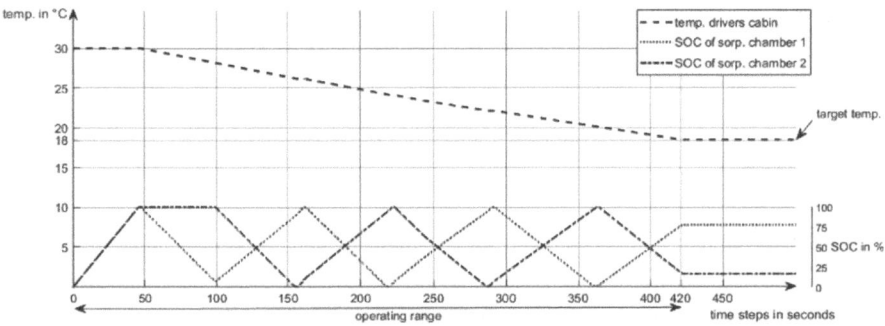

Fig. 8 Temporal course of the driver's cab temperature (temp.) and state of charge (SOC) of both sorption stores (SOC = 100%: battery is fully loaded; SOC = 0%: battery is completely unloaded)

The temperature curve shown in Fig. 8 promises an acceptable process control for the real operation of a prototype with regard to external interferences. In addition, the figure details an adequate cyclic loading and unloading curve for both sorption containers. Once the target temperature has been reached, the auxiliary heating can switch off. The sorption chambers remain at their current state of charge.

5 Outlook

Finally, the developed model of the adsorption air-conditioning system with auxiliary heater represents a potential alternative to conventional air-conditioning systems and thus provides an answer to the considered research question [12]. The calculation of thermodynamics, ad- and desorption processes as well as the loss mechanisms of vehicle temperature control are extremely complex. Further tests are required to investigate unexpected performance losses during real operation. Also, due to strongly varying literature statements about the regeneration time of the zeolite, own experiments are recommended. Due to the wide spectrum of potential disturbances during operation, the development of a control system and the integration of all the components involved can significantly increase the performance potential of the system and most closely resemble the ideal operation described in this paper. Further process and ecological improvements are of course possible, such as the integration of intuitive machine learning. The storage capacity of the sorption reservoirs could also be integrated into public networks when idling in order to collect local waste heat or to cover cooling requirements. Furthermore, we will start building a real pilot prototype to be integrated into an electrified light-utility vehicle. This will validate the values of the design calculations and adapt the performance class of the adsorption machine, the relation of the heating system and the research vehicle. Overall, it is to be expected that adsorption technology will be an effective strategy for the air-conditioning of electric vehicles in the future.

Acknowledgements The authors thank the Biomass Institute for financial support.

References

1. Riess C, Walter MSJ, Weiherer S, Haas T, Haas S, Salceanu A (2018) Heating an electric car with a biofuel operated heater during cold seasons–design, application and test. ACTA IMEKO 7(4):48–54
2. Karle A (2018) Elektromobilität: Grundlagen und Praxis, 3rd edn. Hanser, Munich
3. Vasta S, Freni A, Sapienza A, Costa F, Restuccia G (2012) Development and lab-test of a mobile adsorption air-conditioner. Int J Refrig 35(3):701–708
4. Gassel A (2012) Die Adsorptionskältemaschine: Betriebserfahrungen, thermodynam. Modell und TRNSYS-Simulation. https://www.ib-aton.de/service/fachaufs/Thd-trns.pdf. Last Accessed 12 Feb 2019
5. Abdullah NO, Tan IAW, Lim LS (2011) Automobile adsorption air-conditioning system using oil palm biomass-based activated carbon. Renew Sustain Energy Rev 15(4):2061–2072
6. Gröper M, Riess C, Walter MSJ, Weiherer S (2018) Analysis of electrical energy flow in a battery powered electric vehicle. Comparing the economical mode and the normal mode under real conditions. In: IEEE 2018 international conference and exposition on electrical and power engineering (EPE) Lasi. IEEE, pp 63–68
7. Bathen D, Breitbach M (2001) Adsorptionstechnik, 1st edn. Springer, Berlin
8. Riess C, Walter MSJ, Weiherer S, Gröper M (2018) Evaluation and quantification of the range extension of battery powered electric vehicles in winter by using a separate powered heating unit. In: 2018 international conference and exposition on electrical and power engineering (EPE), Lasi. IEEE, pp 75–80
9. Bonalumi D, Lillia S, Manzolini G, Grande C (2017) Innovative process cycle with zeolite (MS13X) for post combustion adsorption. Energy Procedia 114:2211–2218
10. Storch JCS (2009) Materialwissenschaftliche Untersuchungen an zeolithischen Adsorbenzien für den Einsatz in offenen Sorptionssystemen zur Wärmespeicherung. Dissertation, Technical University Munich
11. Prokopowicz P, Czerniak J, Mikołajewski D, Apiecionek Ł, Ślęzak D (eds) (2017) Theory and applications of ordered fuzzy numbers: a tribute to Professor Witold Kosiński, 1st edn. Springer, Cham
12. Haas S, Weiherer S, Walter MSJ (2020) Design of an adsorption refrigeration machine with an auxiliary heater for CO_2-neutral air-conditioning of e-vehicles. In: Littlewood J, Howlett R, Capozzoli A, Jain L (eds) Sustainability in energy and buildings Smart innovation, systems and technologies, vol 16. Springer, Singapore, pp 651–664

Chapter 20
Sustainable Road Infrastructure in Rural Areas in South Africa—A Preliminary Study

Matthew Ikuabe, Douglas Aghimien, Clinton Aigbavboa, and Ayodeji Oke

Abstract The role of sustainable infrastructures such as roads in the development of any community cannot be overemphasised. Quality and sustainable roads play a vital role in the daily living of individuals. However, roads within the rural areas are in most cases adjudged to be defective and unsustainable, thereby crippling activities within these rural areas. It is based on this notion that this study assessed the possible measures for attaining sustainable road infrastructure within rural communities in Limpopo province, South Africa. The study sought answers from rural dwellers and construction workers within the study area through a questionnaire survey. Data gathered were analysed using percentage, mean item score, standard deviation and one-sample t-test. The reliability of the questionnaire was also tested using Cronbach's alpha which gave an alpha value of 0.948 which indicates the questionnaire used was reliable. Findings of the study revealed that the most significant measures for attaining more sustainable road constructions within the rural areas include using quality materials that will last the expected lifespan of the road, having planned maintenance, proper investment on road projects, and using contractors and skilled workers with the right experience in road construction. It is believed that the findings of this study will help increase the delivery of sustainable road projects within the rural areas in a bid to provide better standard of living for rural dwellers.

Keywords Sustainable roads · Rural areas · Defective roads · Construction · South Africa

1 Introduction

Road infrastructure plays a major role in the livelihood of rural dwellers. When the roads in rural areas are bad, it has significant negative effects that cripple rural

M. Ikuabe (✉) · D. Aghimien · C. Aigbavboa · A. Oke
Department of Construction Management and Quantity Surveying, University of Johannesburg, Johannesburg, South Africa
e-mail: ikuabematthew@gmail.com

dwellers activities in many ways. Unfortunately, according to [1], infrastructure development has been the bane of most African countries with decay in infrastructure being a common sight in these countries. Just like every other African country, South Africa faces issue of poor infrastructure development. According to [2], although the South African road infrastructure seems to be adequate in terms of quantity, the major problem is their unacceptable quality. According to [3], the temporary bridge between Vaalwater and Lephalale that was put in place after the road was destroyed during floods and appears not to be able to withhold the load of all vehicles. In rural areas, the case is worse as the movement of products and people is done over long distances. Fungo et al. [4] noted that road infrastructure is a major element for many rural and urban transport systems. This is because rural transport provides assurance for the supply of agricultural inputs and facilitates the delivery of the farm outputs to the markets in most urban centres. Equally, asset value of the rural dwellers is increased by infrastructure facilities which create benefiting opportunities. This also affects the social and financial growth of citizens.

When roads are defective, farmers will most likely seek better alternative routes in order to ensure the distribution of their farm products. Unfortunately, these alternative routes are in most cases longer and often lead to higher expenditure on transportation cost. In the case of lack of alternative routes, farmers are forced to travel on defective roads. The resultant effect of this includes damage to vehicles used, high maintenance costs, low profit, low earnings for workers and continuous rise of poverty among rural dwellers [5]. In South Africa, Van Heerden et al. [6] observed that the Department of Transport has a comprehensive plan for non-motorised transport. The department also strives to implement a rapid public transport network so that regular transport services can be available to the public in rural areas. Investing in roads creates more employment, and it is an avenue to increase the knowledge of road construction, as well as trainings are being offered. Moeketsi [7] stated that road infrastructure has been unattended to for a long time in some provinces in South Africa, thereby creating significant limitations in the growth and development in rural communities. According to [2], South Africa has a better working road infrastructure network compared to other African countries, which provides a lot in terms of economic and social progress. However, poor quality roads are still evident particularly within the rural settlements. This poor quality becomes more evident during heavy rainfall where materials used on construction roads get eroded easily. Fungo et al. [4] earlier noted that when it comes to quality standards, the roads that are constructed in Africa are far below. Van Heerden et al. [6] noted that the majority of rural areas have no gravel and tar roads and rural dwellers drive in unsafe conditions. It has been stated by [8] that the condition of roads at the provincial and municipal level is disadvantaged by the fact that data collection and management systems have not been updated for numerous roads.

It was indicated by Moeketsi [7] that proper and effective systems must be in place to manage infrastructure cost-effectively and efficiently. Roads authorities should be considerate of the unsealed roads impacts during maintenance strategies developments and maintenance plans implementations. Maintenance of roads targets preventing the waste of finance used in constructing roads and improving the rural areas. Road maintenance should be executed on a constant basis to avoid deterioration that leads to the increased cost of road rehabilitation [8]. Overloading, flood, high traffic volumes cause roads to deteriorate over time while the lifespan of roads is largely determined by the weather conditions, the design of the road and the strength of the materials deployed. Author [8] stated that maintenance of roads managed by municipalities is neglected since most municipalities have to execute this by self-funding. Informal roads in the provinces are usually maintained by a small community, immediate road users who are forced to spend more on vehicle costs and group of farmers who needs safe access to distribute their products. There must be conception, plans and designs to kick start road construction; otherwise, the project will not be successful. Regardless of adequacy of construction and supervision, the absence of proper planning will result in the function of the road project being unsuccessful. There is slow movement on roads that is caused by obstacles cracks, potholes and abandoned vehicles on the roads, resulting in traffic on the roads, and one way of solving decongestion of roads is by removing such obstacles and diverting traffic to other traffic modes [9]. Better road infrastructure can be achieved with the right resources and the right knowledge of road contractors, the appropriate methods of maintenance and constructing it with materials that are lasting and can be easily maintained. Defective road infrastructure causes most rural residents to relocate to the cities, in order to have access to improved infrastructure.

To ensure the improvement of sustainability in rural transportation system, assessing issues surrounding road infrastructure with rural environment is necessary. While this current study recognises the presence of several other existing studies conducted on road infrastructure, the issue of defective road infrastructure is still a major concern in most developing countries around the world and South Africa is no exception. Thus, the motivation for this study lies in the fact that new studies are helpful in discovering the depth of implications and the current issues that the country is facing. Based on this notion, the study assessed the measures needed for improving road infrastructure in rural areas in South Africa with specific focus on Limpopo province. The subsequent part of this paper includes the research methodology adopted, the findings and conclusion. Based on the conclusion, recommendations were made and the areas for further studies were suggested.

2   Research Methodology

In assessing the possible measures for attaining sustainable road infrastructure within rural communities, a quantitative survey approach was adopted. Answers were gotten from rural dwellers and construction workers in Limpopo province of South Africa using a questionnaire. The questionnaire was adopted due to its ease of usage and ability to cover a wide range of respondents [10]. A total of 72 questionnaires were conveniently distributed with 65 retrieved and 60 found fit for data analysis. The remaining five questionnaires were discarded as a result of inaccurate completion and missing vital information. Some of the questionnaires were self-administered while others were done through the help of field agents who read the questions to some of the rural dwellers who were unable to fully understand the questions asked. The questionnaire used was designed in sections; the first section gathered information on the respondent's background. The second section sought answers regarding the possible measures for improving road infrastructure within rural communities in South Africa. Respondents were provided with a list of measures identified from literature to rate based on their level of significance. A Likert scale of 1–5 was employed, with 5 being very high, 4 being high, 3 being average, 2 being low and 1 being very low. Data analyses were done using percentage for data on the background information of respondents. Mean item score (MIS) and standard deviation (SD) were used to rank the identified measures based on their rating with the highest mean value ranking first. However, where two are more measures have the same mean value, the one with the least standard deviation is ranked first [11]. Furthermore, a one-sample t-test was used to identify the level of significance attached to the identified measures by the respondents. The reliability of the questions in this second section was also tested using Cronbach's alpha test which gave an alpha value of 0.948, thus implying high reliability of the questionnaire used as the alpha value is closer to 1.

3 Results and Discussion

3.1 Background Information

The data gathered on the background of the respondents that 52% of population were male while the remaining 48% were female. Forty-five per cent (45%) of these respondents are rural dwellers that operate local businesses such as farming and selling of farm products, petty shop owners, motorist and the likes. The remaining 55% are into construction works within these rural areas. For their academic qualification, result revealed that 5% of the respondents had a below matric qualification, 18% had matric, 13% had National Certificate, 37% had a National Diploma, 20% had a Bachelor's degree and 7% had a Master's degree. Following these results, it can be said that the respondents for the study have considerable understanding of happenings within the rural environment as they reside in these areas.

3.2 Measures for Improving Sustainable Road Infrastructure in Rural Areas

In determining the possible measures for improving road infrastructure within rural communities in South Africa, a total of 11 measures were identified from the review of literature and presented to the respondents to rate. One-sample t-test was then used as method of data analysis to determine the significance of each of the identified measures as rated by the respondents. A null hypothesis which states that a measure is unimportant when the mean value is less than or equal to the population mean (H_0: $U \leq U_0$) was set. The alternate hypothesis set was that a measure was important when the mean value is greater than the population mean (H_a: $U > U_0$). The population mean (U_0) was fixed at 3.0 (the mid-point for the Likert scale adopted), and the significance level set at 95% which is the conventional confidence level [12]. Thus, a measure is said to be important if it has a mean value of above 3.0. Result in Table 1 shows a two-tailed p-value which represents the significance of each identified measure. This significant p-value was further divided by two as shown in Table 2 to get the significant value for a one-tailed test about the test hypothesis (i.e. H_a: $U > U_0$).

From Table 2, it is evident that all the assessed possible measures for improving road infrastructure within rural communities in South Africa have a mean value of well above the 3.0 cut-off point. The implication of this is that the respondents considered all the assessed measures to be important. Similarly, the second to the last column shows the significant p-value of each of the identified measures, and this column reveals that at 95% confidence level, all the 11 assessed measures were considered significant, as a significant p-value of below 0.05 was derived. This result implies that if road infrastructures in rural communities are to improve, all the assessed measures need to be put in place. Among these measures are use of the right quality materials (MIS = 4.27, sig. = 0.000), use of materials will last the expected life span of the road (MIS = 4.27, sig. = 0.000), planned maintenance (MIS = 4.25, sig. = 0.000), government investing in road infrastructure (MIS = 4.22, sig. = 0.000) and using contractors/skilled workforce with the right expertise in road construction (MIS = 4.17, sig. = 0.000).

The findings of the study revealed that in the improvement of road infrastructure in rural areas, the deployment of the right quality material is important and these materials should be expected to last the life span of the road. This is in conformity with the study of [9] which stated that the use of appropriate materials in road construction leads to a longer lifespan and leads to a reduction of the cost of maintenance over time. This is a clear indication that contractors handling the construction of roads should as a matter of importance channel efforts to utilising the appropriate materials that will be durable and long-lasting. Also revealed in the study is that when the construction of the infrastructure is on-going, measures should be taken to accommodate planned maintenance of the road as this is vital in ensuring the continuous improvement of the road. This is in tandem with [7] who stated that proper and effective systems must be in place to manage infrastructure. Planned maintenance will ensure that facility retains the purpose for which it was originally built. The government participation in the construction of road infrastructure is vital to improving road infrastructure as shown in the study. Van Heerden et al. [6] affirmed this by stating that government intervention in the provision of infrastructure has helped in closing the gap in infrastructural deficit. Since government still remains a large financier of infrastructures (roads inclusive) in South Africa, a necessary measure to the improvement of roads is the continuous financial commitment of government to this course.

Table 1 One-sample *t*-test statistics

Measures	Test Value = 3.0					
	T	df	Sig. (2-tailed)	MD	95% confidence interval of the difference	
					L	U
Investing in road infrastructure	8.294	59	0.000	1.217	0.92	1.51
Adequate and thorough planning by municipalities	6.632	59	0.000	1.033	0.72	1.35
The use of right quality materials	8.898	59	0.000	1.267	0.98	1.55
The use of materials that will last the expected life span of the road	8.777	59	0.000	1.267	0.98	1.56
Planned maintenance	9.503	59	0.000	1.250	0.99	1.51
Using contractors/skilled workforce with the right expertise in road construction	8.523	59	0.000	1.167	0.89	1.44
Labour-based technology	4.696	59	0.000	0.667	0.38	0.95
Provision of adequate and appropriate designs	8.435	59	0.000	1.050	0.80	1.30
Decongestion of the roads	2.944	59	0.005	0.467	0.15	0.78
Provision of facilities, such as drainages	7.758	59	0.000	1.050	0.78	1.32
Appropriate sanctions for road failures	5.538	59	0.000	0.817	0.52	1.11

Note MD mean difference,*L* lower,*U* upper

4 Conclusion

The study focused on the measures of improving road infrastructure in rural areas with Limpopo province as the area of study. Through the review of relevant literature, the study identified and outlined the measures for improving road infrastructure in rural areas. Based on the findings, the study concludes that much needs to be done on improving road infrastructure in rural areas as a result of the hindering impacts of bad

Table 2 Summary of t-test showing rankings of the measure for improving sustainable road infrastructure in rural areas

Measures	MIS	SD	SE	t	Sig. (one–tailed)	Rank
The use of right quality materials	4.27	1.103	0.142	8.898	0.000	1
The use of materials that will last the expected life span of the road	4.27	1.118	0.144	8.777	0.000	2
Planned maintenance	4.25	1.019	0.132	9.503	0.000	3
Government investing in road infrastructure	4.22	1.136	0.147	8.294	0.000	4
Using contractors/skilled workforce with the right expertise in road construction	4.17	1.060	0.137	8.523	0.000	5
Provision of adequate and appropriate designs	4.05	0.964	0.124	8.435	0.000	6
Provision of facilities, such as drainages	4.05	1.048	0.135	7.758	0.000	7
Adequate and thorough planning by municipalities	4.03	1.207	0.156	6.632	0.000	8
Appropriate sanctions for road failures	3.82	1.142	0.147	5.538	0.000	9
Labour-based technology	3.67	1.100	0.142	4.696	0.000	10
Decongestion of the roads	3.47	1.228	0.159	2.944	0.000	11

Note MIS mean item score,*SD* standard deviation,*SE* standard error

roads on the socio-economic well-being of rural dwellers. In achieving an improvement in the state of road infrastructure in rural areas, the study makes the following propositions: the use of quality materials that is expected to last the expected life span of the road should be deployed when road constructions are being done. Also, planned maintenance of erected road infrastructure should be considered with utmost seriousness as this would go a long way in ensuring that the infrastructure retains the purpose for which it was originally constructed. Government's role in the provision of road infrastructure is prominent as such projects are capital intensive. Also, construction contractors handling road projects should as a matter of importance deploy skilled personnel or workforce with appropriate expertise when undergoing road construction projects.

The scope of this study was limited to Limpopo province in South Africa. Since the country is made up of nine provinces, further research can be carried out in these other provinces. By so doing, a wider reach and a more balanced perspective from a national point of view can be ascertained.

References

1. Jerome A, Ariyo A (2004) Infrastructure reform and poverty reduction in Africa. Paper presented at African development and poverty reduction: the macro-micro linkage Forum. Somerset, South Africa
2. Mamabolo MA (2016) Provision of quality roads infrastructure in South Africa: rural villagers' perceptions, Polokwane municipality in Limpopo province. J Public Adm Development Alternat 1(2):28–44
3. No money to fix roads and bridges-Northern News, Available from: https://www.noordnuus. co.za/articles/news/32274/2015-07-30/no-money-to-fix-roads-and-bridges. Last Accessed 08 March 2018
4. Fungo E, Krygsman S, Nel H (2017) The role of road infrastructure in agricultural production. In: 36th annual southern African transport conference. South Africa
5. Tunde AM, Adeniyi EE (2012) Impact of road transport on agricultural development: A Nigerian example. Ethiopian J Environ Stud Manag 5(3):232–238
6. Van Heerden H, Burger M, Coetsee M, Mahlangu N, Naude K (2015) The current infrastructure conditions and the problems relating to it. In: 3rd global virtual conference. Zilina, Slovakia
7. Moeketsi AKW (2017) The relationship between road infrastructure investment and economic growth in South Africa. Unpublished M.Sc dissertation submitted to North-West University, South Africa
8. Department of Transport: Roads Branch, https://www.transport.gov.za/web/department-of-tra nsport/roads. Last Accessed 03 May 2018.
9. Okigbo N (2012) Causes of highway failures in Nigeria. Int J Eng Sci Technol 4(11):4695–4703
10. Tan P (2011) Towards a culturally sensitive and deeper understanding of "rote learning" and memorization of adult learners. J Stud Int Edu 15(2):124–145
11. Field AP (2005) Discovering statistics using SPSS, 2nd edn. Sage, London
12. Pallant J (2005) SPSS survival manual: A step by step guide to data analysis using SPSS for Windows (Version 12). Open University Press, Maidenhead

Part VIII

Chapter 21
Optimising Offsite Manufacturing of Timber-Frame Roof Trusses for UK Housing

V. L. Moorhouse, John R. Littlewood, P. Wilgeroth, and E. Hale

Abstract This study has been undertaken in response to drivers from the Welsh Government to increase the number of houses and quality and as such increase the offsite manufacturing of modern methods of construction and also to meet Wales' low- to zero-carbon agenda, launched in March 2019. The need to increase the quality and operational energy use of buildings has been acknowledged since 2019 when it was recognised that the UK's climate change targets were impossible to meet without almost 100% elimination of greenhouse gas emissions from UK buildings. This paper discusses and presents the first of three case studies undertaking time and motion and value stream mapping studies by one of Wales' (UK) largest offsite manufacturers (referred to as the company hereafter) of timber-frame construction systems, in order to evaluate optimisation opportunities in the offsite manufacture of the company's modern methods of construction within roof component production, such as trusses. A time and motion matrix to capture the live manufacturing data is presented and discussed. The preliminary results from the time and motion and value stream mapping assessments of the company's manufacture of a number of roof truss case studies conducted between May and September 2019 are presented and highlight opportunities for quick win refinement to their operational processes with the aim to increase production efficiency, reduce waste and close the performance gap, therefore increasing the quality and thermal performance of offsite manufactured timber-frame buildings resulting in reduced operational energy usage and therefore minimising greenhouse gas emissions. This paper will be useful for academics, timber-frame manufacturers, offsite manufacturers, building contractors, estimators, housing developers and clients.

V. L. Moorhouse (✉) · J. R. Littlewood · P. Wilgeroth
Cardiff Metropolitan University, the Sustainable & Resilient Built Environment Group, Cardiff CF52YB, UK
e-mail: vlmoorhouse@cardiffmet.ac.uk

E. Hale
Unit 2, Milland Road Industrial Estate, Neath SA111NJ, UK

Keywords Timber frame · Roof truss · Value stream mapping · Time and motion studies · Offsite manufacturing · Dwellings · Wales · UK

1 Introduction

This paper introduces a knowledge transfer partnership (KTP) project undertaken between Cardiff Metropolitan University and a timber-frame manufacturer (the company) in South Wales, UK. Context to the KTP project is given, which includes developing innovative practices for timber-frame products, designed and manufactured by the company for the UK construction market, thereby refining their operational performance and thus with a target to increase their market share in Wales and the UK, so pioneering expansion of prefabricated components and transitioning into modular construction. The suitability of value stream mapping including time-motion studies, to examine the manufacturing process of roof trusses, is discussed and illustrated, in order to identify any waste from materials or operative labour, and opportunities for improvement to the company's operational processes are discussed.

2 The Need for Offsite Manufacturing of Modern Methods of Construction

2.1 Context to the KTP Project

A KTP project enables researchers to be based in a company with support from a university and their staff (Cardiff Metropolitan University and the two-second paper authors, in the project discussed in this paper) and managers and directors from the company, and in this case, the project is three years in duration (February 2019 to January 2022). The KTP is funded by Innovate UK, the Welsh Government and part by the company. One of the aims of the KTP is to assist the transition from a manufacturer of traditional timber-frame building fabric componentry, such as roof trusses and floor joists to one that specialises in high-quality, high-performance and energy-efficient prefabricated and modular buildings for different sectors, such as residential, education, retail, fire and rescue services, medical and specialised dwellings. The two key tools which help to achieve the KTP's overall aim are to embed an innovation management system, with a holistic feedback and optimisation loop for transparent cost appraisal/reduction, and product performance enhancement, and these tools form two golden threads throughout the three-year KTP work plan. The main challenges for the KTP team are:

- To combine the expertise in the multi-disciplinary team (architectural engineering, building performance assessment, manufacturing management and product design) to develop a new approach for creating high-quality buildings.

- To develop an innovation management system as a framework and a holistic feedback loop with a toolkit of methods for costing, performance testing and refinement to provide the most efficient processes.
- To apply to buildings ethnographic research methods and analysis for refined manufacture and advanced product design assembly methodologies, normally only used in the design of high-quality consumer products.
- To pioneer the expansion of prefabricated components and transitioning into modular construction.

2.2 Context to the Need for OSM of MMC

It has been recognised since 2016 in the UK that there is an increased need to drive efficiencies in housing supply and also performance with the current and future skills shortages in construction, that is such encouraging a shift to offsite construction, known as offsite manufacturing (OSM) and modern methods of construction (MMC) [1]. In 2020, the Welsh Government's target for new affordable dwellings is 20,000 (not yet realised in January 2020), and traditional forms of construction with brick/block exterior walls are both unsustainable from a materials and climate change perspective, but are also slow to construct compared with OSMMMC [2]. OSM for housing requires the construction industry to combine design practices from both architectural technology and manufacturing, which is less than apparent in Wales [3]. The potential advantages of OSM over conventional construction of housing include quicker completion, greater quality of finish, less defects and better integrated building services. Therefore, the operational performance of buildings tends to be much nearer to design aspirations than conventional construction techniques [4].

OSM systems in the UK typically use either timber, steel or aluminium as the frame for the fabric such as exterior walls, floors or roof structures. Of these materials, only timber is a renewable and natural material. There are many organisations in the UK that supply timber open panel and to a lesser extent closed panel exterior systems [5]. In Wales, modern timber construction is predicted to reach 32% of new houses in 2018 [6]. In Scotland, timber construction in 2017 accounted for over 80% of the new-build housing market [7]. However, Wales imports at least 85% of the timber used in construction from outside the UK, but a strong timber OSM sector in Wales using home-grown timber would reduce dependence on imports and help to catalyse a substantial increase in forestry planting, and there are 35 OSMs of open or closed panel timber exterior wall systems based in Wales [8]. This KTP project as discussed in section one above is driving some of the innovation for timber-frame OSMs in Wales.

3 Methodology

3.1 The Company

The company is one of the largest timber-frame manufacturers in Wales and in 2019 started expanding from their existing factory circa 1000 m² (m²) to a new factory that they are refurbing of circa 23,000 m². The intention for the company is to double their workforce and triple production, to manufacture prefabricated fabric elements such as exterior walls, roof trusses and floor cassettes and complete modular pods and buildings for up to 3000 homes per year [9]. In addition, to creating a dedicated research and development hub within the company's new factory, to build entire buildings in the factory to test their performance and quality standards. This significant increase in space and equipment has allowed the company to improve the manufacturing of their timber pre-fabricated elements used in buildings, such as roof trusses. As such from 2019, this has enabled the company to tender for projects to provide single components, such as roof trusses, rather than multiple fabric components, so creating deeper penetration into the timber prefabricated market in Wales. In their existing factory, pricing for single-component projects, rather than whole buildings, such as roof trusses was impossible. This was previously an economically inefficient process when placed in comparison with timber-frame systems for whole buildings (this would generally include trusses, loose joists or floor cassettes, closed or open timber-frame panels, vapour control layers, breather membranes, insulation and in some instances factory fitted windows). When pricing timber-frame elements for complete buildings, trusses would be manually priced on a £/m² figure using the company's pricing system, as it would not be time efficient to design an entire roof for a more accurate quote. Instead, if more complex roofs (such as vaulted or hip style roofs or roofs that include multiple valleys or dormers) a smaller section of the roof would be designed to base the cost (£/m²). Whereas the company has only had the capacity to quote and manufacture for roof truss only projects since 2019, each truss kit is fully designed within the MiTek Pamir software [10] and consequently priced using the MiTek MBA software [11].

3.2 Ethnographic Observation During Time and Motion and Value Stream Mapping Methods

Since one of the central foci for the KTP is to optimise costs and as a result increase margins, it was agreed with the company that the research team would conduct both a T&M study and VSM exercise on the roof truss manufacture, in the new factory between May and September 2019. This was in order to identify any opportunities for cost savings and changes to the manufacturing and design processes, as a pilot study for the other timber-frame products that the company manufactures.

A KTP project uses by its very nature an ethnographic approach to engaging with the industry partner that is co-funding the project, whereby the associate (the lead author) is immersed within an empirical setting for long periods, i.e. the company's offices and operational facilities [12]. During this time, the researcher's experience, in terms of his or her observation at the research site, is used to generate a narrative-based interpretation of the events that took place. In the case of this paper, the interpretation is based on a T&M study and VSM activity of 23 truss projects [ibid].

The T&M data collection of operative processes in the manufacture of roof trusses from sawn timber to OSM MMC will provide a direct comparison of real-time production output, with anticipated production output—using MiTek's UK industry-standard pricing brackets [13]. This should give a good indication of expected profit/loss margins based on the current state of production in 2019/20 and the company's competitiveness within the UK market. The results will also present a comparison of the company's pricing estimates and MiTek's Business Application Software (MBA) estimates (MBA is a *"fully customisable, SQL database driven, project and client management system"), and* this will determine the accuracy and appropriateness of pricing trusses on an m^2 basis for complete build projects [11]. The T&M study was undertaken at the same time as the VSM exercise to aid the development of manufacturing processes, in order to optimise efficiency and consequently increase competitiveness of the company's pricing.

Value stream mapping is a lean manufacturing tool developed to track the flow of materials and information, from customer order through to delivery [14]. Value stream mapping illustrates value adding and non-value adding processes, allowing informed refinement of inefficient practices[ibid]. Toyota use value stream mapping as a means of eliminating Muda (waste) which includes *'Over processing, rework, transportation, inventory, waiting and motion'* [15]. Typically, maps produced by value stream mapping are first drafted during a walkthrough of the operating space using a pencil and paper and then refined and developed using several common value stream mapping symbols. Once a current value stream map has been produced (not discussed within this paper), an ideal or future state value stream map is drafted, and steps to achieve this are mapped out and implemented [16]. The company were invited by the Welsh Government in 2019 to take part in training, delivered by Toyota, UK, to learn about their expertise in lean manufacturing and then apply these lessons (with the support of Toyota) to their design and manufacturing practices [17].

The KTP's time and motion and value stream mapping pilot activities have been undertaken during the first stage of the company engaging with Toyota, UK [ibid].

3.3 Methodology for Value Stream Mapping Assessment of the OSM of the Truss MMC Production Line

In order to understand, in depth, the process in question, an extensive observational ethnographic study was conducted following Dey's (2002) [12] recommendations,

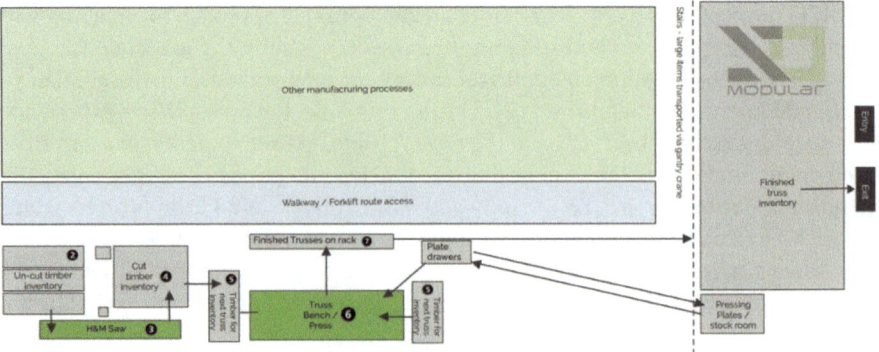

Fig. 1 Factory floor map highlighting the layout and route of material of the truss production process. Image created by author 2019

whereby every step of the roof truss manufacture process was identified to create a shop floor map (developed by the first author). The shop floor map highlights the route of material/product transfer from start to finish, including inventory and machinery placement. The map was used as an initial visualisation tool, which helped to identify any waste, such as over conveyance of people and stock, see Fig. 1.

3.4 Time and Motion

Traditionally, truss pricing was determined using a board footage methodology [18], (a board foot refers to a volume of timber 12 inches (30 centimetres (cm)) wide, 12 inches (30 cm) long and 1 inch (") thick) [19] which used the assumption that the size and weight of a truss had a direct correlation to manufacturing time and effort [18]. Walters et al. [20] suggest that Houlihan has since proved board footage an imperfect practice, instead Houlihan used the science of time and motion studies combined with the understanding that productivity and efficiency are reliant on variables related to people and machines, which could be refined through a systemised and scientific approach resulting in increased outputs with equal inputs [21]. Houlihan's [20] time and motion studies led to the development of realistic expectancies, often known as time standards or Man Minutes [20]. Realistic expectancies were calculated using the factory's maximum capacity production time and deducting unavoidable human factors that will increase production time. This approach was not only more accurate than board footage, but also it was popular with factory operators as it set realistic and achievable time standards [ibid].

Walters et al. [20] stated that Houlihan understood how company culture changes and policy adoption was equally as paramount to the success of his methodology and therefore implemented an hourly operator progress update, which allowed supervisors to calculate live efficiency percentages. If this percentage fell below 90%, investigations would take place into the reasons why and then recommendations made for improvements [ibid]. The study documented in this paper as part of the KTP applies the theories stated above through T&M studies, designed to calculate total realistic expectancies (or Man Minutes) in truss production in order to compare against MiTek's MBA software which also reflects Houlihan's methodology as stated by Walter et al. [20].

3.5 Understanding MiTek's Pricing System

With a huge array of roof trusses available to the UK market in different sizes, shapes and complexities designed to fulfil a variety of structural and aesthetics require-ments, Gang-Nail (now owned by MiTek) developed their own unique standard for modelling truss production costs [13]. The MiTek term for modelling truss costs is known as the Equivalent Fink (EFINK) costing system. First developed in Australia, *'the EFINK is also known as an EqA or Equivalent A truss, where an A truss is a Fink truss'* [13]. This pricing methodology forms the basis of MiTek's Business Application (MBA) software in order to predict labour costs and assist production scheduling[ibid]. The EFINK system utilises many of Houlihan's [20] theories on realistic expectancies/Man Minutes; however, MiTek has adapted and refined to be specific and unique to truss manufacture [ibid]. An EFINK is simply the amount of work required to assemble one standard Fink truss [22]. Note, a fink roof truss is one of several types of roof trusses known as a trussed rafter, see Fig. 2. Trussed rafters have been used and manufactured in the UK since 1964, and Fink trusses are a simple webbed design providing the most economical roof solution for roof structures, where roof loads are transferred down each heel to the exterior wall plate [23].

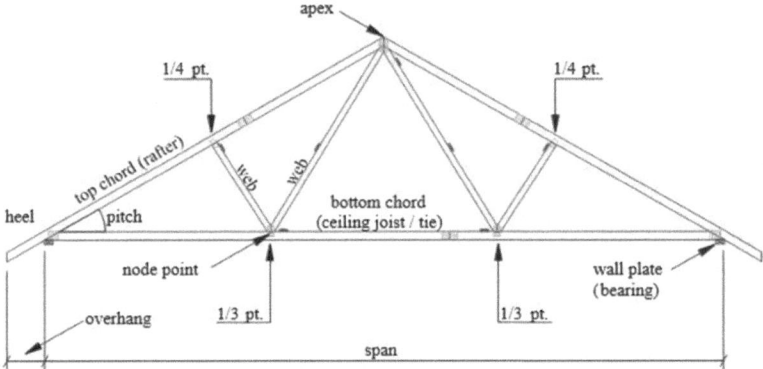

Fig. 2 Standard fink roof truss [24]

The most common fink roof truss that the company designs and manufactures has a timber length (top chord) of eight metres and manufactured at a pitch of 30°.

All other work is then valued in this EFINK unit, i.e. the number of EFINKs required to set a truss up (prepare the materials and truss bed for assembling a truss) is equal to the number of standard fink roof trusses that could have been assembled during the set-up time [13]. It is claimed by MiTek [ibid] that a key element in MBA's success is the software's ability to distribute set-up time across a group of trusses, meaning 20 trusses with the same set-up time will cost less per unit than two trusses with the same set-up time, which is reflective of the reality of manufacturing trusses at the company.

Figure 3 shows a screenshot of the MBA software upon calculation of a Fink truss [11]. Notice the total set-up cost (£16.37) is divided by the number of trusses in the run (9), to give an individual set-up cost per truss (£1.82); therefore, the more trusses produced, the more cost-effective the set-up process.

The concept of price reduction in direct correlation to quantity is far more consistent and reliable than other methods which use a fixed per truss price, and it also helps to indicate less profitable work enabling management to make informed decisions when committing to a tender [13].

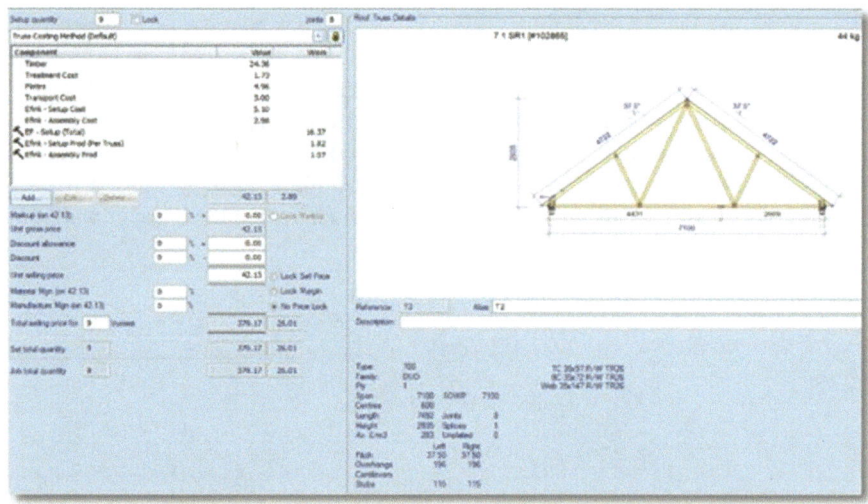

Fig. 3 Screenshot of the MBA pricing software, showing the breakdown of cut, set-up and assembly time [25]

3.6   Methodology for Time and Motion Assessment of the Company's OSM Roof Truss MMC Production Line

Following the value stream mapping exercise, a thorough time and motion study was carried out over two, two week periods, during working hours (07:45 to 16:30). The aim of the time and motion study conducted was to determine the average labour costs associated with the three main processes involved in roof truss production (cutting, set-up and assembly) in order to compare to that of the MiTek MBA system [11], which MiTek claim to be based on UK industry best practice, but altered to suit each organisation [13]. The study aimed also to uncover any possible optimisation and efficiency modifications to the current process which could result in reduced labour costs and increased competitiveness.

3.6.1   Production Line Processes for Roof Trusses

Cutting. The cutting operation refers to the time taken for one operative to cut the timbers needed in one truss type, such as a Fink truss, which could be as little as one truss or as many as 80 (or more) of the same truss type. There is always one operator on this process.

   Recording cut time. In order to track this time, the truss manufacturing supervisor was given a T&M record sheet. The start and end time of each truss type is recorded, giving an accurate figure for the time taken to cut each element of the truss.

   Set-up. This refers to the time taken between the last trusses of one set being placed on the truss rack, to when the first truss of the next job/truss type is placed on the truss rack (the wheeled truss rack enabled easy tying up and transportation of finished trusses across the factory). Set-up always includes the first truss of a set and also (Fig. 4):

- Tying up and labelling the finished set of trusses, this secures a set of trusses together, meaning they can be transported as a pack with the appropriate identification label;
- Moving the completed truss/set of trusses (and rack) to another part of the production line in the factory where a gantry crane is positioned. The gantry crane is used to lift each roof truss and place on one level below the production line, where the transport vehicles are located;
- Collection of an empty truss rack from the transport level using the gantry crane, ready for the next step/truss manufacture;
- Collection of connector plate drawers. These drawers (as shown in Fig. 5) were designed and manufactured in house to assist in the organisation and stock control of the metal connector plates [26], and each labelled drawer contains a different plate size and is pre-filled by stock control with all the plates necessary to complete on set of trusses:

Fig. 4 (Left)—Connector plate drawers

Fig. 5 (Right)—Completed truss being lifted off the press bed and onto the rack

- Clearing/cleaning the truss bed ready for the truss next set-up (removing stops and guides);
- Setting up of the truss press bed (the surface where trusses are laid out and pressed), which includes measuring and placement of stops and sticky plate locators. Stops are positioned using measurements from the design drawings to act as a jig for securing the cut timbers in the correct and consistent position for each truss. Two sticky plate locators are placed on the press bed underneath the timbers at every join point, this acts as a jig to enable quick and accurate positioning of each connection plate, and they also prevent the plates from moving during the pressing process.
- Movement (by hand) of top chords from the stack of cut timber lengths for the next project. These are moved to the ground next to the press bed in order to access the bottom chords first;
- Movement of bottom chords to appropriate position on the truss bench;
- Positioning of all chords, webs and plates of the first truss between pre-set stops so that the truss can be checked to ensure it meets the design drawings (please see image below for reference), and if any issues are identified, they can be resolved;
- Pressing plates of the first truss, which secures the timber elements together;
- Movement of the first completed truss to the rack, this is shown in Fig. 5.

Assembly. This refers to the time taken between one truss within a set of trusses being manufactured and placed on the truss rack, to the next truss of the same job being placed on the truss rack. In essence, the time taken to assemble, press and stack one truss. In groups of trusses, this time should be consistent. Common activities within this time include:

- Positioning of all elements of the truss, which includes the bottom chords, top chords, webs (see Fig. 2) and connection plates in place between the pre-set stops (see Fig. 6);

Fig. 6 Chords being placed into position between pre-measured stops during the set-up process

– Checking that the timber elements used in the manufacture of each truss are straight and not twisted or buckled before securing together;
– Occasionally, webs will need trimming manually by an operative with a hand saw, as they are sometimes not sized correctly after the cutting process;
– Pressing of plates;
– Movement of completed truss to the rack (please see Fig. 5).

Recording set-up and assembly time. The set-up and assembly time were recorded using a digital stopwatch. This time was then logged on the truss set-up and assembly time and motion sheet, in addition to the number of operators working on the operations at any time.

4 Results—Value Stream Mapping and Time and Motion Assessment of the Company's OSM of the Truss MMC Production Line

4.1 Results of T&M of Roof Truss Manufacture

Data gathered from the time and motion study is presented in a matrix (see Fig. 7) highlighting the actual Man Minutes for the truss cut, set-up and assembly, in comparison with the Man Minutes allowed in by the MiTek MBA software that was used to price the job. In Fig. 7, the red and green colour key gives a quick indication as to whether the company's manufacture of roof trusses was over or under the target hours allocated by MiTek's MBA time standards (red is over and green is under).

Overall, throughout 23 case studies (between May and September 2019) of truss manufacture that were observed by the authors of different sizes, the company was under target by 352 Man Minutes which equates to +£64.55 in labour; however, eight of the twenty-three (34.7%) studies were over target, meaning eight sets of

Truss Type	Qty	Size	Plates	Cuts	System/Per Truss		Actual / Per Truss		Difference (£/Truss)	Difference (MM/Truss)	TOT Difference	Tot MM for JOB	Tot Cost for JOB
Attic Truss		H: 3194			Cut		Cut	£ 2.77					
		L: 8611			Set Up	£ 3.11	Set Up	£ 3.65	-£ 3.31	-18.09			
T1	20		18	14	Assembly	£ 4.02	Assembly	£ 3.95	£ 0.07	0.38	-17.70	-354.10	-64.80
Finks		H: 2753			Cut		Cut	£ 1.17					
		L: 7612			Set Up	£ 2.70	Set Up	£ 1.50	£ 0.03	0.16			
T1	17		16	22	Assembly	£ 2.90	Assembly	£ 2.13	£ 0.77	4.21	4.37	74.32	13.60
Finks		H: 2753			Cut		Cut	£ 1.22					
		L: 7092			Set Up	£ 3.79	Set Up	£ 0.95	£ 1.62	8.85			
T2	13		16	23	Assembly	£ 3.16	Assembly	£ 2.65	£ 0.51	2.79	11.64	151.31	27.69
Mono		H: 3450			Cut		Cut	£ 2.92					
		L:7744			Set Up	£ 4.17	Set Up	£ 4.03	-£ 2.78	-15.19			
T2	11		10	28	Assembly	£ 4.59	Assembly	£ 4.20	£ 0.39	2.13	-13.06	-143.66	-26.29
Mono		H: 3451			Cut		Cut	£ 2.82					
		L:7910			Set Up	£ 5.43	Set Up	£ 5.09	-£ 2.48	-13.55			
T4	11		20	27	Assembly	£ 4.18	Assembly	£ 3.64	£ 0.54	2.95	-10.60	-116.61	-21.34

Fig. 7 Example of truss T&M data comparison matrix (image created by author)

trusses took longer than the MBA software suggested they should. Initially, overall this would be portrayed as a positive result, as the hours that were negative were compensated by faster production in the manufacture of other trusses.

However, upon consultation with MiTek advisors (the software that is used in the design and manufacture process for the trusses), it was noted that the company's design office was not using the 'match group' function within the software. A match group, groups all trusses that could use the same set-up, often differences in a run of 10 T1s and 10 T2s which is minor (a T1 would be the first type/group of trusses in a project and a T2 would be the second type, etc.), such as there could be on an overhang or nib detail (see Figs. 8 and 9 for reference), meaning they will use the same jig and set-up, but as they are technically 'different', they are classed as different truss types (T1 & T2) and so are given their own set-up time, which in reality should not be used, please see figures below for an example.

MiTek states that by utilising the match group function within their software, this will often significantly reduce Man Minutes assigned to the manufacture of each truss, which could if used by the company, and lead to more competitive pricing, plus reduce the target Man Minutes for each truss. This is a finding that the company's technical

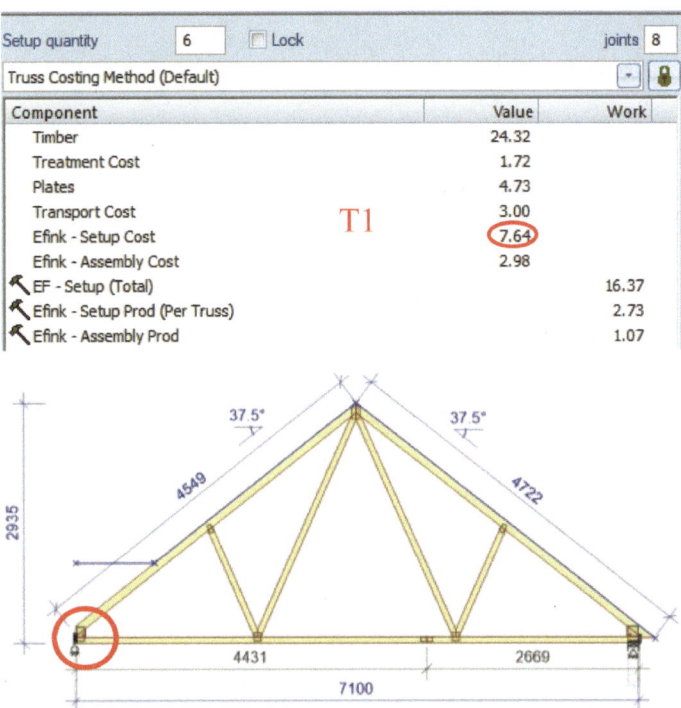

Fig. 8 Combined screenshot of T1 in MiTek's MBA software [11]. The circle on the left highlights the set-up cost for each T1. The circle on the right highlights the nib detail which differs to T2 (below) (screenshot from MiTek MBA software)

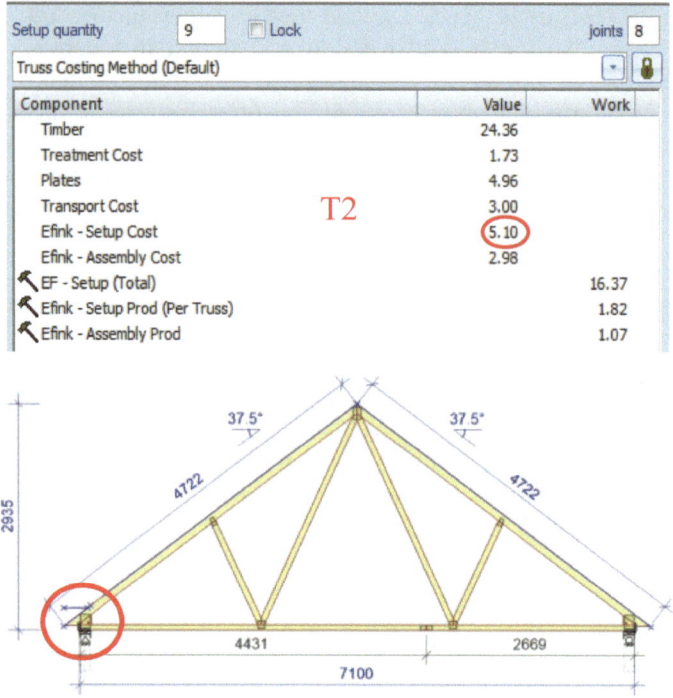

Fig. 9 T2 which is slightly different to T1 as it has an overhang detail rather than a nib (circled); however, the rest of the truss is the same and so uses the same set-up as T1; therefore, the match group function could have been utilised to reduce set-up costs (screenshot from MiTek MBA software)

team are investigating further to pass on savings to their clients. From the data gathered, it is found that if match groups were utilised, there could be concern that the current state of production would not meet the reduced time standards associated with the match group function (there would be less set-up time) and would therefore create a loss on truss labour costs. In contrast, if match groups are not utilised, there would be concern that the company are not manufacturing at MiTek's perceived industry best practice, which could impact upon their pricing of trusses. As such, following these findings, the company's technical team are investigating further in order to make enhancements to their design and manufacturing processes in order to pass on savings to their clients. The implementation of efficiencies identified within this study will speed up the truss production process in order to meet/exceed match group and industry standards to provide competitive truss pricing within the market. This is one example of work from the KTP providing a benefit to the company and contributing to meeting the project aim, which is to optimise operational performance.

Further analysis of the time and motion results emphasised another potential error within MiTek's MBA software for the company. The set-up and assembly allowance within MBA have appeared significantly too high for flat, kings post and un-standard trusses (such as attic trusses with a raised BC designed for a particular client), as well as valleys. This could be due to issues within the initial set-up of the MBA software and is undergoing investigation with MiTek software specialists. This over prediction currently allows a 'safety net' should another Truss's take longer than planned which is not correct practice. Modification to the pricing of these truss types will contribute to bringing overall quotation costs down and therefore placing the company more competitively within the market.

Another issue highlighted was within the production of simpler trusses (such as mono trusses or attic trusses), with Man Minutes consistently not meeting MiTek's target production hours. This could be due to a number of factors. Firstly, it is believed to be due to the 'over-manning' of these simpler trusses; generally, three operators work on the assembly bench at one time, irrespective of the size or complexity of truss in production. This has proved an inefficient use of manpower, often leaving the 3rd member of the team redundant, when they could be better utilised in the preparation of the next truss or within another process in the manufacturing suit. This could also be due to the unfamiliarity of the setting up and assembly of these less common trusses, indicating additional operative training may be required something which the company are investigating.

The company's technical team are using these findings from the KTP project to investigate the production time being outside of MiTek target hours, for these other truss projects, and are as such implementing T&M on a regular basis for all roof truss manufacture to ensure that any problems are identified and solutions implemented. So that cost savings can be realised, waste minimised and benefits passed onto their clients.

4.2 Results of the VSM of Roof Truss Manufacture

Results from the VSM of the cutting, set-up and assembly and the design processes for roof trusses are discussed and illustrated below.

4.2.1 Cutting

The company's production line operatives use a Randek SP720 saw [27] to cut the timber used in the production of roof trusses, and the saw relies on manual input through the operator, determined from a cutting list produced by the MiTek Pamir software. Another finding of the VSM exercise is that the lengths of the timber specified on the cutting list by the MiTek Pamir software as a means of optimisation to reduce timber waste are not yet routinely stocked by the company due to the manufacture of roof trusses as a singular component being a new product offering as a result of the company's rapid expansion. This sometimes means that the operative undertaking the cutting operation has to use longer timber lengths than specified, and as consequence, this can lead to offcuts of timber that are classified as waste. The company are looking to start stocking all timber lengths specified by the software in order to reach maximum operational efficiency for their clients. Currently, instead, the company ensure offcuts of timber can be reused elsewhere within the factory, such as for noggins or where smaller timber elements are needed in the manufacture of a roof truss or other timber product. The company's technical team are investigating this finding further, to identify which additional timber sizes could also be stocked on the production line for roof trusses, to ensure that unnecessary waste timber is not created and production costs could also be refined. Waste minimisation will also reduce capital costs for truss manufacture. In addition, the company are looking into the purchase of a semi-automated saw which receives information directly from the design software, meaning lengths and angles would no longer require manual input which is a time-consuming process open to error.

Other findings from the value stream mapping activity have identified that with simple adjustments, waste reduction and cost savings can be further enhanced related to the use of webs, offcuts, and also how materials are delivered and scheduling of processes.

Webs (which join the top and bottom chords together–see Fig. 2 [24]) were occasionally slightly too large (6–9 mm) at the assembly stage, and this resulted in operators on the truss bench re-cutting timbers using a hand saw. After investigation, the source of the problem was partly due to the angle calibration of the Randek saw, which could be up to one to five degrees out of alignment, depending on timber widths, and this resulted in webs that were one to three mm too long. Investigation into further causes of this issue is ongoing (Figs. 10 and 11).

Offcuts of timber are placed adjacent to the cutting saw ready for possible re-use, whilst offcuts with dimensions typically less than 600 mm are placed in a recycling bin adjacent to the operative undertaking the cutting process. Once the offcut recycling bin is full, the operative transfers the offcuts and stacks them on a pallet, so that they can be transferred by pallet vehicle to the existing factory and used as biomass

Fig. 10 (Left) Operator cutting a web on the press bed with a hand saw, as it is too long for the truss set out (authors image 2019)

Fig. 11 (Right) Angle
inaccuracy by 3 mm (authors
image 2019)

in the boiler to generate heat (see Fig. 12). This is a temporary measure until the new factory has its biomass boiler installed.

This finding related to offcuts has led the company to improve its processing, by considering how offcuts could be minimised or reduced through improvements in design management, design to manufacture and manufacturing management, particularly bearing in mind the Toyota philosophy to minimise waste [15].

Delivery of timber to operative cutting. The operative is sometimes left waiting for up to 30 min for the pallet truck operator to deliver the correct lengths of timbers to correlate with the optimisation of the cutting list, and this is often due to the

Fig. 12 Offcuts/waste timber neatly stacked ready to be burnt in the biomass boiler (authors image 2019)

driver already being occupied with another task. One option which the company are evaluating (following this finding from the VSM exercise) is to provide project scheduling for all production line operators via smartphone or tablet, which would allow pallet truck drivers to anticipate areas due for re-stocking and plan their day accordingly and avoid bottleneck areas so that other operatives are not waiting. In addition, the company are also investigating RFID tagging on components used in the manufacture of each timber product, such as roof trusses, as one way of tracking and ensuring just-in-time delivery to the production line process. RFID tagging has been used since 2019 to track felled trees in the USA, and as such, the company are exploring whether the same technology can be used on processed timber members that they stock/use in their factories and production lines [28].

4.2.2 Set-up and Processing

The operative cutting the timber for use in each roof truss stacks all timbers in neat piles after cutting, with the longer timbers such as the bottom and top chords at the bottom of the stack and smaller timbers (webs) on top of the stack for stability, and these stacks are usually 0.6–1.2 m high. Figure 13 shows these stacks of cut timber lengths. However, when assembling a truss, the assembly operatives position the bottom and top chords first on the truss bench. This requires each timber stack to be unpacked in order to access the bottom and top chords. One minor change to this stacking operation would be to stack the bottom and top chords in one stack and the webs in another stack. This is also an improvement to the production line process that the company's production team are investigating, to save time and any resulting impact on production costs.

End of shift operative activity. It was noted that some of the operatives working on the roof truss production line were consistently finishing 20 min early at the end of their shift to dispose of waste or recycle materials. The VSM exercise also identified

Fig. 13 Stacks of cut timbers arranged into truss types ready for assembly. The smaller timbers (webs) can be seen on the top of the stack, whereas the larger timbers (TCs and BCs) which are accessed first can be seen on the bottom of the stack (authors image 2019)

an alternative period during the production cycle that could save these 20 min each day. The process of pressing the connector plates to join timber members creates a few minutes wait time for two operators on every truss, whilst a third operative is operating the plate press. If this time was utilised for tidying/waste disposal/material recycling throughout each shift, then it would not be necessary to finish 20 min early; a couple of minutes should be sufficient.

Operative idleness. Similar to operative cutting timber wasting time waiting for timber to be delivered for procession, other operatives on the truss production line are sometimes idle.

Due to nature of construction industry, occasionally the company have to prioritise a particular project over their production schedule in order to prevent late charges or to accommodate for their clients' urgent requirements. This disruption to the production schedule can cause truss assembly operators to be left idle whilst waiting for a high priority job to finish being cut on the Randek saw, despite there being a surplus of cut jobs ready for assembly. This is something currently out of the company's control due to the unpredictability of the construction industry. But, this issue is something the company hope to work on with the assistance of Toyota beginning from the 5th February 2020, to streamline the production process, which will be of benefit to them and their clients.

4.2.3 Timber Element Fixing

The VSM exercise has also highlighted opportunities for refinement to the fixing process for timber elements used as part of roof trusses. Occasionally, the metal plates used to secure together the different timber elements in a roof truss are too large for purpose, which requires additional timber components to be cut specifically for this purpose and pressed into the plate overhang (known as 'packers'), see Fig. 14.

Fig. 14 Highlighted dashed box illustrates the timber packers to compensate for the plate which is too large for fixing timber elements together on the roof truss (author image 2019)

Incorrect sized plates that are larger than required for joining the truss components are more expensive than correctly sized plate; labour time is increased as the operative cutting timber has to process the timber for the packers, and other operatives have to fit and secure the packers in place. The company are investigating the reason why overly large plates are specified occasionally, and once the error is identified, the necessary refinement will be made to the design and production processes. Therefore, reducing the labour and material costs further for their roof truss manufacture could lead to further optimisation of production costs and ultimately product costs for the market.

5 Discussion

This paper set out to discuss the context and main aims of the KTP between Cardiff Metropolitan University and the company. The focus for the paper has been to discuss and illustrate one of the first KTP targets, which aim to optimise production of roof components (such as trusses, gable ladders and valleys) though VSM exercises and T&M studies in order to increase the quality of components and also create a cost saving for both the company and their clients.

The findings of the time and motion study and value stream mapping exercise on 23 case studies of roof truss manufacture between May and September 2019 have proved extremely valuable for the company. In particular, the findings have identified ways in which materials, processes and labour hours can be optimised, and software can be tailored to the company's individual requirements. This has already resulted in faster production, reduced waste and consequently increased the company's competitiveness within the industry, with further changes still planned.

The key findings from the T&M study, which the company are investigating further and refining their processes to reduce material waste and labour costs, include the refinement and correct use of MiTek's MBA software that the company currently use as part of the design and manufacture of roof trusses. This includes the realisation and implementation of the match group function, as well as alterations to some pre-set units within the software set-up, and a reduction in labour time allocated to flat, kings post and unique trusses (such as the attic truss with raised BC) and also valleys. Results highlighted issues within the production of mono trusses and attic trusses taking longer than targeted hours. It was determined that this is likely due to the over-manning of less complex and sometimes smaller trusses.

The key findings from the VSM exercise, which the company are also investigating further and refining their processes to reduce material waste and labour costs, include ensuring that the most appropriate materials for purpose are stocked for use (such as correct timber lengths and connecting plates) meaning reduced waste from components that are larger than necessary. If possible, any unavoidable waste produced should be utilised within other manufacturing processes within the factory; if unavoidable, smaller offcuts should be recycled or disposed of efficiently through well-thought-out conveyance, minimising distance travelled (and therefore time) or people and vehicles.

Additionally, production scheduling should be streamlined yet accommodating for the unpredictability of the construction industry preventing unnecessary operator wait times.

The limitations of the study documented in this paper are that only 23 case study projects were observed (345 trusses), out of an estimated 7500 roof trusses manufactured per annum by the company. The time of year and subsequent temperature would likely have had an effect on the speed of work. This study was undertaken throughout the summer months when the production facility was very hot due to external air temperature combined with the facilities largely glass roof which amplifies the sun's rays. For more accurate results, if possible future studies should be carried out during alternate seasons as a true reflection of annual productivity. Finally, it is likely that the factory-based observational methodology could have had inaccurate results due to operators feeling under pressure and therefore working faster or in a way that is not reflective of a 'normal' day. If possible, in future observational studies should be done without operators knowing, such as through CCTV footage to ensure work patterns are not altered as a result of the presence of an individual collecting data. This study was conducted on one UK timber-frame manufacturer out of hundreds and so is not reflective of the industry as a whole. Some UK manufacturers currently use less sophisticated technology, and others use more sophisticated/state-of-the-art equipment, and therefore, this study will only be useful either for those utilising the same of similar MMC or for companies looking to upscale to MMC where they have previously not.

The next steps in the KTP project and with support of the company's technical team and directors are to undertake a time and motion study and value stream mapping exercise on the production lines for their exterior timber closed wall panel systems and floor cassettes, where the former studies will be documented in a paper for the 12th Sustainability in Energy and Buildings International Conference [29]. Furthermore, following the findings presented in this paper time and motion and value stream mapping are being regularly included in activities by the company's production manager in order to improve the company's quality control, waste minimisation, operational costs and savings to their clients and also identify when machinery should be improved and additional training provided to their staff.

6 Conclusion

As populations increase and housing demands grow in unity with the Welsh Governments low- to zero-carbon agenda, the construction industry is experiencing a drive towards offsite timber-frame manufacture utilising MMC as an efficient and low carbon alternative to traditional buildings. Whilst giving context to the KTP project, this paper documents and compares methods for optimising MMC within timber-frame truss manufacture through results from time and motions studies and value stream mapping, as well as investigating the best methods for accurate costing of such a complex and versatile process.

Through investigation into the use of MiTek's MBA software supporting evidence suggests the Equivalent Fink theory behind MiTek's pricing system is the most appropriate methodology for accurate truss labour costing, however the importance of tailoring and thoroughly checking the software's algorithms against individual companies requirements is priceless, and ensuring match groups which are being utilised will reduce consumer costs significantly, and it is particularly important to check the pricing strategy produced by MBA for non-standard truss types that may be unique to a specific project or company as the software sometimes overcompensates for these. As standard, the MBA software displays cut and set-up time as one item and then assembly time as another, and this created a challenge when converting and comparing the time and motion data from the cut and set-up time as separate items. After consultation with MiTek, the company was able to split the cut and set-up times, meaning future time and motion study data will be faster and simpler to compare.

Low-cost and easily implemented alterations to the company's current manufacturing processes have been identified to reduce waste in relation to Man Minutes, materials and, therefore, money. These include solutions for minimising the over conveyance of waste such as timber offcuts through alternative disposal methods and reducing unnecessary movement during the cutting and set-up process, such as altering the method for stacking cut top and bottom chords. The benefit of ensuring initial designs is checked to ensure the correct products/materials which are used were highlighted through the observation that incorrect plate sizes are often used, causing time, material and monetary wastes.

Limitations including the appropriateness of physical observational study have been discussed. Further studies should look to conduct observations in a non-intrusive manor, such as through CCTV to prevent affecting workflow and therefore impacting the reliability of results. If possible, studies should take place throughout the year to represent the pace of work in relation to the internal environmental conditions of the workplace.

The next steps have been discussed which are related to implementing time and motion and value stream mapping of the company's closed panel timber-frame exterior wall system and floor cassettes. Identified process efficiency changes will first be trialled on a low cost and low-risk basis, and if trials are successful, full implementation will take place. After every change made to production, new time and motion studies will be performed in order to understand the impact, the change has had on production speed and cost, and this should then be reflected directly into the company's pricing system. Therefore, the findings of the time and motion studies and value stream mapping exercises undertaken will be significant and invaluable moving forward, to the company and their clients (existing and new), the wider timber-frame market in the UK and also the academic and scientific community.

Acknowledgements The authors wish to thank the company, Innovate UK and the Welsh Government who are co-funding and supporting the research as part of a three-year KTP project undertaken by staff from Cardiff Metropolitan University.

References

1. Farmer M (2016) The farmer review of the UK construction labour model. Online access to official document developed by Cast Consultancy, published by the Construction Leadership Council (CLC). Cited at: http://www.constructionleadershipcouncil.co.uk/wp-content/upl oads/2016/10/Farmer-Review.pdf, (available). Accessed on 7 May 2019
2. Anon (2017) Offsite manufacturing. Cited at: https://chcymru.org.uk/uploads/events_attach ments/Offsite_Maufacturing_-_A_Real_Solution.pdf, (available). Accessed 1 Jan 2020
3. Offsite Construction Hub (2018) Offsite Hub, Cited at: https://www.offsitehub.co.uk/home/, (available). Accessed 1 Jan 2020
4. Hetherington D (2016) Delivering new homes—a future off-site. Cited at: https://www. northern-consortium.org.uk/2016/04/29/delivering-new-homes-a-future-off-site/, (available). Accessed 1 Feb 2018
5. STA (2018) Project Gallery. Cited at: http://www.structuraltimber.co.uk/project-profiles, (available). Accessed 1 Jan 2020
6. Savills (2017) Demand for timber primed for growth. Cited at: http://www.savills.co.uk/res earch_articles/141557/215817-0, (available). Accessed 1 Jan 2020
7. STA (2017) Annual survey of UK structural timber markets. Cited at: http://www.forestryscot land.com/media/370371/annual%20survey%20of%20uk%20structural%20timber%20mark ets%202016.pdf, (available). Accessed Jan 2020
8. Littlewood ZF, Lancashire JR, Newman R, Hedges GD An evaluation of offsite timber frame manufacturers in Wales, UK. Chapter 61. In: Smart innovation, systems and technologies, vol 163, pp 723–733
9. Gregory R (2019) Iconic metal box factory in neath starts production again after Hiatus. Cited at: https://wales247.co.uk/iconic-metal-box-factory-in-neath-starts-production-again-after-hia tus/, (available). Accessed Jan 2020
10. MiTek UK and Ireland (2020) PAMIR Truss & joist design & manufacturing software. MiTek, UK. Cited at: https://www.mitek.co.uk/software/pamir/, (available). Accessed 23 Jan 2020
11. MiTek UK and Ireland (2019) MBA. MiTek UK and Ireland. 2020 cited at: https://www.mitek. co.uk/software/mba/, (available). Accessed 19 Jan 2020
12. Dey C (2002) Methodological issues: The use of critical ethnography as an active research methodology. Account Audit Account J. Cited at: https://www.emerald.com/insight/content/ doi/10.1108/09513570210418923/full/html, (available). Accessed 1 Jan 2020
13. MiTek (2020) The MiTek EFINK Truss costing system. MiTek
14. Irani S, Zhou J (2019) Value stream mapping of a complete product. The Ohio State University, Department of Industrial, Welding and Systems Engineering. Cited at:http://www.lean-man ufacturing-japan.com/Value%20Stream%20Mapping%20of%20a%20Complete%20Product. pdf, (available). Accessed Oct 2019
15. Toyota Production System Glossary. The Official Blog of Toyota GB. 2013.Cited at:https:// blog.toyota.co.uk/toyota-production-system-glossary, (available) Accessed 29 October 2019
16. Bradley J (2015) Improving business performance with Lean, 2nd edn. Business Expert Press, New York
17. Lean Training Courses and Workshops 2020 cited at: https://tlmc.toyotauk.com/our-services/ training/, (available). Accessed 22 Jan 2020
18. Drummond T (2007) Estimating labour with averages not sufficient. Struct Build Comp Mag 38–40. Cited at:https://www.sbcmag.info/sites/default/files/0708_estimating.pdf, (available). Accessed Oct 2019
19. Shepherd J, Shepherd J (1996) Be your own contractor and save thousands, 1st edn. Real Estate Education Co., Chicago, IL, p 79
20. Walters L, Houlihan AS (2004) Recipe for production success. structural building components magazine 2004 Cited at: https://www.sbcmag.info/Archive/2004/dec/0412%20Houlihan.pdf, (available). Accessed Nov 2019

21. Hershey B (2017) 4Ward Consulting Group, LLC. Houlihan, Efficiency, Lean & the 5 M's. Component Manufacturing Advertiser 2017. Page 57. 10216:56–58 cited at: https://issuu.com/componentadvertiser/docs/1710216, (available). Accessed 1 Nov 2019

22. Donohue T (2019) Do you know your real costs? Gang-Nail Guidelines—MiTekcited at: http://www.mitek.com.au/Publications/GN-Guidelines/GN-Guideline-39, (available). Accessed Sept 2019

23. Wolf Systems (2004) Trussed rafter technical manual. Cited at: https://www.djr-roof-trusses.co.uk/images/Trussed-Rafter-Tech-Manual.pdf, (available). Accessed 1 Jan 2020

24. Figure Two (2020) Fink roof truss structure. Cited at: https://mbctimberframe.co.uk/fink-truss/, (available). Accessed 20 Jan 2020

25. Figure 3. Screen shot of the MBA pricing software, showing the breakdown of cut, set up and assembly time. MiTek UK and Ireland, 2020. MBA. Cited at: https://www.mitek.co.uk/software/mba/. Accessed 19 Jan 2020

26. Connector Plates (2020) Cited at: https://www.mitek-us.com/products/Connector-Plates/, (available). Accessed 23 Jan 2020

27. Randek (2020) House Production Technologies—Cut Saw SP720. Cited at: http://www.randek.com/en/products/cut-saws/crosscut-saw-spl728, (available). Accessed 25 Jan 2020

28. Swedberg C (2019) New RFID staple tag for rough wood tracks timber. Cited at: https://www.rfidjournal.com/articles/view?18541, (available). Accessed 20 Jan 2020

29. Moorhouse VL, Littlewood JR, Hale E A pilot study evaluating offsite manufacturing of timber frame panels using lean manufacturing principles for dwellings, to be publilshed in KES International Sustainability in Energy and Buildings Research Advances, Vol tbc, Page tbc